HYDRAULIC ENGINEERING III

PROCEEDINGS OF THE 3RD TECHNICAL CONFERENCE ON HYDRAULIC ENGINEERING (CHE 2014), HONG KONG, 13–14 DECEMBER 2014

Hydraulic Engineering III

Editor

Liquan Xie

Department of Hydraulic Engineering, Tongji University, Shanghai, China

CRC Press
Taylor & Francis Group
Boca Raton   London   New York   Leiden

CRC Press is an imprint of the
Taylor & Francis Group, an **informa** business

A BALKEMA BOOK

CRC Press/Balkema is an imprint of the Taylor & Francis Group, an informa business

© 2015 Taylor & Francis Group, London, UK

Typeset by V Publishing Solutions Pvt Ltd., Chennai, India
Printed and bound in Great Britain by CPI Group (UK) Ltd, Croydon, CR0 4YY

Published by: CRC Press/Balkema
    P.O. Box 11320, 2301 EH Leiden, The Netherlands
    e-mail: Pub.NL@taylorandfrancis.com
    www.crcpress.com – www.taylorandfrancis.com

ISBN: 978-1-138-02743-5 (Hbk)
ISBN: 978-1-315-71598-8 (eBook PDF)

Hydraulic Engineering III – Xie (Ed)
© *2015 Taylor & Francis Group, London, ISBN 978-1-138-02743-5*

Table of contents

Hydraulic Engineering III – Xie (Ed)
© *2015 Taylor & Francis Group, London, ISBN 978-1-138-02743-5*

Preface

Hydraulic research is developing beyond traditional civil engineering to satisfy increasing demands in natural hazards, structural safety assessment and also environmental research. In such conditions, this book embraces a variety of research studies presented at the 3rd Technical Conference on Hydraulic Engineering (CHE 2014), 2014 Structural and Civil Engineering Workshop (SCEW 2014) and the 4th Workshop on Environment and Safety Engineering (WESE 2014), held in Hong Kong during December 13–14, 2014. The series of conferences was conceived and organized with the aim to promote technological progress and activities, technical transfer and cooperation, and opportunities for engineers and researchers to maintain and improve scientific and technical competence in the field of hydraulic engineering, environment and safety engineering, and other related fields.

62 technical papers were selected for publication in the proceedings. Each of the papers has been peer reviewed by recognized specialists and revised prior to acceptance for publication. The papers embody a mix of theory and practice, planning and reflection participation, and observation to provide the rich diversity of perspectives represented at the conference. This book review recent advances in scientific theories and modelling technologies that are important for understanding the challenging issues in hydraulic engineering and environment engineering. Some excellent papers were submitted to this book, and some highlights include:

– river engineering and sediment transport, e.g. unsteady open-channel flow over rough bed, sparse planted open channel flow, bed load deposition in vegetated area of a stream, flows with non-submerged vegetation in depth-averaged 2D scheme, seabed sediment at Leizhou Bay of Guangdong (china), sand ridges in Liaodong Bay in northern Bohai Sea.
– waterway engineering, e.g. measures for the waterway of middle Yangtze river, deep-water channel regulation at the lower reach of the Yangtze river.
– flood control, irrigation and drainage, e.g. levee breach process by overflow, rain adaptative landscape in Fujian coastal cities of China, "99.6" Meiyu front rainstorm in China, fractal dimension in the evolution of drainage networks, decadal wind velocity in Henan province of China, hydrological responses of the upper reaches of Yangtze river to climate change.
– hydraulic structures, e.g. thin-plate rectangular weir, wall-jets for hydraulic engineering applications, pier base of hydroturbine with ring beam and columns.
– geotechnical aspects, e.g. slope stability analysis by the improved simplified Bishop's method considering the difference of inter-slice shearing force, instability of composite subgrade in a railway project under construction.
– safety analysis in engineering, e.g. earthquake response of self-anchored suspension bridges, seismic response of PC continuous box-girder bridge with corrugated steel webs, risk of accident investigation and correlation analysis of the foundation pit, key construction technologies for long-span aqueduct with reinforced concrete box arch, fire detection in the condition of low air pressure, risk assessment of domino effect in LPG tank area.
– water resources and water treatment, e.g. index system of water resources allocation, evaporation duct prediction models, characteristics of a heavy rain event in Guilin (china), diatomaceous earth precoat filtration for drinking water treatment, effect of Sanosil on removing algae.
– environmental fluid dynamics, e.g. measured wave in Andaman Sea (Myanmar), particulate organic matter (POM) with sediment at vegetated area in rivers.

- waste management and environmental protection, e.g. environment risk research on Yongjiang river basin in Guangxi province (China), risk assessment of heavy metals in the sediments of Jiulong river estuary in Fujian province (China).
- pollution and control, e.g. adsorption of cu^{2+} from aqueous solution by aminated ephedra waste.
- modelling technology in hydraulic engineering, e.g. mathematical model for closure of a cofferdam in Caofeidian Harbour (China), elementary physical model to study the monogranular cohesive materials, Hoek–Brown disturbance factor for analyzing an axisymmetrical cavern.
- mechanics in engineering, e.g. mesoscopic mechanics theory for concrete in hydraulic engineering, mechanical structure system of straight arch with two articulations by break point.
- numerical software and applications, e.g. tidal current modeling of the yacht marinas of double happiness island in Xiamen (China) by Mike21 series of FM module, shoreline change on Beidaihe New District Coast (China) by model-GENESIS based on one-line theory, construction safety of Beijing subway.
- civil engineering, e.g. reinforced concrete beam by high titanium blast furnace slag, connection design of concrete-filled twin steel tubes column, blind bolted end-plate connection on structural square hollow section.
- sedimentary records and climate environment evolution, e.g. Qinghai lake (China).
- other aspects, e.g. early warning system for groundwater hazard in coal mine, heat storage of solar-ground concrete pile, temperature characteristics for concrete box girder, comprehensive evaluation of cracking characteristics of box girder high performance concretes, land use changes in the coastal cities in China.

At the very time we would like to express our deep gratitude to all authors, reviewers for their excellent work, and Léon Bijnsdorp, Lukas Goosen and other editors from Taylor & Francis Group for their wonderful work.

The 3rd technical conference on hydraulic engineering

Hydraulic Engineering III – Xie (Ed)
© 2015 Taylor & Francis Group, London, ISBN 978-1-138-02743-5

Bed roughness boundary layer and bed load deposition in vegetated area of a stream

Ho-Seong Jeon, Makiko Obana & Tetsuro Tsujimoto
Department of Civil Engineering, Nagoya University, Japan

ABSTRACT: Depth-averaged analysis of flow and sediment transport has become familiar and powerful means in river management. When it is applied to a stream with vegetation, form drag due to vegetation is taken into account, but the change of frictional resistance law is not properly treated. Though the flow with non-submerged vegetation, the velocity is constant except only the thin layer near the bed where the shear flow appears due to bed roughness to govern the frictional resistance law. In this paper, the bed roughness boundary layer is investigated for reasonable estimation of shear velocity and subsequently bed load transport process.

1 INTRODUCTION

Flood mitigation and ecosystem conservation are simultaneously required in recent river management, and understanding and analysis of flow and river morphology in a stream with vegetation have become important topics in river hydraulics. Recently, the depth averaged 2-dimensional model has become familiar with the analysis of river morphology. The drag coefficient is the key in determining the drag related with vegetation and also to understand the vertical distribution of the velocity. The applicability is fine and sometimes it is available to describe the outline of fluvial process there (Tsujimoto 1999). In depth averaged model, the resistance law to relate the depth averaged velocity U to the shear velocity u^* for uniform flow is introduced. If the logarithmic law (Eq. 1) is applied as velocity profile $u(z)$, Keulegan's equation obtained by integration the velocity profile along the depth is employed.

$$\frac{u}{u^*} = \frac{1}{\kappa}\ln\left(\frac{z}{k_s}\right) + B_s(R_{e*}) \quad \frac{U}{u^*} = \frac{1}{\kappa}\ln\left(\frac{h}{k_s}\right) + B_s(R_{e*}) - \frac{1}{\kappa}$$ (1)

where κ = Karman's constant; h = depth; k_s = equivalent sand roughness, $B_s(Re^*)$ = function of roughness Reynolds number $Re^* = u^*k_s/v$; and v = kinematic viscosity.

In vegetated area, form drag is predominant and velocity profile is uniform along the depth only except the thin layer near the bed where the boundary layer is developed to bring a shear flow (see Fig. 1).

In the conventional depth analysis for streams with vegetation, form drag for vegetation is introduced in addition to the bed friction treated as similarly as that in non-vegetated area. However, depending on the velocity profile shown in (Fig. 1), a proper resistance law should be applied in vegetated area. As shown later, the resistance law is not sensitive for the calculation of depth and depth-averaged flow, but it brings underestimation of the shear velocity and subsequently it may not bring a reasonable analysis of sediment transport and subsequent fluvial process. In this paper, we discuss the bed roughness boundary layer in vegetated area and deduce a reasonable relation between U and u_* in vegetated area to proceed to an analysis of sediment transport.

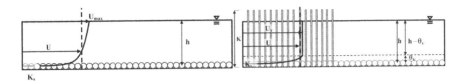

Figure 1. Vertical distribution of velocity in vegetated area.

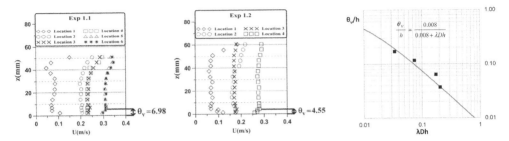

Figure 2. Velocity profile in vegetated area measured by Liu et al.

Figure 3. Relation between θ_v/h and λDh.

2 BED ROUGHNESS BOUNDARY LAYER IN VEGETATED AREA

2.1 *Bed roughness boundary layer thickness in vegetated area*

Figure 2 shows elaborate measurement of vertical distribution in vegetated area by (Liu et al. 2008), where group of piles arranged staggered pattern was utilized as non-submerged vegetation; D = diameter; λ = number density of piles; h = depth; U_v = characteristic velocity in vegetated area: and θ_v = bed roughness boundary layer thickness. The characteristic velocity in vegetated area is given as follows by equating the gravity component and the form drag, where g = gravitational acceleration; I_e = energy gradient, and C_D = drag coefficient.

$$U_v = \left[2gI_e/\lambda DC_D\right]^{\frac{1}{2}} \qquad (2)$$

The bed roughness boundary layer thickness is subjected to the characteristics of vegetation and dimensional analysis propose the relation between θ_v/h and λDh. By considering θ_v may decrease with the vegetation density and tends to the flow depth with sufficiently disperse density, the following relation is proposed as a favorable formula to estimate the bed roughness boundary layer thickness (see Fig. 3).

$$\frac{\theta_v}{h} = \frac{0.008}{0.008 + \lambda Dh} \qquad (3)$$

2.2 *Velocity distribution in bed roughness boundary layer in vegetated area*

Velocity distribution in bed roughness boundary layer is investigated, and logarithmic law is expected to be applied.

$$\frac{u(z)}{u^*} = \frac{1}{\kappa}\ln\left(\frac{z}{k_s}\right) + B_s(R_{e^*})\,(z < \theta_v) \qquad (4)$$

Though the number of the data is small for each run, the shear velocity in the vegetated area, u^*, is evaluated by fitting the logarithmic law for each run, then the data of the all

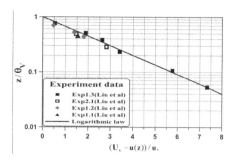

Figure 4. Defect law expression of velocity distribution in bed roughness boundary layer.

runs are plotted in the defect law expression in (Fig. 4). Defect law expression is written as follows.

$$\frac{U_v - u(z)}{u^*} = -\frac{1}{\kappa}\ln\left(\frac{z}{\theta_v}\right)(z < \theta_v) \tag{5}$$

According to this figure, it is recognized that the velocity profile follows the logarithmic law though the number of the measured data for each run is very few.

2.3 Resistance law in vegetated area

In order to obtain the resistance law to be applied to the vegetated area, velocity distribution is integrated from the bottom to the free surface as follows:

$$\frac{U}{u^*} = \frac{(h - \theta_v)}{h}\frac{U_v}{u^*} + \frac{\theta_v}{h}\left[\frac{1}{\kappa}\ln\left(\frac{\theta_v}{k_s}\right) + B_s(R_{e*}) - \frac{1}{\kappa}\right] \tag{6}$$

In vegetated area, the shear velocity to govern sediment transport is evaluated by using equation 6 from the obtained depth-averaged velocity in horizontal 2D flow analysis, and it is applied to evaluate bed load transport rate, entrainment flux of suspended sediment and so on.

3 FLUME EXPERIMENT AND CERTIFICATION OF MODEL IMPROVEMENT

3.1 Laboratory experiment

To observe the sediment transport in the vegetation area during flood stage. In the laboratory experiment, a model vegetation made by a group of cylinders made of bamboo arranged in staggered pattern ($D = 0.25$ mm, $\lambda = 0.25/\text{cm}^2$) was set in the interval of 5.0 m in a flume 20 m long and 0.5 m wide with the constant slope. The bed slope was rigid.

Firstly, the flow measurements (U and h) were conducted along the centerline of the flume (Obana et al. 2012). Then, sand ($d = 0.5$ cm, $\sigma/\rho = 2.65$; d = diameter, σ, ρ = mass density of sand and water) was fed at 1.0 m upstream of the vegetated area with constant volume along the width. The supplied sediment rate was 0.047 cm²/s. After 20 min of sediment supply, water is stopped and then deposition of sand in the vegetation area was measured along the centerline in the vegetated zone.

3.2 Simulation of depth-averaged flow and comparison with flume experiment

The measured data and the results of 2D depth averaged model longitudinal changes of depth and depth-averaged velocity are compared in figure 6. The calculation was conducted

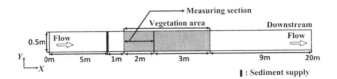

Figure 5.   Experimental Flume.

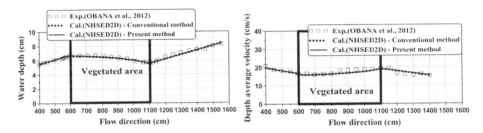

Figure 6.   Comparison between measured and calculated results of depth and depth-averaged velocity.

by using a program developed for horizontal 2D depth averaged flow where the bed friction and the form drag due to vegetation are taken into account by the following equation.

$$\tau_x = \rho C_f U \sqrt{U^2 + V^2}, \quad \tau_y = \rho C_f \lambda V \sqrt{U^2 + V^2} \tag{7}$$

$$F_x = \frac{1}{2} C_D \lambda U \sqrt{U^2 + V^2}, \quad F_y = \frac{1}{2} C_D \lambda V \sqrt{U^2 + V^2} \tag{8}$$

where U, V = longitudinal and lateral component of depth-averaged velocity, C_f = friction coefficient defined as $(U/u^*)^2$. In the conventional way, Keulegan's equation (Eq. 1) is employed to both non-vegetated and vegetated areas. While, when the present method is applied, equation 5 is employed in the vegetated area. As shown in figure 6, Conventional way and present way give good agreements between experimental data and calculated data about depth-averaged velocity and water depth. Figure 7 shows the shear velocity may be appreciably underestimated if the conventional way is employed.

3.3   *Bed load transport and deposition in vegetated area*

Bed load transport can be described by the formula proposed by (Ashida & Michiue 1972), and written as follows.

$$q_{B*} \equiv \frac{q_B}{\sqrt{(\sigma/\rho - 1)gd^3}} = 17\tau^{*3/2}\left(1 - \frac{\tau^*_c}{\tau^*}\right)\left(1 - \sqrt{\frac{\tau^*_c}{\tau^*}}\right) \tag{9}$$

where q_B, q_{B*} = bed load transport rate; $\tau^* = u^{*2}/[(\sigma/\rho - 1)gd]$ = Shields number; and τ^*_c = dimensionless critical tractive force. If bed load transport formula by (Ashida & Michiue 1972) is applied, the supplied sediment in the present experiment is around equivalent in the upstream of the vegetated area but excessive in the vegetated zone. Thus, bed load sediment deposit just upstream of the vegetated area and the upstream part of the vegetated area.

6

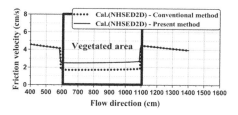

Figure 7.   Calculated results on shear velocity.

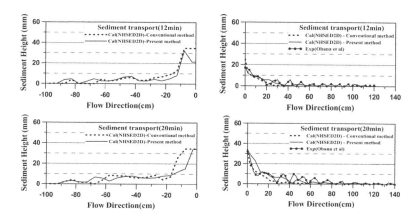

Figure 8.   Deposition profile of bed load sediment.

In figure 8, longitudinal profile of bed load deposition with time is depicted with the meas-ured profile in the vegetated area after 12 min. and 20 min. As for the deposition of bed load in the upstream of the vegetated zone, we did not carried the measurement in the flume experiments.

As for the calculated results, those employed the conventional resistance law and the pres-ently proposed one for the vegetated area are compared with the measured data. The present model can describe the deposition profile with steeper downstream slope with faster migra-tion because of the higher value of the shear stress and it shows better conformity with the experimental result. Thus it is concluded that introduction of the present proposal of the resistance law based on the concept of bed roughness boundary layer in vegetated area.

4   CONCLUSION

Recently, 2D horizontal depth-averaged flow model becomes familiar to be recognized as powerful means of stream with vegetation by adding the form drag of vegetation. Though it is expected to apply fluvial process of streams with vegetation, the shear stress may be underestimated and fluvial process may not be properly described. In this study, we focused on the bed roughness boundary layer in the vegetated area to deduce the resistance law in the vegetated area.

The modification of the resistance law by introducing the bed roughness boundary layer brings less changes in flow calculation represented by depth and depth-averaged veloc-ity but significant correction of shear velocity and subsequently sediment transport. The proposed modification will affect other aspects in fluvial processes, which will be clarified successively.

REFERENCES

Ashida, K. & M. Michiue. 1972. Hydraulic resistance of flow in an alluvia bed and bed load transport rate. *Proc., JSCE* 206: 59–69 (in Japanese).

Liu, D., P. Diplas, J. Fairbanks & C. Hodges. 2008. An experimental study of flow through rigid vegetation. *Journal of Geophysical Research* 113(F4).

Obana, M, T. Uhida & T. Tsujimoto. 2012. Deposition of Sand and particulate organic matter in riparian vegetation. *Advances in River Engineering., JSCE:* 47–52 (in Japanese).

Tsujimoto, T. 1999. Fluvial processes in streams with vegetation. *Journal of Hydraulic Research* 37(6): 789–803.

Hydraulic Engineering III – Xie (Ed)
© 2015 Taylor & Francis Group, London, ISBN 978-1-138-02743-5

Levee breach process by overflow using a small scale experimental model

M. Arita
Graduate School Student, The University of Tokushima, Tokushima, Japan

Y. Muto
Professor, Department of Civil Engineering, The University of Tokushima, Tokushima, Japan

T. Tamura
Associate Professor, Department of Civil Engineering, The University of Tokushima, Tokushima, Japan

ABSTRACT: The process of three-dimensional levee breach by overflow has not well been clarified in past studies. It is very important to elucidate this process for disaster prevention. The authors conducted small scale experiments on levee breach caused by overflow. Special attentions are paid to the effect of the levee material on levee breach process. As a result, relationship between characteristics of levee material and widening process is indicated.

1 INTRODUCTION

River levee is designed as a safe structure under the condition below the estimated high water level. In recent years, Water level above the estimated high water level were frequently recorded by typhoons and torrential rains and then damage caused by levee breaches may be substantial. More than 80% of levee destruction were caused by overflow but the process of levee breach by overflow has not been clarified in past studies (Muramoto 1981). It is very important to elucidate of this process or disaster prevention. For example, Hokkaido Regional Development Bureau and Civil Engineering Research Institute for Cold Region conduct a three-dimensional experiment on levee breach by overflow (Simada et al. 2011). A large scale experiments takes a long time and require the cost of considerable, but these problems are not in a small scale experiments. The purposes of this study are difference in levee materials affect the levee breach process by a small scale experiments.

2 EXPERIMENT OF OUTLINE

The configuration of the experimental flume and the location of the hydraulic observation position are shown in Figure 1. The Experimental flume has a length of 4 m, a width of 0.8 m and the bed slope was set up approximately 1/1000. Both of the height and crown width of the levee was designed to 0.1 m. The length of the breach section was 1.8 m from the upstream, and there were a notch of 0.01 m in depth to trigger a breach. In addition, the other parts except breach section made of wood. A damming facility was set at the downstream end to maintain a proper water level. The flow rate and levee shape were constant for all the cases.

The soils used for experiments are shown in Table 1, the grain size distribution of these soils are show in Figure 2, and case conditions for the experiments are shown in Table 2. The average grain size was almost the same except for DL clay.

To capture the collapse time of the levee, a video camera was placed over the levee, and recorded the time when change of the collapse of the levee. In addition, the levee collapse

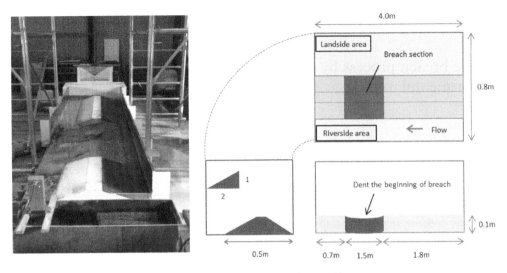

Figure 1.　View of experiment flume and hydraulic observation position.

Table 1.　Soils for experiments.

Soil name	Mean diameter (mm)	Uniformity coefficient
Weed control soil	0.97	15.11
Decomposed granite soil	1.43	5.91
Sea sand	0.48	2.7
Soil for horticultural	1.12	7.5
Antibacterial soil	0.71	5.28
Vermiculite	1.42	2.65
DL clay	0.022	6.25
Silica sand	1.4	1.66

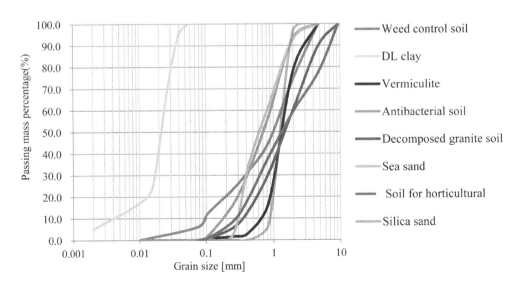

Figure 2.　Distribution of levee material.

Table 2. Case conditions for experiments.

Levee material	Case
Silica sand:Weed control soil = 4:1	W-1
Silica sand:Weed control soil = 2:1	W-2
Silica sand:Weed control soil = 1:1	W-3
Sea sand:Weed control soil = 4:1	W-4
Decomposed granite soil	D-1
Decomposed granite soil:Vermiculite = 1:4	D-2
Sea sand	S-1
Sea sand:Soil for horticultural = 4:1	S-2
Antibacterial soil	A-1
Silica sand:DL clay = 4:1 2)	Q-1

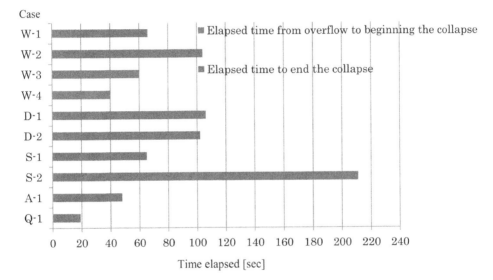

Figure 3. Levee collapse time in each case.

start time was defined as the time collapse of the toe of back slope began, and the levee collapse termination time was as the time when collapse time had reached the outer of the riverside.

3 EXPERIMENTAL RESULTS

3.1 Levee collapse time

The levee collapse time in each case were shown in Figure 3. In cases W-1, W-2, and W-3, changing the distribution amount of weed control soil, the more weed control soil blended, the longer levee collapse time is, and case W-3, blending weed control soil up to half, didn't collapse. In addition, case W-4, blending sea soil with weed control soil, nor did collapse. This is because the average particle size of the sea sand is smaller than silica sand, and possibly has it viscous properties. Next, in cases D-1 and D-2, using decomposed granite soil, the average particle size of the silica sand and decomposed granite soil is almost unchanged, but the levee collapse time were relatively long. On the other hand, in case S-1, using sea sand, even though the average grain size of the sea sand was less than the decomposed granite soil, the levee collapse time was shorter than case D-1 and case D-2.

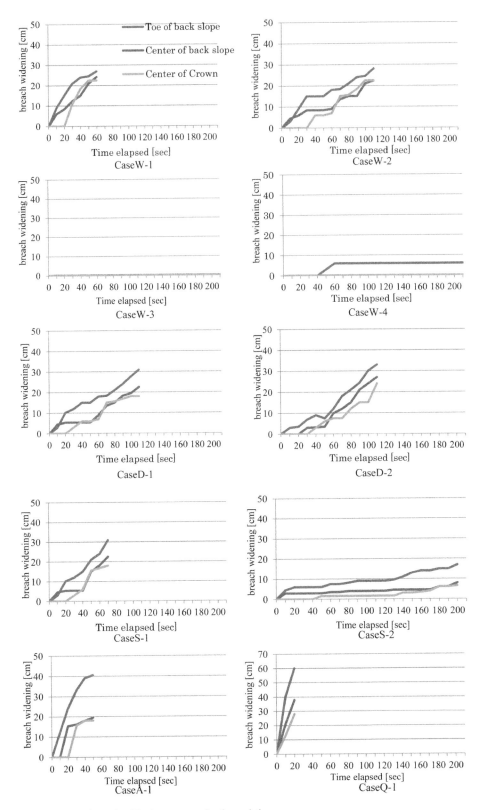

Figure 4.   Levee breach widening process in elapsed time.

However, the levee collapse time in the case S-2, blending sea sand with soil gardening, was the longest in the case that the levee had collapsed. The reason can be attributed to sawdust contained in soil gardening. The levee collapse time of the case A-1, using antibacterial soil, and case Q-1, blending silica soil with DL clay, were shorter than the others. The reason is that antibacterial soil was washed away in an early stage because it was mainly composed of pumice. Whereas, case Q-1 was considered that adhesive strength was small because of a small portion of DL clay.

3.2 *Widening process of the levee breach*

Figure 4 shows widening processes of the levee breach. Widening of the levee breach begins from the toe of back slope, followed by center of back slope and center of crown in all cases. In case W-1 and case W-2, widening of toe of back slope, center of back slope and center of crown began to progress gradually, but in the case of case W-3 and case W-4, not collapsed, breach widening hardly developed at 3 points, namely toe and center of back slope and center of crown. Widening process of case D-1, case D-2 and case S-1 was similar to the case W-1 and case W-2. In case S-2, not only the widening process itself but also reaching the levee collapse to the crown was slow, and the final breach width was the narrowest in all cases. The reason can be attributed to sawdust contained in soil for horticultural had preventing the breach progress. In the case of short collapse time such as case A-1 and case Q-1, collapse of the toe of back slope begins at the same time as over-flow begins and widen in a short time. In addition, when widening process of toe of back slope progressed in a short time, both center of back slope and of crown also progressed in a short period.

4   CONCLUSIONS

Construction of levee material used by this study, were no major differences in the grain size distribution. However, differences the collapse time and the breach widening process. In the future, it is necessary to study characteristics of the soil in more detail such as adhesive strength. It is important to examine the scale effect of the model by comparing with the other researchers results in a field or large scale experiment.

REFERENCES

Kakinuma et al. 2013. Levee breach observation at Chiyoda experimental flume. ISBN 978-1-138-00062-9.
Muramoto et al. 1981. Studies on Safety Evaluation of River Levee and Flood Fighting Technology.
Simada et al. 2008. Basic Study on Sediment Behavior in Chiyoda Experimental Channel. *Journal of applied mechanics JSCE* Vol.11:699–707.
Simada et al. 2009. Cross-Levee Breach Experiment by Overflow at the Chiyoda Experimental Channel. *Annual Journal of Hydraulic Engineering, JSCE* Vol.53:871–876.
Simada et al. 2010. Levee Breach Experiment by Overflow at the Chiyoda Experimental Channel. *Annual Journal of Hydraulic Engineering, JSCE* Vol.54:811–816.
Simada et al. 2011. Experiment for the Destructive Mechanism of the Overflow Levee and Flooding Area at the Chiyoda Experimental Channel. *Annual Journal of Hydraulic Engineering, JSCE* Vol.55:841–846.

Hydraulic Engineering III – Xie (Ed)
© *2015 Taylor & Francis Group, London, ISBN 978-1-138-02743-5*

Velocity distribution and characteristics in unsteady open-channel flow over rough bed

Ai-Xing Ma
Nanjing Hydraulic Research Institute, State's Key Laboratory of Hydrology Water Resources and Hydraulic Engineering Science, Nanjing, China

Yong-Jun Lu
Nanjing Hydraulic Research Institute, Nanjing, China

Min-Xiong Cao
Nanjing Hydraulic Research Institute, Construction Headquaters of Deep-water Channel of the Yangtze River below Nanjing City, Nanjing, China

Xiu-Hong Wang & Ting-Jie Huang
Nanjing Hydraulic Research Institute, Nanjing, China

ABSTRACT: The synchronous real-time vertical velocity is measured with Particle Image Velocimetry in unsteady open-channel flow over fixed rough bed. This paper discusses the vertical velocity distribution and velocity variation in unsteady open-channel flow. Test data analysis shows that the mean velocity distribution in unsteady open-channel flow over rough bed still obeys log-wake law, while parameter B_r and wake-strength parameter Π vary with the hydrograph, differing from those in uniform flow conditions. Parameter B_r varies in the range of ±13% on the basis of uniform flow value 8.88. The test range of wake-strength parameter Π is 0.11~0.18, and the average value is 0.00~0.05. For the velocity magnitude in outer region, the deviation from the log-law is not obvious. Velocity difference Δu at different heights from bed can be divided into three sections according to relative distance y/h_b which takes 0.045 and 1.0 as dividing points. The flow velocity reaches its maximum value before the water depth does in unsteady flow, causing the velocity in rising branch larger than that in falling branch under the same water depth condition. The time gap between velocity peak and water depth peak increases with the enhancement of the unsteadiness parameter P.

1 INTRODUCTION

Under natural conditions, water flow in open channel shows unsteadiness, especially the flood-wave affected by heavy rains during flood season, and daily regulated unsteady flow affected by power generation of hydropower stations. The unsteadiness of flow is strong, thus river sediment movement is different from that in steady flow conditions (Sutter 2001, Kuhnle 1992, Ma 2012). Study of the vertical velocity distribution and variation of unsteady flow in open channel flow is the base to determine the friction factor and bed shear stress of unsteady flow, and the base to predict sediment incipient motion, transport and riverbed evolution under the action of unsteady flow as well. However, since the measurement of unsteady flow requires more advanced instrument and more complex experimental methods, research on the unsteady flow in open channel is more difficult than that in steady flow, therefore, research progress relatively lags behind. The mean velocity distribution in steady open-channel flow obeys log-wake distribution. With the increase of Reynolds number, velocity distribution in outer region deviates from the log-law distribution, and the deviation value can be expressed by Coles' wake function (Nezu 1995). Nezu and Nakagawa use two-component

LDA to measure the velocity distribution of unsteady flow on smooth bed, and they confirm velocity distribution in rising and falling period still obeys the log-law distribution, and explore the change rule of relevant parameters in the formula. Liu C.J. also conduct similar research on velocity distribution of unsteady flow on smooth bed. Tu and Graf, Song and Graf use micro-propellers and Acoustic Doppler Velocity Profiler respectively to measure velocity distribution of unsteady flow on gravel bed surface, and find that velocity distribution of unsteady flow can be expressed by log-law distribution or wake function. Instruments used in previous studies can only measure single point, so each point in vertical line cannot be measured synchronously. Meanwhile, rough bed surface has the problem of theoretical zero, and its velocity distribution is more complicated than that of smooth bed. In this paper, the synchronous real-time vertical velocity in unsteady open-channel flow over fixed rough bed is measured with PIV (Particle Image Velocimetry) and its variations are discussed as well.

2 A BRIEF INTRODUCTION TO SYNCHRONOUS TEST OF UNSTEADY FLOW AND FLOW CONDITIONS

Vertical velocity measurement of unsteady flow is carried out on slope flume which is 40 m long, 0.8 m wide and 0.8 m deep. The test section is 15 m long, about 15 m from sink entrance on the front, and 10 m from the tail-gate. Uniform gravel particles with size of 3.2 mm are pasted closely within the test section. The same gravel particles are also pasted on upstream and downstream fixed bed of test section, which can ensure that water flow develops fully before entering the test section and reduce the influence of backwater on test section in tail gate. Five automatic water level meters (W1–W5) are laid along the flume to observe change process of water level. PIV system is laid 0.3 m away from the middle of test section (W3) on upstream to measure velocity changes. In the experiment, bottom slope of the flume keeps unchanged at 2.65‰, and grille of tail-gate remains open. Flume test arrangement is shown in Figure 1.

Synchronous test system of unsteady flow is constructed to measure real-time changes of hydraulic elements synchronously. Water flow in flume entrance can be controlled by adjusting openings of electric actuators; discharge value can be measured by ultrasonic flowmeter; water level and water depth can be measured by automatic water level meter; real-time vertical velocity can be measured by PIV synchronously. PIV system adopts two-dimensional PIV velocimeter made by Dantec Dynamics in Denmark, which can record the relevant information of the whole flow filed at the same time, and has the characteristics of high spatial resolution, access to large amount of information, and continuous measurement without interference. When PIV works, laser shoots vertically into water from water surface. Through the test, the system can generate stable and repeatable process of unsteady flow, and collect the changes of velocity and water level (or depth) synchronously and instantaneously at the same time.

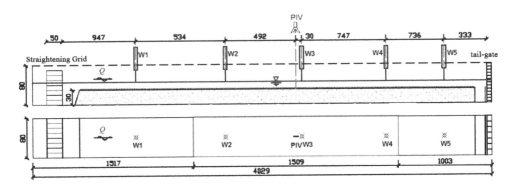

Figure 1. Layout of flume experiment (unit: cm).

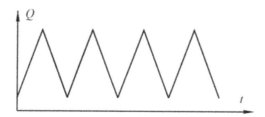

Figure 2. Triangular hydrographs of experiment.

Table 1. Hydraulic conditions for unsteady flow experiments.

Test	Wave shape	T (s)	h_b (cm)	B/h_b	Q_b (l/s)	V_b (m/s)	R_{eb} (10^5)	F_{rb}	h_p (cm)	B/h_p	Q_p (l/s)	V_p (m/s)	R_{ep} (10^5)	F_{rp}	P
					Base flow							**Peak flow**			
UF1	Continuous	72	11.0	7.3	43.9	0.50	1.68	0.48	14.8	5.4	118.6	1.00	4.54	0.83	0.43
UF2		80	10.8	7.4	44.2	0.51	1.69	0.50	13.4	6.0	103.8	0.97	3.97	0.85	0.23
UF3		88	10.3	7.8	44.0	0.53	1.68	0.53	12.4	6.4	89.7	0.90	3.43	0.82	0.18
UF4		96	10.8	7.4	43.7	0.51	1.67	0.49	12.7	6.3	80.2	0.79	3.07	0.71	0.16
UF5		60	10.6	7.6	44.6	0.53	1.71	0.52	13.4	6.0	103.2	0.96	3.95	0.84	0.36
UF6		70	11.7	6.8	44.2	0.47	1.69	0.44	14.4	5.6	102.8	0.89	3.93	0.75	0.28
UF7		90	11.5	6.9	44.2	0.48	1.69	0.45	14.5	5.5	103.5	0.90	3.96	0.75	0.26
UF8	Single	60	8.8	9.1	43.4	0.61	1.66	0.66	13.1	6.1	103.6	0.99	3.96	0.88	0.79

Note: h is water depth, B is flume width, Q is discharge, V is average velocity of section, R_e is Reynolds number ($=4hV/v$), v is kinematic viscosity, F_r is Froude number ($=V/(gh)^{1/2}$), T_r is the duration (s) in rising period of unsteady flow, index b and p is base flow and peak flow, P is unsteadiness parameter (Ma 2012) proposed by the author, and the expression is as follows:

$$P = \left[1000BT_r/Q_p - Q_b\right]\left(h_p - h_b/T_r\right)^2 Q_p \neq Q_b \qquad (1)$$

The actual hydrographs (the changes of discharge Q with time t) in open channel unsteady flow is relatively complex, and flow change rate in rising and falling periods (dQ/dt) changes with the time. In order to simplify the research problem, dQ/dt in rising or falling period keeps constant for the same unsteady flow waveform, i.e., triangular waveform (Fig. 2) is adopted in the entrance of the unsteady flow process. The test mainly uses continuous triangle wave, and a set of single-wave experiments have been carried out for comparison. 8 pairs of trials are conducted, flow condition for each test is listed in Table 1, the ratio of width and depth for peak flow B/h_p is 5.4~6.4, and vertical velocity distribution is not affected by flume wall.

3 THE VELOCITY DISTRIBUTION OF UNIFORM FLOW ON ROUGH BED

When the roughness of the $k_s^+(= k_s U_*/v)$ is more than 70, velocity distribution is only affected by roughness height of the bed k_s, and this is called rough bed. For the inner region of rough bed ($y/h < 0.2$) velocity distribution obeys the log-law:

$$\frac{u(y)}{U_*} = \frac{1}{\kappa}\ln\frac{y_T + y_0}{k_s} + B_r \qquad (2)$$

where, y is the distance from test point to theoretical zero, $y = y_T + y_0$, y_T is the distance from test point to the top of the grains of the bed, y_0 is the distance from theoretical zero ($u = 0$) to the top of the grains of the bed, $u(y)$ is the average velocity of test point, U_* is friction

17

Table 2. Hydraulic conditions for uniform flow experiments.

Parameter	Test						Average value
	SF1	SF2	SF3	SF4	SF5	SF6	
h (cm)	6.90	8.79	10.98	12.41	13.85	15.15	/
Q (l/s)	29.8	44.2	60.3	74.7	88.2	103.4	/
V (m/s)	0.54	0.63	0.69	0.75	0.80	0.85	/
S_f (‰)	2.69	2.61	2.62	2.65	2.72	2.74	/
R_e (10^5)	1.14	1.69	2.31	2.86	3.38	3.96	/
F_r	0.66	0.68	0.66	0.68	0.68	0.70	/
U_* (cm/s)	3.94	4.37	4.76	4.82	4.92	5.71	/
B_r	8.79	9.12	8.27	9.31	9.18	8.58	8.88
Π	0.050	0.054	0.021	−0.018	0.111	−0.066	0.025

Note: S_f is energy slope.

velocity, k is karman constant (0.41), k_s is roughness height of the bed, uniform gravel usually takes its size, B_r is constant.

For outer region ($y/h > 0.2$), it is generally believed that velocity deviates from log-law distribution and wake region exists, its velocity distribution can be expressed as follows:

$$\frac{u}{U_*} = \frac{1}{\kappa} \ln \frac{y_T + y_0}{k_s} + B_r + \frac{2\Pi}{\kappa} \sin^2 \left(\frac{\pi}{2} \frac{y_T + y_0}{h} \right) \qquad (3)$$

where, Π is wake flow strength coefficient.

Velocity distribution parameter B_r and Π in log-law wake distribution obtained in uniform flow test can be seen in Table 2. Obviously, velocity distribution for uniform flow B_r is 8.27~9.31, and the mean value is 8.88; Π is −0.066~0.111, and the average is 0.025. Judging from Π, the measured velocity deviation from log-law distribution is not obvious in outer region. For fixed rough bed, B_r in Keulegan's experiment is 8.5, and B_r is 8.47 in Kironoto and Graf's gravel bed experiment with k_s=2.3 cm. B_r value for rough bed in this experiment is consistent with that of Keulegan, and Kironoto & Graf, which confirms the reliability of test methods and instrument.

4 VELOCITY DISTRIBUTION OF UNIFORM FLOW ON FIXED ROUGH BED

In each test, the minimum rough scales k_s^+ ($= k_s U_* / v$) are greater than 70, which shows the bed surface is rough. Figure 3 shows the vertical velocity distribution for each featured point in rising and falling period gathered in test UF5. t/T_r=0 and 1 are the velocity distribution for base flow and peak flow respectively. In order to show velocity distribution for different time clearly, velocity in rising period increases by 0.05, 0.1, 0.15, …, 0.45 m/s from the original value respectively, and velocity in falling period increases by 0.45, 0.40, 0.35, …, 0.05 m/s in turn. Based on the viewpoint that velocity distribution for inner region ($y/h < 0.2$) in uniform water flow obeys the log-law distribution, the log law matches inner region velocity distribution for different time in unsteady flow (Fig. 3), and the correlation coefficients are greater than 0.99. It is obvious that in the inner region of unsteady flow, velocity distribution in rising and falling periods still obeys log-law distribution, but parameters B_r and Π changes with the rising and falling process, which differ from those in uniform water flow.

Figure 4 shows part of test parameters B_r change with rising and falling process. Judging from the overall trend, B_r decreases gradually with the increase of water level, reaches the bottom at water peak level; B_r increases gradually to the original value at the beginning of the rising period with the decrease of water level; Judging from the values, vertical velocity distribution parameter B_r changes in a certain range: B_r changes from 7.99 to 10.57 in

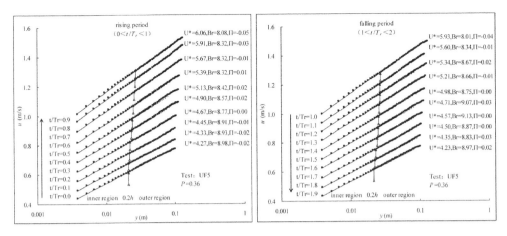

Figure 3.　Velocity distribution for experiment UF5 at rising and falling branch.

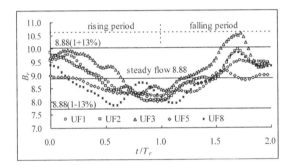

Figure 4.　B_r against normalized time t/T_r.

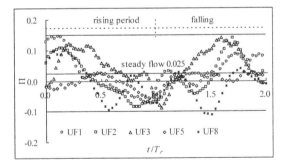

Figure 5.　Π against time t/T_r.

the test, which is 13% more or less than the value of B_r in uniform water flow (8.88), i.e. $B_r=8.88(1 \pm 13\%)$.

Figure 5 reveals that wake strength coefficient Π changes along with rising and falling process. It can be seen that the value and change range of wake strength coefficient Π obtained in unsteady flow tests are small: Π changes from −0.11 to 0.18 in each test, and the average of Π is 0.00 to 0.05. In the outer region, velocity deviation from the log-law distribution is not obvious. Test shows that in the rising period, wake flow coefficient Π has a peak value, reaches the bottom at water level peak, fluctuates in falling period, and its changes are rather complicated.

5 VELOCITY VARIATION OF UNIFORM FLOW

Figure 6 shows mean velocity from varying heights above the bed changes with time in test UF5, and the number is the height above the bed (mm). It is clear that velocity is small near bottom, but the velocity gradient is large, whereas velocity is larger away from bed surface, but the velocity gradient is smaller; The trend of velocity variation is consistent for the hydrograph at different depth in the same unsteady flow, and reach the velocity peak synchronously, but earlier than peak of water depth. Through further analysis on changes of velocity difference at different heights Δu ($= u(y)_{max} - u(y)_{min}$) along with the relative distance from bed surface y/h_b in a unsteady flow cycle (Fig. 7), it can be seen that, velocity changes gradually with small difference because of the influence of rough bed near the bottom; after leaving the bottom for a certain distance, generally $y/h_b > 0.045$, influence of rough bed decreases quickly, velocity difference increases rapidly, that is, the gradient of Δu near the bottom is relatively large, then Δu increases gradually; when $y > h_b$, since the location is in water depth changing region where part of the water did not reach the area, Δu decreases rapidly. Overall, changes of Δu along with y/h_b can be divided into three sections (Fig. 7): the first section ($y/h_b < 0.045$) near the bottom, the second section ($0.045 < y/h_b < 1.0$), the third section ($y/h_b > 1.0$) in water level changing area, changes of Δu along with y/h_b in three sections conform to log-law.

Figure 8 shows the relationship between the vertical average velocity and water depth, it can be seen vertical average velocity in rising period is greater than that in falling period under the same water depth condition, i.e., velocity arrives at the peak ahead of water depth. In the rising and falling period of unsteady flow, water depth and velocity show the relationship of counterclockwise loop curve. From the width of the noose, UF8 (P=0.79) with strong unsteadiness is wide, while UF3 (P=0.18) with weak unsteadiness is narrow. The width of noose relationship between water depth and mean velocity distribution reflects strength of asynchrony between two peak values. Further analysis on the relationship between relative time $\Delta t/T_r$ (time that peak value of velocity arrives ahead of peak value of water depth) and unsteadiness parameter P (Fig. 9) shows that $\Delta t/T_r$ increases along with the increase

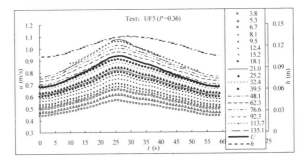

Figure 6.   Time-variation of velocity at different heights from bed (UF5).

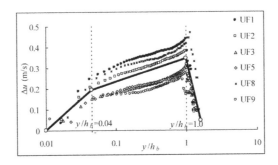

Figure 7.   Velocity difference Δu at different heights.

20

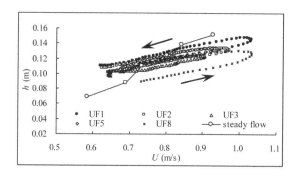

Figure 8.   Loop property of water depth and velocity.

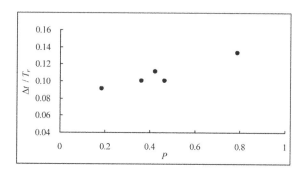

Figure 9.   Relative time gap $\Delta t / T_r$ against P.

of P, namely the stronger the unsteadiness of unsteady flow, the earlier velocity arrives at the maximum than water depth, and the wider the loop curve between water depth and velocity.

6   CONCLUSION

Hydraulic elements of unsteady flow in open channel change with time and space, and PIV is applied to measure synchronously and instantaneously vertical velocity on fixed rough bed in rising and falling process.

1. The experiment reveals that velocity distribution of rough bed in unsteady flow in open channel still obeys the log-wake law, but in the formula, parameter B_r and wake strength coefficient Π change along with the rising and falling process, different from those in uniform flow conditions. Parameter B_r decreases gradually as the water level rises, and reaches the lowest point in the peak of water level; along with the falling of water level, B_r increases to the value at the beginning of rising period, the range for B_r in the test is 7.99~10.57 and B_r=8.88 (1 ± 13%); the range for wake strength coefficient Π is −0.11~0.18 with the mean value of 0.00~0.05, indicating that velocity deviation from the log-law distribution in outer region is not obvious.
2. The trend of vertical velocity variation is consistent at different height above bed surface, and the velocity reaches the maximum synchronously.
3. Velocity difference Δu can be divided into three sections by relative distance y/h_b=0.045 and y/h_b=1.0. Changes of Δu in three sections along with y/h_b all obeys log-law variation.
4. Under the influence of unsteady flow, velocity in rising period is greater than that in falling period at the same water depth, that is, velocity peak arrives ahead of peak of water depth, and time lag increases with the intensity of unsteady flow.

REFERENCES

De Sutter R. & Verhoeven R & Krein A. 2001. Simulation of sediment transport during flood events: laboratory work and field experiments. *Hydrological sciences journal* 46(4):599–610.

Keulegan G.H. 1939. Laws of turbulent flow in open channels. *Journal of the Franklin Institute* 227(1):119–120.

Kironoto B.A. & Graf W.H & Reynolds. 1994. Turbulence characteristics in rough uniform open-channel flow. *Proceedings of the ICE—Water Maritime and Energy* 106(4):333–344.

Kuhnle R.A. 1992. Bed load transport during rising and falling stages on two small streams. *Earth Surface Processes and Landforms* 17(2):191–197.

Liu C.J. 2004. Experimental study on unsteady open-channel flow and bed-load transport properties. *Tsinghua University*.

Ma A.X. & Liu S.E. & Wang X.H. 2012. Near-dam reach propagation characteristics of daily regulated unsteady flow released from Xiangjiaba Hydropower Station. *Proceedings of the 2012 SREE conference on hydraulic engineering*: 103–110.

Ma A.X. & Lu Y. & Lu Y.J. 2012. Advances in velocity distribution and bed-load transport in unsteady open-channel flow. *Advances in Water Science* 23 (1):134–14.

Nezu I & Nakagawa H. 1995. Turbulence measurements in unsteady free-surface flows. *Flow measurement and instrumentation* 6(1):49–59.

Nezu I & Rodi W. 1986. Open-channel Flow Measurements with a Laser Doppler Anemometer. *Journal of Hydraulic Engineering* 112(5):335–355.

Song T. & Graf W.H. 1996. Velocity and turbulence distribution in unsteady open-channel flows. *Journal of Hydraulic Engineering* 122(3):141–154.

Tu H. & Graf W.H. 1992. Velocity distribution in unsteady open-channel flow over gravel beds. *Journal of Hydroscience and Hydraulic Engineering* 10(1):11–25.

Hydraulic Engineering III – Xie (Ed)
© 2015 Taylor & Francis Group, London, ISBN 978-1-138-02743-5

Experimental study of Manning's roughness in sparse planted open channel flow

Yan Hong Li
School of Naval Architecture, Ocean and Civil Engineering, Shanghai Jiaotong University, Shanghai, China

Li Quan Xie
Department of Hydraulic Engineering, College of Civil Engineering, Tongji University, Shanghai, China

ABSTRACT: The plant branches with multi-leaf were used in an indoor flume to simulate the submerged plant in open channel flow. The flow velocities were measured for 27 tests under sparse branch densities. Manning's roughness coefficient and the plant's blockage effect on it were estimated based on the Cowan's empirical formula. The dependence of Manning's n on the composite plant blockage factor, being defined as a function of the projection area and the submerged volume of plants, was analysed. The result shows that Manning's roughness coefficient presents logarithmic relationship with this blockage factor.

1 INTRODUCTION

Aquatic plants result in significant increase of resistance in open channel flows due to the blockage effect of plant stands and the skin friction effect of their leave. However, how much dose the plant contribute the boundary roughness cannot be exactly calculated due to the complexity of the interaction between the plant and the water flow. For predicting flow discharge and water level in design of man-made drainage channels, management of biological rivers and flood prediction of lowland area, Cowan's empirical adjustment factors (Cowan, 1956) for estimating Manning's n-values have been widely used since 1960's.

The well-known Manning's equation can be written as:

$$V = \frac{1}{n} R^{2/3} S^{1/2} \tag{1}$$

where V is the cross-sectional averaging flow velocity (m); R is hydraulic radius (m); and S is the friction slope (dimensionless).

The n-value can be determined after the V, R, and S are known. For uniform, steady and smooth channel, there have been quite accurate empirical n-values for researchers and engineer to select. For rough channel flow, however, the variables in Eq. (1) are affected by some physical factors and processes derived from fluid mechanics (Yen, 1991, 2002). On the one hand, in site descriptions with photographs of the main physical factors and processes, including size and type of the bed and bank materials and their distribution, longitudinal variation in bed profile and cross-sectional shapes (form resistance) were given based on a large amount of data and picture collection (e.g. Barnes, 1967; Hicks and Mason, 1998; Yen, 1991, 2002; Soong, et. al. 2012). On the other hand, these physical factors and processes were organized as parameters representing the roughness characteristics (e.g. Cowan, 1956). Cowan (1956) organized the roughness characteristics in Manning's n formula through 5 adjustment factors as follows:

$$n = (n_0 + n_1 + n_2 + n_3 + n_4)m_5 \tag{2}$$

where n_0 is a basic n-value for a straight, uniform, smooth channel;

n_1 is the adjustment factor for the effect of surface irregularity;

n_2 is the adjustment factor for the effect of variation in shape and size of the channel cross section;

n_3 is the adjustment factor for obstruction;

n_4 is the adjustment factor for vegetation;

and m_5 is a correction factor for meandering channels.

Magnitude of each adjustment factor in Eq. (2) for various classified groups of roughness has also been given with examples (descriptions) by Cowan (1956). Manning's n-values can be obtained by comparing method, i.e. by comparing the roughness characteristics observed during measurements and the adjustment factors in available manuals.

Variations in n-values with flow discharge or velocity increase the complexity of comparing method. Multiple n-values are usually available for the same site at varying flow discharges or velocities especially for the vegetated river channels. It is more difficult to determine the Manning coefficient for vegetated streams than for open channel flow (Chow, 1959). Ebrahimi et al. (2008) conducted experiments at low channel slope to demonstrate variation of Manning's n-value and shear velocity with flow velocity for low submerged vegetation. The n-value and shear velocity were shown to be very sensitive at low Froude numbers. Vegetation on the bed is the main roughness feature in shallow river channel for that it produces more turbulence and drag than normal concrete, clay and sediment bank. For the submerged condition, many experiments have been conducted (e.g. the U.S.A Department of Agriculture) to estimate vegetation roughness in small channels and crop furrows. The result of the work was the construction of a series of n-VR curves to relate the Manning roughness coefficient (n) to velocity (v) and hydraulic radius (R) for submerged crops (Kouwen, 1992). Wu (Wu F.C., and Shen H.W., 1999) experimentally studied the variation of Manning's n-value with stage for an artificial form of vegetation. Many studies show that Manning's n-value tends to converge with the increase of VR value (e.g. Kim, et al., 2010). But these results can hardly be used to another situation because vegetation's complex shape, distribution, different bending degree and wavy motion with the varying flow discharge or velocity. Fathi-Moghadam, et al. (2008) used a dimensional analysis supported by experimental results to develop a relationship to estimate Manning's n-value for the submerged vegetative zones of rivers and flood plains under variable flow and vegetation conditions. Their model is able to account for effects of stream flow velocity and depth, as well as density and flexibility of vegetation. However, there remain uncertainties using analytical methods. From a practical viewpoint, water level and discharge as variables computed by numerical model are influenced by uncertainty in estimating the roughness coefficient (Kim, et al., 2010).

This study conducts experiments in an indoor flume aiming to evaluate the relationship between the Manning's n-value and the roughness parameters in vegetated open channel flow. The volume of the submerged vegetation, frontal area and the horizontal projection area of the vegetation are considered as the main roughness parameters. The relationship between these parameters and the Manning's n-value are analyzed.

2 LABORATORY EXPERIMENTS AND DATA ANALYSIS

The experiments were conducted in a 60 m long by 1 m wide indoor circulating flume (as schematically shown in Fig. 1). The sidewalls of the flume compose of steel frames and embedded glass windows for observation. The deck of the flume was made of concrete and its slope was fixed. The experimental zone was above the deck. And beneath the deck was a rectangular corridor for circulating the experimental water flow. These two parts that are above and beneath the deck have the same length and width. The water flow in the flume was driven by two water pumps located in the circulating corridor beneath the deck.

The branches with green leave of a kind of evergreen trees were selected as an alternative of plastic model of vegetation. The tree branches were fixed to the steel bars and then put on the deck of experimental flume and immersed by the water within 2 hours after being cut

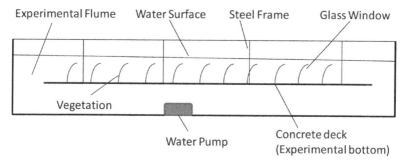

Figure 1. Schematic figure of experiment flume.

Figure 2. Photo of the experiment.

from the trees. It was assumed that the branches and leaves didn't wither during each set of test which lasted no more than 8 hours (Fig. 2).

All the tests were conducted in under-critical flow conditions (Fr <1). 3 water depths were tested. For each water-depth condition, several flow discharges or cross-sectional averaged flow velocities were obtained through adjusting the power of the driving water pumps. Flow velocity was measured by the propeller velocimeter. For each test, the water pumps kept at a constant power value so that the flow can be considered steady. Several stem densities of vegetation were used for each flow depth and velocity condition.

3 DATA ANALYSIS

In this study, the base factor of Manning's n formula expressed in Eq. (2), n_0 was tested for each water depth and condition on bare flume bottom. The adjustment factors n_1 and n_2 were considered to be zero. The adjustment factor n_3 for obstruction roughness were considered as the steel bars' contribution and tested before each set of test for every flow depth, flow velocity and stem density of vegetation. The correction factor for meandering channels m_5 was considered to be zero. Thus Eq. (2) can be written as:

$$n = n_0 + n_3 + n_4 \qquad (3)$$

Manning's n-value was determined from Eq. (1) for vegetated water flow. The adjustment factor n_4 can be obtained from Eq. (3).

To study the effect of vegetation element on water flow roughness, the frontal area of vegetation A_f, the parietal area of vegetation A_p on per unit area of the flume bottom, and the ratio of the volume of vegetation to the water column V_v were selected as the main roughness parameters. The experimental results showed that n_4 and the composite factor, $\alpha V_v^{\beta(A_f/A_p)}$

25

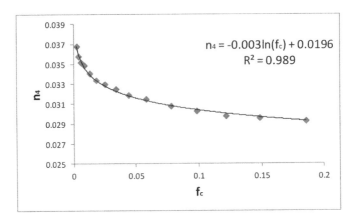

Figure 3. Relationship curve of n_4—f_c.

(α and β are two rational factors which need to be determined according to the experiments; in this study $\alpha = 1$, and $\beta = 1$) presented nearly a logarithmic relationship (Fig. 3).

Figure 3 shows that the adjustment factor of vegetation, n_4 depends only on the composite factor $\alpha V_v^{\beta(A_f/A_p)}$ for all water depths and flow velocities. It is implicated that the variation in flow resistance brought by the flow depths and flow velocities may be balanced by the deformation of flexible vegetation. The streamlined deformation of the vegetation leads to a reduction of turbulence and thus the flow resistance.

ACKNOWLEDGEMENT

The writer would like to thank the National Science Foundation Council of China for the financial support under grant 51109130.

REFERENCES

Barnes, H.H., 1967. Roughness characteristics of natural channels; U.S. Geological Survey Water-Supply Paper 1849, 213 p.

Chow V.T., 1959. Open channel hydraulics, McGrawHill, New York.

Cowan, W.L., 1956. Estimating hydraulic roughness coefficients: Agricultural Engineering, 377, 473–475.

Ebrahimi, N.G., Fathi-Moghadam, M., Kashefipour, M., Saneie, M., and Ebrahimi, K., 2008. Effects of flow and vegetation states on river roughness coefficients, Journal of Applied Sciences and Environmental Management, 8(11), 2118–2123.

Green J.C., 2005. Modelling flow resistance in vegetated streams: review and development of new theory, Hydrological Processes 19, 1245–1259.

Hicks, D.M., and Mason, P.D., 1998. Roughness characteristics of New Zealand rivers: National Institute of Water and Atmospheric Research Ltd.. Water Resources Publications, LLC, 329 p.

Kim J.S., Lee C.J., Kim W., And Kim Y.J., 2010. Roughness coefficient and its uncertainty in gravel-bed river, Water Science and Engineering, 3(2), 217–232.

Kouwen N., 1992, Modern approach to design of grassed channels. Journal of Irrigation and Drainage Engineering, ASCE, 118(5), 733–743.

M. Fathi-Moghadam, 2007. Characteristics and mechanics of tall vegetation for resistance to flow, African Journal of Biotechnology, 6(4), 475–480.

Soong D.T., Prater C.D., Halfar T.M., and Wobig L.A., 2012. Manning's Roughness Coefficients for Illinois Streams, Report.

Wu F.C., Shen H.W., Chou Y.J., 1999. Variation of roughness coefficients for unsubmerged and submerged vegetation, Journal of Hydraulic Engineering 125(9), 934–942.

Yen, B.C., 1991. Hydraulic resistance in open channels, in Yen, B.C., ed., Channel flow: centennial of Manning's formula: Water Resources Publications, Littleton, Colo. 1–135.

Yen, B.C., 2002. Open channel flow resistance, Journal of Hydraulic Engineering, ASCE, 128(1), 20–39.

Hydraulic Engineering III – Xie (Ed)
© 2015 Taylor & Francis Group, London, ISBN 978-1-138-02743-5

Lateral mixing of flows with and without non-submerged vegetation in depth-averaged 2D scheme

Tetsuro Tsujimoto, Ho-Seong Jeon & Makiko Obana
Department of Civil Engineering, Nagoya University, Japan

ABSTRACT: Recently depth-averaged 2D analysis of flow and fluvial process has become popular. In vegetated area, the form drag is taken into account, but frictional resistance law is conventionally assumed to the same to flow without vegetation. This paper introduces the concept of bed roughness boundary layer in vegetated area, and based on this concept, not only the velocity distribution but also the suspended sediment concentration profile are discussed. Then, the frictional resistance law for flow in vegetated area is deduced to evaluate the frictional velocity. Furthermore, the kinematic viscosity and the diffusion coefficient of suspended sediment, and the depth-averaged concentration of suspended sediment are discussed to prepare depth-averaged 2D analysis for flow and fluvial processes in streams with vegetation.

1 INTRODUCTION

When flow and fluvial process in a stream are discussed, vegetation is a key from the view point of rive ecosystem as well as flood management. Recently, 2D depth averaged analysis has become popular where vegetation is regarded as dispersive obstacles represented by form drag, and it is expected to apply to description of fluvial process in a stream with vegetation (Tsujimoto 1999).

Usually the flow has diffusivity depending on the bed shear, and even in lateral mixing is governed by shear velocity. In flow of stream with vegetation, flow velocity varies spatially and it brings horizontal mixing in momentum and sediment suspension, and depth-averaged 2D analysis has advantage to apply to such situations.

However, it is pointed that the introduction of the concept of bed roughness boundary layer in vegetated area is necessary to evaluate the shear velocity reasonably when it is applied to description of sediment transport and subsequent fluvial process (Jeon et al. 2014). Flow in area with non-submerged area shows the characteristic velocity U_v determined by balance of form drag and longitudinal component of the gravity, and only a thin layer near the bed it shows a shear flow and it is related to the bed shear stress. It suggest the frictional resistance law represented by the ratio of the depth-averaged velocity and the shear stress $C_f^{1/2}$ in the vegetated area must be quite different from that derived by the logarithmic law (Keulegan's equation), though in a conventionally it has been applied to vegetated area as well as non-vegetated area. Jeon et al discuss the thickness of the boundary layer to propose the frictional resistance law in the vegetated area and the proposed model gives a reasonable estimation of the shear stress and bed load transport rate.

2 VELOCITY DISTRIBUTIONS AND SUSPENDED SEDIMENT CONCENTRATION PROFILES UNDER EQUILIBRIUM IN AREAS WITHOUT AND WITH NON-SUBMERGED VEGETATION

2.1 *Velocity distribution in bed roughness boundary layer in vegetated area*

Flow without vegetation can be described by the logarithmic law, and the vertical distribution of velocity $u(z)$ and the frictional resistance law known as "Keulegan's equation" is written as follows:

$$\frac{u(z)}{u_*} = \frac{1}{\kappa} \ln\left(\frac{z}{k_s}\right) + B_s(R_{e*}) \tag{1}$$

$$\frac{U}{u_*} = \frac{1}{\kappa} \ln\left(\frac{h}{k_s}\right) + B_s(R_{e*}) - \frac{1}{\kappa} \tag{2}$$

where $u(z)$: vertical distribution of velocity; $u_* =$ shear stress; $U =$ depth averaged velocity; $h =$ depth; $\kappa =$ Karman's constant; $k_s =$ equivalent sand roughness, $B_s(R_{e*}) =$ function of roughness Reynolds number $R_{e*} = u_*k_s/v$. and $v =$ kinematic viscosity.

While, the vertical distribution of velocity of flow in the area with non-submerged vegetation is written as follows according to the recent study by (Jeon et al. 2014).

$$\frac{u(z)}{u_{*v}} = \frac{1}{\kappa} \ln\left(\frac{z}{k_s}\right) + B_s(R_{e*}) \quad (z < \theta_v); \quad U_v = \sqrt{\frac{2gI_e}{C_D \lambda D}} \quad (\theta_v < z < h) \tag{3}$$

where vegetation is represented by a group of piles $D =$ diameter of a pile; $\lambda =$ number density of piles; $\theta_v =$ bed roughness boundary layer thickness; $g =$ gravitational acceleration; $I_e =$ energy gradient, and $C_D =$ drag coefficient. θ_v is related to the vegetation characteristics, and Jeon et al. formulates according to the measured data by (Liu et al. 2008) in the following equation.

$$\frac{\theta_v}{h} = \frac{0.008}{0.008 + \lambda Dh} \tag{4}$$

And, subsequently the following frictional resistance law is proposed (Jeon et al. 2014):

$$\frac{U}{u_*} = \frac{(h - \theta_v)U_v}{h} + \frac{\theta_v}{u_*} + \frac{\theta_v}{h}\left[\frac{1}{\kappa} \ln\left(\frac{\theta_v}{k_s}\right) + B_s(R_{e*}) - \frac{1}{\kappa}\right] \tag{5}$$

Based on the above velocity distributions, the kinematic eddy viscosity is discussed. For flow without vegetation, it distributes along the depth and its depth-averaged value v_t is as follows:

$$v_t = \frac{\kappa}{6} u_* h \tag{6}$$

In the vegetated area, the turbulence diffusivity exists only in the bed roughness boundary layer, and the kinematic eddy viscosity will be formulated as follows:

$$v_t = \frac{\kappa}{6} u_* \theta_v \quad (0 < z < \theta_v); \quad v_t = 0 \quad (\theta_v < h) \tag{7}$$

The apparent depth-averaged value of the kinematic eddy viscosity is written as follows:

$$v_t = \frac{\kappa u_* \theta_v^2}{6h} \tag{8}$$

2.2 Suspended sediment concentration profile

When the diffusion coefficient of suspended sediment ε_s is expressed as βv_t, the equilibrium suspended sediment concentration distributes along the depth in the area without vegetation and the profile $c(z)$ is written as follows (Lane & Kalinske 1941) by substituting Equation 6.

$$\frac{c(z)}{C_B} = \exp\left(-\frac{6w_0}{\beta \kappa u_*} \frac{z}{h}\right) \tag{9}$$

28

where w_0 = settling velocity of sand; and β = reciprocal of the Schmidt number, which is evaluated by (Tsujimoto 1984) as follows by comparing the diffusion theory with the stochastic model.

$$\beta = 1 + k_\beta \left(\frac{w_0}{u_*}\right)^2 \tag{10}$$

where $k_\beta = 1.54$. Then, the depth averaged suspended sediment concentration C is written by

$$\frac{C}{C_B} \equiv \frac{6 w_0 h}{\beta \kappa u_*} \left[1 - \exp\left(-\frac{6 w_0 h}{\beta \kappa u_*}\right)\right] \tag{11}$$

On the other hand, suspended sediment distributes only in the bed roughness boundary layer $(0 < z < \theta_v)$ because the turbulent diffusivity never exist in the upper layer $(\theta_v < z < h)$ as follows:

$$\frac{c(z)}{C_B} = \exp\left(-\frac{6 w_0}{\beta \kappa u_*} \frac{z}{\theta_v}\right) \quad (0 < z < \theta_v); \quad \frac{c(z)}{C_B} = 0 \quad (\theta < z) \tag{12}$$

And, the layer-averaged concentrations, C_θ for roughness boundary layer and $C_{h-\theta}$ for the upper layer, are written as follows.

$$\frac{C_\theta}{C_B} \equiv \frac{6 w_0 \theta_v}{\beta \kappa u_*} \left[1 - \exp\left(-\frac{6 w_0 \theta_v}{\beta \kappa u_*}\right)\right] \quad (0 < z < \theta_v); \quad \frac{C_{h-\theta}}{C_B} = 0 \quad (\theta < z) \tag{13}$$

The depth-averaged concentration in the vegetated area is written as follow:

$$\frac{C}{C_B} \equiv \frac{6 w_0 \theta_v^2}{\beta \kappa u_* h} \left[1 - \exp\left(-\frac{6 w_0 \theta_v}{\beta \kappa u_*}\right)\right] \tag{14}$$

The equilibrium bottom concentration, C_{Be}, is related to the shear velocity.

3 LATERAL MIXING BETWEEN FLOW AND SUSPENDED SEDIMENT BETWEEN AREAS WITHOUT AND WITH NON-SUBMERGED VEGETATION

3.1 *Lateral mixing of momentum and lateral distribution of depth-averaged velocity*

The governing equations of flow in depth-averaged 2D scheme are written as follows:

$$\frac{\partial h}{\partial t} + \frac{\partial h U}{\partial x} + \frac{\partial h V}{\partial y} = 0 \tag{15}$$

$$\frac{\partial}{\partial t}(h U) + \frac{\partial}{\partial x}(h U U) + \frac{\partial}{\partial y}(h U V) = -g h \frac{\partial}{\partial x}(h + z_b) + D_x - F_x h - \frac{\tau_{bx}}{\rho} \tag{16}$$

$$\frac{\partial}{\partial t}(h V) + \frac{\partial}{\partial x}(h U V) + \frac{\partial}{\partial y}(h V V) = -g h \frac{\partial}{\partial x}(h + z_b) + D_y - F_y h - \frac{\tau_{by}}{\rho} \tag{17}$$

$$\tau_{bx} = \rho C_f U \sqrt{U^2 + V^2}; \quad \tau_{by} = \rho C_f V \sqrt{U^2 + V^2} \tag{18}$$

$$F_x = \frac{1}{2} C_D \lambda U \sqrt{U^2 + V^2}; \quad F_y = \frac{1}{2} C_D \lambda V \sqrt{U^2 + V^2} \tag{19}$$

where x, y = longitudinal and lateral directions, U, V = depth-averaged velocity in x and y direction; z_b = bed elevation; τ_{bx}, τ_{by} = bed shear stress in x and y directions; D_x, D_y = diffusion terms in x and y directions; F_x, F_y = x and y components of form drag due to vegetation; C_f = friction coefficient given by $(U/u_*)^2$ (Equation 2 for area without vegetation and Equation 5 for area with vegetation). There are some programs based on the above governing equations, but in conventional approaches, the friction resistance law for flow without vegetation is applied even in the vegetated area though the form drag is taken into account for the vegetated area. The authors have derived a reasonable equation of frictional resistance law based on the concept of bed roughness boundary layer in the area with non-submerged vegetation. Applying this law can bring a reasonable evaluation of the shear velocity in the vegetated area and subsequently equilibrium bottom concentration of suspended sediment (entrainment rate is $C_{Be}w_0$).

In lateral mixing, the horizontal diffusion written in Equation 16 is subjected to the eddy kinematic viscosity. Conventionally, the vertical diffusion coefficient v_t which is related to the vertical velocity distribution is applied by simply multiplying a constant even in horizontal mixing as v_{th}, and furthermore, the value given for flow on the area without vegetation, as follows:

$$v_{th} = \alpha u_* h \tag{20}$$

where α is empirically determined in the range 0.1~0.5.

Depth-averaged kinematic eddy viscosity is modified as expressed by Equation 8, and thus the following equation is reasonably employed for flow with vegetation instead of Equation 20.

$$v_{th} = \alpha u_* \theta_v^2 / h \tag{21}$$

3.2 Lateral mixing of suspended sediment and lateral distribution of depth-averaged concentration of suspended sediment

The spatial distribution of depth-averaged concentration is described by the flowing equation:

$$\frac{\partial hC}{\partial t} + \frac{\partial}{\partial x}\left(hCU - \beta v_{th} h \frac{\partial C}{\partial x}\right) + \frac{\partial}{\partial y}\left(hCV - \beta v_{th} h \frac{\partial C}{\partial y}\right) = \Xi_\Theta \tag{22}$$

where Ξ_Θ = difference between deposition and entrainment from the bed to be expressed by $(C_{Be}-C_B)w_0$ and it represents the bed deformation; and the horizontal diffusion coefficient of suspended sediment is assumed to be βv_{th}.

4 LATERAL DISTRIBUTION OF DEPTH-AVERAGED VELOCITY FOR EQUILIBRIUM FLOW IN A STRAIGHT STREAM WITH VEGETATION ZONE

As the simplest example of flow to be treated in 2D scheme, the flow in a straight channel with a zone of non-submerged vegetation is investigated. Far from the boundary between the zones with and without vegetation, the flows with and without vegetation described in Chapter 2 appear without mutual interactions. Without any separation wall to prevent from mixing between two zones, there appears a mixing zone where velocity and suspended sediment concentration changes laterally to connect the characteristic values of respective zones.

In the conventional approach, α is assumed to be appreciably larger than $\kappa/6$ (same to the vertical mixing). With the larger value of α, the wider the mixing width becomes as shown in Figure 1(A). And, the experimental data (Tsujimoto 1996) can be described well by larger value of α (around 0.3) as shown in Figure 1(B). According to the elaborate flow

30

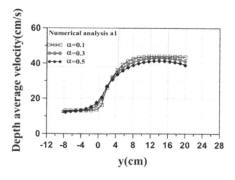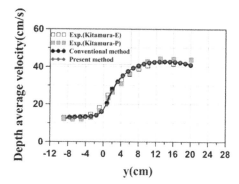

Figure 1. Lateral distribution of depth-averaged velocity: (A) Sensitivity due to α value; (B) Comparison with flume experiment (Tsujimoto 1996).

measurement, migration of horizontal vortex along the interface between two zones is recognized by detecting low-frequency fluctuation of horizontal components of flow (Tsujimoto 1996), and it may cause an increase of horizontal diffusivity.

5 CONCLUSION

In this paper, fundamental differences between vegetated and non-vegetated zones are discussed and in particular the characteristics of flow and sediment suspension are investigated based on the introduction of the concept of bed roughness boundary layer in the vegetated area.

Recently numerical analysis on 2D depth-averaged flow and fluvial process has become familiar, the treatments of flow and sediment transport should be carefully prepared based on the above discussion. Discussion in this paper is limited to 2D behavior of flow and suspended sediment concentration in a straight channel with vegetation zone, further discussion should be developed in more general 2D analysis under more complicated conditions.

REFERENCES

Jeon, H.S., M. Obana & T. Tsujimoto. 2014. Bed roughness boundary layer and bed load transport in vegetated area. *Proc. Conf. on Hydarulic Eng.*, Hong Kong.
Lane, E.W. & A.A Kalinske. 1941. Engineering calculations of suspended sediment in open channels. *Trans. AGU* 22: 307–603.
Liu, D., P. Diplas, J. Fairbanks & C. Hodges. 2008. An experimental study of flow through rigid vegetation. *Journal of Geophysical Research* 113(F4).
Tsujimoto, T. 1986. Sediment transport by turbulence—Diffusion coefficient of suspended sediment. *Proc. 30th Annual Conf. Hydraul.*, *JSCE*: 637–642 (in Japanese).
Tsujimoto, T. 1996. Coherent fluctuations in a vegetated zone of open-channel flow: Cause of bedload lateral transport and sorting, Coherent Flow Structures in Open Channels. edited by P.J. Ashworth, S.J. Bennett, J.L. Best & S.J. McLelland, John Wiley & Sons: 375–396.
Tsujimoto, T. 1999. Flow and fluvial processes in streams with vegetation. *Journal of Hydraulic Research* 37(6): 789–803.

Hydraulic Engineering III – Xie (Ed)
© 2015 Taylor & Francis Group, London, ISBN 978-1-138-02743-5

Transportation and deposition of Particulate Organic Matter (POM) with sediment on vegetated area in a river

Makiko Obana, Ho-Seong Jeon & Tetsuro Tsujimoto
Department of Civil Engineering, Nagoya University, Japan

ABSTRACT: The transport and deposition of Particulate Organic Matter (POM) in river streams has recently received much attention as one of important ecological process in rivers. We focused on interacted behaviors of bed load and POM in vegetated area on sand bars. The purpose of this study is to clarify the characteristics of deposition of POM with suspended sediment on sandbars with riparian vegetation through field observation and laboratory experiments. The main results of this study are that ripples are formed by bed load that was deposited in riparian vegetation. And Coarse POM (CPOM) is captured by trough of ripples formed by wave action. It was clearly showed that the behavior of bed load and that of POM have significant differences under the same condition. Based on these results, the mechanism of POM deposition with formation of ripples in vegetated area was also modeled.

1 INTRODUCTION

River landscapes are characterized by an interrelating system of flow, sediment transport, morphology and vegetation. It is known that suspended sediment and Particulate Organic Matter (POM) transported by flood are captured and deposited on sandbar with riparian vegetation. Capture of POM there must be significant in ecosystem and it is important to understand how POM drifts are different from sediment behavior in riparian vegetation, because it influences vegetation productivity and supports diversity through riverine bio-geochemical processes. The purpose of this study is to clarify the deposition mechanism of POM in consideration of the influence of sediment transport in riparian vegetation. Thus field observation and laboratory experiment were conducted, and then necessary modeling in numerical calculation was also conducted.

2 CHARACTERISTICS OF CPOM BASED ON FIELD OBSERVATION

An investigation was conducted on the Yahagi River (Chubu region, Japan) during normal flow stage. To evaluate the deposition characteristics of various POM with sand on sandbar with riparian vegetation in a field, we selected a conspicuous island sandbar covered with vegetation. The longitudinal length of sandbar was around 300 m, the cross sectional length was around 100 m and the average bed slope was 1/150. The bed was composed mainly cobbles and boulders as basement rock. An herbaceous plant occupied the whole sandbar.

Deposition characteristics of POM with sand was assumed by measuring the topography of sandbar and the thickness of sediment deposition layer, observing the spatial distribution of vegetation, analyzing the grain size distribution of deposited sediments and the quantity of POM contained the deposited sediments. The thickness of deposition layer was measured the height from the top of basement rock to the top of fine sand layer by using a soil auger at 15 locations in the target sandbar (see Fig. 1).

POM content ratio included in the sediment through longitudinal direction of sandbar was distributed from 0.5% to 1.5% at each point. In addition, specific gravity of POM was

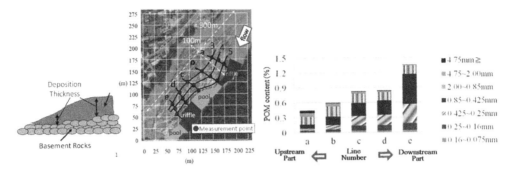

Figure 1. Site location and spatial distribution of POM content ratio according to grain size.

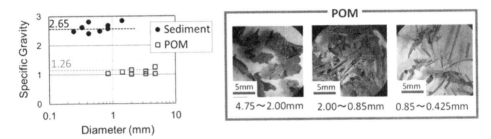

Figure 2. Measurement of specific weight of POM.

observed in a laboratory. Figure 2 shows that it is distributed from 1.02 to 1.26 compared to sediment distribution are around 2.65. The movement of POM in settling process is completely different from sediment. Sediment was settled rapidly in a group however POM moved flutteringly and settled slowly from top to the bottom of water. It is clear that the settling process of sediment and POM is differed even if they have same diameter.

3 DEPOSITION PROCESS OF POM TRANSPORTED WITH SAND

3.1 *Flume experiment*

Flume experiment was conducted to understand the deposition mechanism of POM, especially CPOM such as vegetation seeds, litters was selected as our target in this study. In the laboratory, a model vegetation made by a group of cylinders made of bamboo arranged in staggered pattern ($D = 0.25$ mm, $\lambda = 0.25/\text{cm}^2$) was set in the interval of 5 m in a flume 20 m long and 0.5 m wide with the constant slope as shown in Figure 3. The bed was rigid.

Firstly, the flow measurements (U and h) were conducted along the centerline of the flume (Obana et al. 2012). Shear velocity (u_*) is also important factor for an analysis of sediment transport. Tsujimoto (1999) pointed out that form drag is predominant and velocity profile is uniform along the depth only except the thin layer near the bed where the boundary layer is developed to bring a shear flow. Thus, after measuring the velocity near the bed with/without vegetated area respectively, shear velocity of vegetated area (u_{*v}) was deduced based on the assumption under the similarity of velocity near the bed (u_b) and the shear velocity between with and without vegetated area (see Table 1). The results of this method is confirmed with the theoretical discussion on bed roughness boundary layer (Jeon et al. 2014).

Then, sand ($d = 0.025$ cm, $\sigma/\rho = 2.65$; d = diameter, σ, ρ = mass density of sand and water) and POM was fed at 1 m upstream of the vegetated area with constant volume along the width. As for CPOM model, we selected PVC (polyvinyl chloride) controlled the specific weight ($d = 0.15$ cm, $\sigma/\rho = 1.26$) in reference to examination as mentioned above. Each supplied sand and POM rate

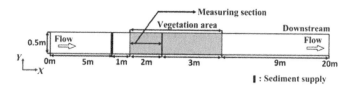

Figure 3. Plan view of experimental channel.

Table 1. Experimental condition.

Q (cm³/s)	I_b	I_f	h_0 (cm)	u_{*0} (cm/s)	u_{*v} (cm/s)	u_{b0} (cm/s)	u_{bv} (cm/s)
6940	1/150	1/185	3.5	4.3	2.3	14.0	7.4

	d_{50}(cm)	σ/ρ	W_0 (cm/s)
Sand	0.025	2.65	3.13
CPOM (using PVC)	0.15	1.26	5.5

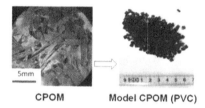

CPOM   Model CPOM (PVC)

Figure 4. Detail of each experimental sample.

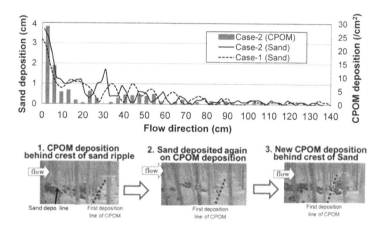

Figure 5. Comparison of deposition state between Case-1 (Sand) and Case-2 (Sand + CPOM) after 20 min.

was 0.047 cm²/s, 0.06/cm². After 20 min. of sediment supply, water is stopped and then deposition of sand and POM in the vegetation area was measured along the centerline in the vegetated zone. Two kinds of experimental cases were run to observe each fluvial process of sand and POM. Case-1 was provided only sand, the other Case-2 was provided both of sand and CPOM. Both sand and POM were transported as bed-load under this hydraulic condition.

3.2 Deposition process of POM transported with sand

Figure 5 shows comparison of deposition state of sand thickness and CPOM between Case-1 and Case-2 after 20 minutes from starting this experiment. We observed that sand deposition was

Table 2. Comparison of wave length, height and propagation velocity of ripple (x=40–60 cm).

x=40–60 cm	Case-1	Case-2
Wave length: L (cm)	10	9
Wave height: H (cm)	0.65	0.45
Propagation velocity: U_w (cm/s)	0.021	0.019

occurred at the beginning of vegetation area in both of cases; ripples were formed by sand and propagated with time progress. In contrast, CPOM had never deposited when it was transported by itself in vegetated area. However it was deposited with sand by the both of interactions.

The CPOM deposition mechanism with sand is as follows; 1) firstly fine sand formed ripples, 2) CPOM is deposited behind the crest of ripple, and then, fine sand is deposited on CPOM deposition, 3) finally new CPOM is coming and deposited behind crest of ripple. These steps are then repeated. In Figure 5, we can confirm that CPOM deposition is increasing and propagating with the development of ripples. By measuring each parameter of ripples at measuring section (x = 40–60 cm), the averaged wave length of ripple in Case-1 was about 10 cm ($L/d = 400$), wave height was 0.65 cm. And it was clear that each parameter of case-2 was decreased because of CPOM which was reduced the development process of ripples by disturbing sand supply from upstream as shown in Table 2.

4 MODELING OF CPOM DEPOSITION WITH RIPPLES

4.1 Model concept

The flume experiment made it clear that CPOM was captured behind the crest of ripple by separation vortex, and was deposited. The CPOM particles transported to the rearward of crest of ripple is presented by the accumulation of supplied number density v_g, particle velocity u_g and capture ratio ψ. The supplied number density v_g and particle velocity are written as follows (Ashida & Michiue 1972).

$$v_g d^2 = \frac{1}{A_3 \mu_R}(\tau_* - \tau_{*c})$$ (1)

$$\frac{u_g}{u_*} = \phi_d\left(1 - \sqrt{\frac{\tau_{*c}}{\tau_*}}\right)$$ (2)

where A_3 = geometrical coefficient of sand; μ_R = friction coefficient of sand (=0.4); ϕ_d = 6.8 (constant); τ_* = dimensionless tractive force; τ_{*c} = critical tractive force.

During the 1st step of time progress (Δt), $\psi v_g u_g \Delta t$ of CPOM is captured by the trough of ripple, and number of ripple will be increased toward downstream part ($t_k = kL/U_W = k\Delta T$, k: number of ripple) according to the sand deposition. They are buried and deposited along the longitudinal direction (Δx) during the time needed for form the one wave length of the ripple. Thus, CPOM is deposited on a straight line which is connected with the trough of ripple (see Fig. 6).

As mentioned above, we can calculate the temporal change of CPOM deposition every time step at representative point in the vegetated area. Accumulation quantity of CPOM (in number) N_{kj} from x_1 to x_j along longitudinal direction according to the temporal change t_k could be calculated as shown in Figure 7.

4.2 Simulation of CPOM deposition and comparison with flume experiment

Ripples are formed under the hydraulic condition in this study. As for the sand deposition, 2D simulation had already improved based on the concept of the resistance law based on

36

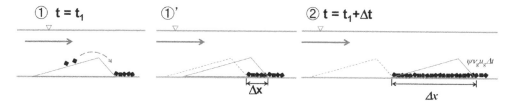

Figure 6.   Concept of deposition process of CPOM with ripple.

Figure 7.   Model concept of CPOM deposition process with ripple.

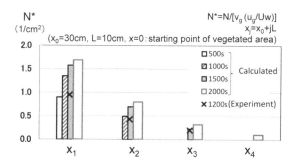

Figure 8.   Comparison between measured and calculated results of CPOM deposition.

the bed roughness boundary layer in vegetated area (Jeon et al. 2014.). The CPOM deposition was evaluated by superimposing the scale of ripple and propagation time ($L = 10$ cm, $U_w = 0.02$ cm/s) by our experiment and simulation result of sand deposition.

As for the numerical calculation of CPOM, the temporal change of N^* on the basis of assumption of $\psi = 0.9$, $\beta = 0.5$. Our model showed that ripples were formed every 500 s, and propagated to the downstream. After 1500 s, there are 3 numbers of ripples formed. In addition, CPOM deposition N^* after 1200 s through the flume experiment was compared with calculation result N^* considering the following value; $U_w = 0.02$ cm/s, $u_g = 5$ cm/s (estimation by Equation (1)), $v_g = 0.06$/cm² calculated by using supplied POM ratio and u_g. The calculated result is explained the experimental result in detail as shown in Figure 8.

5   CONCLUSION

In this study, our efforts have highlighted on evaluation of POM capturing on sandbar which is characterized by vegetated area and fine sand deposition there. We investigated the deposition mechanisms of sediment and POM through a field observation and a laboratory experiment. After the basic deposition characteristics were extracted by the field observation, we observed the deposition mechanisms by the laboratory experiment. The conclusions of this study are summarized that ripples are formed by fine sediment that was deposited in riparian vegetation. And CPOM is captured by trough of ripples formed by wave action.

In contrast, their multifunction of movement of both sample were decreased the propagation velocity of wave height and length of ripples. Then, **POM** behavior with ripples was modeled. The proposed model will affect other aspects in ecosystem management based on fluvial processes.

REFERENCES

Ashida, K. & M. Michiue 1972. Hydraulic resistance of flow in an alluvia bed and bed load transport rate. *Proc. JSCE*, No. 206, pp. 59–69 (in Japanese).

Jeon, H.S., M. Obana & T. Tsujimoto 2014. Bed roughness boundary layer and bed load deposition in vegetated area of a stream, *Proc. Conf. on Hydraulic Eng.,* Hong Kong (submitted).

Liu, D., P. Diplas, J. Fairbanks & C. Hodges 2008. An experimental study of flow through rigid vegeta tion, *Jour. Geophys. Res.*, 113, F04015.

Obana, M, T. Uchida & T. Tsujimoto 2012. Deposition of Sand and particulate organic matter in riparian vegetation, *Advances in River Eng.*, JSCE, pp. 47–52 (in Japanese).

Tsujimoto, T. 1999. Flow and fluvial processes in streams with vegetation, *Jour. Hydraul. Res.*, IAHR, Vol. 37, No. 6, pp. 789–803.

Hydraulic Engineering III – Xie (Ed)
© 2015 Taylor & Francis Group, London, ISBN 978-1-138-02743-5

Research on Baimao Sand Shoal waterway deep-water channel regulation idea in the lower reach of the Yangtze River

Hua Xu
Nanjing Hydraulic Research Institute, Nanjing, China
College of Harbor, Coastal and Offshore Engineering, Hohai University, Nanjing, China

Yun-Feng Xia, Xiao-Jun Wang & Shi-Zhao Zhang
Nanjing Hydraulic Research Institute, Nanjing, China

ABSTRACT: Baimao Sand Shoal waterway is the key navigation-obstructing waterway first encountered in the project of 12.5 m deep-water channel of the Yangtze estuary extending upward to Nanjing. This article utilizes the hydrographic and sediment topography data of Baimao Sand Shoal reach for years to study the riverbed evolution characteristics, navigation-obstructing features and development trend of Baimao Sand Shoal reach. The research shows that, in recent years, Baimao Sand Shoal has retreated because of the impact on the head, with the addition of washout in the lower sand body of Baimao Sand Shoal, the inlet riverbed of south waterway is developing toward wide and shallow riverbed unfavorably, and the hydrodynamic force has dispersed and weakened with sediment siltation, forming a navigation-obstructing shoal at the transition section of branch. Based on the above riverbed evolution characteristics, the research presents the regulation idea of fixing shoal and stablizing channel.

1 INTRODUCTION

The Phase III deep-water channel regulation of Yangtze estuary has been completed at the end of March 2010, with the channel depth rising to 12.5 m. In Jan. 2011, the 12.5 m deep-water channel of the Yangtze estuary has extended upward to Taicang, Jiangsu. However, it's difficult to extend upward to Nanjing port. Baimao Sand Shoal waterway is the key navigation-obstructing waterway first encountered in the upward extension of the deep-water channel of Baimao Sand Shoal waterway. So far, the maintenance depth is 10.5 m. Baimao Sand Shoal reach is shaped as a micro-bended branching river, with wide surface and interlaced shoals and troughs. Under the interaction of runoff and tide, the bed sediment is very active, with complicated water and sediment conditions as well as riverbed evaluation. Therefore, an in-depth study on the riverbed evolution characteristics and navigation-obstructing features of Baimao Sand Shoal reach is necessary, with the addition of presenting the regulation idea of deep-water channel.

2 BASIC INFORMATION

2.1 Overview of river reach

The south branch of the Yangtze River stretches from Xuliujing to Wusongkou, with total length about 75 km. Bounded by Qiyakou, the south branch is divided into the upper south branch and lower south branch, and bounded by Baimao estuary, the upper south branch may be divided into Xuliujing reach and Baimao Sand Shoal branching segment. Xuliujing reach stretches from Chengtong to Baimao Sand Shoal branching segment in the south branch of the Yangtze River. From Xupu to Baimao estuary, the total length is 15 km. Below the Baimao estuary, it's the broadening branching segment, and the Yangtze River is divided into south and west branches here, wherein, the north branch is the sub-branch, the south branch is the main branch. The mainstream of Yangtze River flows through the artificial narrow segment of Xuliujing into the south waterway of Baimao Sand Shoal, with diversion ratio accounting for about 70%. After converging with the north waterway of Baimao Sand Shoal near Qiyakou, it enters the lower south branch of Yangtze River, and enters the sea from the south port and north port after multi-branch diversion (Xia et al. 2000).

2.2 *Hydrology and sediment conditions*

1. Runoff
   Datong Station—the last hydrologic station in the lower reaches of Yangtze River is about 460 km away from Tongzhou Sand Shoal reach. According to the flow statistics during 1950~2011, the maximum flow is 92,600 m^3/s; the minimum flow is 4,620 m^3/s; the average flow is 28,500 m^3/s; the average flow in flood season is 39,700 m^3/s; and the average flow in dry season is 16,500 m^3/s. From May to October in each year is the flood season. The downward runoff flow of Yangtze River is 70.7%.

2. Tide and tidal current
   The Yangtze estuary is a tidal estuary with moderate strength, belonging to informal semi-diurnal tide. According to the tide level statistics of Xuliujing station, the maximum tidal range is 4.01 m, and the mean high tide level is 2.07 m (85 Yellow Sea base level, similarly hereinafter), and the mean low tide level is about 0.03 m. Baimao Sand Shoal reach is characterized by reversing current. During the flood season, the internal flood current of main channel is weak, with flood current of certain intensity in the sub-branch and beach face only. The ebb current is the main impetus to shape riverbed. The maximum ebb in the flood season can reach more than 2.5 m/s; the flood current of sub-branch and beach face in the dry season are basically the same as or slightly stronger than the ebb current. The maximum flood of spring tide in dry season can reach more than 1.2 m/s.

3. Sediment
   The river sediment in the project mainly comes from basins. According to the statistical analysis of Datong Station in the upper reaches during 1951~2011, the maximum sediment concentration is 3.24 kg/m^3, the minimum sediment concentration is 0.016 kg/m^3, and the mean sediment concentration is 0.428 kg/m^3. After the Three Georges Reservoir started to reserve water in 2003, the mean sediment runoff over the years has reduced from 0.14 billion tons before water reserving to 0.14 billion tons after water reserving. Before and after water reserving, the sediment concentration has reduced by about 50%. The bed load is fine sand, with 0.10~0.25 mm median particle diameter; the median particle diameter of suspended sediment is about 0.01 mm.

3 RIVERBED EVOLUTION CHARACTERISTICS OF BAIMAO SAND SHOAL REACH

3.1 *Historical evolution*

Baimao Sand Shoal reach stretches from Tongzhou Sand Shoal reach to the main channel of south branch. In the history, due to the unstable river border, the plane shape of watercourse was easy to change, thereby causing the changes such as mainstream swing, sandbar change as well as the rise and decline of branches. Due to that there's a lack of node segment control between two adjacent reaches, the change of the upstream riverbed may cause the severe change of riverbed in the lower reaches. And there's a strong evolution correlation between two adjacent reaches.

In history, Baimao Sand Shoal has undergone the alternate development process of south and north waterways more than once. And the change rule was directly influenced by the change of river regime in the upper reaches such as Rugao Sediment Shoals waterway and Tongzhou Sand Shoal waterway. See Table 1. When the mainstream in the upper reaches took the north waterway of Rugao Sediment Shoals, the mainstream of Tongzhou Sand Shoal waterway turned from the east to west waterway, and accordingly, Baimao Sand Shoal waterway turned from the north to the south waterway; when the mainstream took the south waterway of Rugao Sediment Shoals, the mainstream of Tongzhou Sand Shoal waterway turned from the west to the east waterway, and accordingly, Baimao Sand Shoal waterway turned from the south to the central (north) waterway. In the 1950s, after the mainstream of the upper reaches was stabilized in Liuhai Sand Shoal waterway, there's no shift between the main branch and sub-branch in Tongzhou Sand Shoal and Baimao Sand Shoal waterways, and the mainstream was stabilized in the east waterway of Tongzhou Sand Shoal and the south waterway of Baimao Sand Shoal (Xia et al. 2000, 2001 & Cao et al. 2000).

3.2 Recent evolution

3.2.1 Change of branching distributary

In the recent years, the diversion ratio of north branch remains 2–5% and huge volume of water flow is drained by the south branch; the diversion ratio of south waterway of Baimao Sand Shoal is always greater than that of the north waterway; the main reason why the distributary of south waterway has increased constantly in recent years is the scour area and open area of its inlet enlarge while the sedimentation area of inlet of the north waterway becomes narrowed and the flow capacity goes down, thus the distributary situation of south waterway better than the north is further strengthened.

3.2.2 Major shoals change

1. Change of sand body of Baimao sand shoal

   The change of sand body of Baimao sand shoal can be seen in Figure 1. Baimao sand shoal is a U-shaped central bar at the middle of river channel and it was formed in the middle of the 19th century. In its development history, Baimao sand shoal has gone through the periodical evolution process of appearance, northward movement, bank consolidation and reformulation in which Baimao sand shoal encountered flood and its sand body was

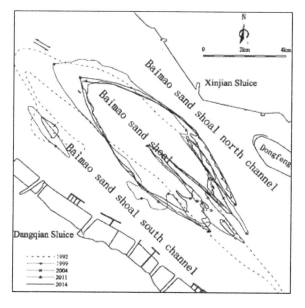

Figure 1.   Change of −5 m-line of sand body of Baimao Sand Shoal in recent years.

41

apt to be cut and dispersed by it. In late 1990s, the changes of sand body were mainly displayed in the scour and retrogression of sand head and its retrogression has retarded in recent years. The retrogression of sand body leads to the broadening of inlet of the south waterway and the riverbed develops toward wide-and-shallow type. The change of cross section of sand head is as shown in Figure 2. However, the broadening of inlet of the waterway goes against the maintenance and development of the entrance channel conditions of the south waterway.

2. Change of sand body of small Baimao sand shoal

The change of sand body of small Baimao sand shoal can be seen in Figure 3. Small Baimao sand shoal, which was formed in the 1970s by the incision of Xuliujing Side Shoal, constitutes of the upper and the lower sand bodies. The upper sand body is comparatively stable while the lower one is active. After 1999, being impacted by the decrease of upstream sand, reclamation along Taicang and human factors, the lower sand body shrank year by year. By the end of 2008, the washout of the lower sand body further broadened the width of entrance main channel of the south waterway; however, it was not conducive to the development of the entrance channel conditions.

3.2.3 *Change of main channel*

The periodical change of Baimao sand body induces the adjustment of water power distribution of the waterways at both sides. In the initial stage of siltation of Baimao Sand Shoal, the siltation at the entrance of the south waterway is usually shallow and a huge volume of water is often drained via the north waterway, causing the scouring development of the north waterway; while in the process of moving down and northward movement of Baimao Sand

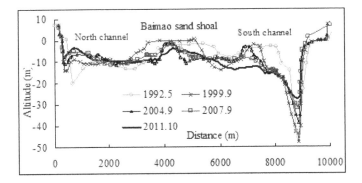

Figure 2.   Change of cross section of sand head of Baimao Sand Shoal in recent years.

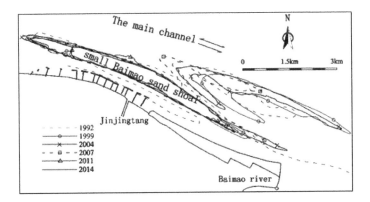

Figure 3.   Change of −5 m-line of sand body of small Baimao Sand Shoal in recent years.

Shoal, the south waterway develops and the north waterway shrinks correspondingly. During such change, there existed a period of integrated sand body and appropriate position as well as mutual development of the south and north waterways, with favorable water depth condition of deep channel. In the 1990s, the sand body attained its maximum size with complete shape, thus the water depth condition of the north waterway was good, and the volume ratio, which was less than 0.2 in early 1980s, of the 10 m-deep channel between the north and the south waterways was about 0.4. After the 1990s, the situation of "south stronger than north" between the south and north waterways has become severe gradually. After 2007, the volume ratio of the 10 m-deep channel between the north and the south waterways was less than 0.2 again. Now, the south waterway mainly focuses on the scour development and the channel width has broadened continually.

The changes of −10 m main channel of Baimao Sand Shoal waterway are as shown in Figure 4. It is thus clear that, under the control of Xuliujing constriction, the plane modality of deep channel of Xuliujing reach gradually heads for stability, while the deep channel below Baimao estuary gradually moves toward the south; being influenced by the upstream river regime and the change of partial swales, the entrance deep channels of the south and north waterways are unstable and the siltation of entrance deep channel of the north waterway is shallow in recent years; the situation of "south stronger than north" is further reinforced. The plane modality of main channel in the downstream of Xinjing River of Baimao Sand Shoal south waterway seldom changes, maintaining the function of main deep channel.

3.2.4 Main influence factors and development tendency of riverbed evolution

1. Main influence factors

   The riverbed evolution of Baimao Sand Shoal reach is mainly pushed by the coming water and sediment, flood and human activities. The flood factor thereof generally leads to a great change of river regime. Due to the impoundment of Three Gorges Reservoir, the coming sediment decreases, bottomland siltation retards and the riverbed is on the scour tendency, thus it becomes harder to recover the circularly changed sand body and the evolution cycle of swale will be prolonged. In recent years, the human activities have impacted greatly on the riverbed evolution and the illegal sand excavation will bring an adverse impact on the river regime and the channel stabilization.

2. Development tendency

   Through years' natural evolution and manual governance, the overall river regime of Baimao Sand Shoal segment gradually tends toward stabilization and has laid a solid foundation for the deep-water channel regulation. The tendency of "south stronger than north" will be remained; the mainstream of Baimao Sand Shoal reach will go on moving

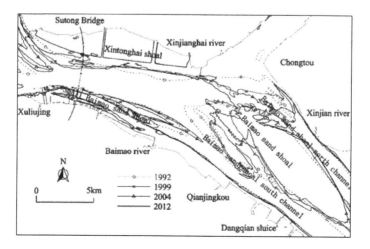

Figure 4. Changes of −10 m main channel of Baimao Sand Shoal waterway.

to the south bank in a long run and the two branches will coexist and the mainstream will flow along the south waterway. Although the sand head position has changed slowly in recent years, the sand head will be likely to recede because of being scoured when encountering flood, hence, the south rim of the sand body will be lashed, the entrance of the south waterway will be broadened further, the riverbed will develop toward wide-and-shallow type and the channel conditions will move in an adverse direction. Being influenced by its great on-way resistance and the back flow of water and sediment of north branch, the development of Baimao Sand Shoal north waterway is restricted; the upper opening of deep channel has been separated with that of the Xuliujing reach and the situation of "south stronger than north" will be intensified.

4 NAVIGATION-OBSTRUCTING CHARACTERISTICS OF BAIMAO SAND SHOAL WATERWAY

4.1 *Navigation-obstructing characteristics*

The navigation-obstructing shoal at the entrance of the south water channel of Baimao Sand Shoal is a navigation-obstructing shoal at the transition section of branch. The width of river channel in the lower Xuliujing has been gradually broadened and the lower reaches of Baimao Sand Shoal estuary is separated into the south and the north waterways by Baimao Sand Shoal; because of hydrodynamic dispersion, the sediment silted in the main deep channel of Xuliujing and at the transition section of the entrance of the south wateway of Baimao Sand Shoal.

The lower reaches of Xinjingkou of the south waterway of Baimao Sand Shoal are good reaches with favorable water depth, thus a natural channel of deep-water upper channel of Yangtze Estuary is formed. But because of the instability of the lower sand body of small Baimao Sand Shoal and the sand head of Baimao Sand Shoal which are at the border of both sides of the channel, and the influence of upstream river regime and back flow of water and sediment of north branch, the inlet water depth condition of the south water channel of Baimao Sand Shoal is unstable with shallow siltation sometimes. In recent years, the head of Baimao Sand Shoal has still been lashed and receded; due to the washout of the lower sand body of small Baimao Sand Shoal and the further broadening of entrance transition section, the water depth of navigation channel is developing in an adverse direction (Xu et al. 2011).

4.2 *Variation tendency analysis of channel conditions*

The 12.5 m isobath changes of Baimao Sand Shoal waterway is seen in Figure 5. The water depth condition of the south waterway of Baimao Sand Shoal is now developing in an adverse direction; the upper opening and the lower reaches of the north waterway are apt to be silted, thus the navigation channel is less stable comparatively. With the sustained retrogression of the head of Baimao Sand Shoal caused by lashing, in the upper reaches of Baimao estuary, the scouring of sand body of small Baimao Sand Shoal will keep on going and if the main channel is further broadened, the stabilization of entrance navigation channel of the south waterway and the water depth of navigation channel will be influenced. Now the sand body of Baimao Sand Shoal is relatively tall and integrated; the water depth of the south waterway is a little deeper than 12.5 m and the width of 12.5 m-isobath is greater than 500 m. However, because of not being protected, the head of Baimao Sand Shoal recedes constantly and the channeling develops on sand body, the small Baimao Sand Shoal is washed out constantly, the south waterway entrance is broadened and the riverbed develops toward the wide-and-shallow type (the minimum water depth decreases from 16.1 m to 13.4 m, and the riverbed develops from "V-shape" toward "U-shape"). Now, three's a shallow area with a top water depth less than 10 m at the north entrance of the south waterway of Baimao Sand Shoal, the deep channel moves toward the south and the forefront water area of the dock is adjacent to the south, influencing the ship navigation safety (Xu et al. 2011).

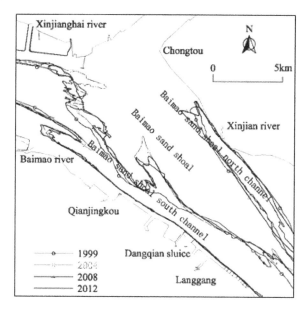

Figure 5.   12.5 m isobath change of Baimao Sand Shoal waterway.

5   RESEARCH ON THE REGULATION IDEA OF DEEP-WATER CHANNEL OF BAIMAO SAND SHOAL

The water depth conditions of Baimao Sand Shoal reach are able to meet the requirements of deep-water channel construction now. The main existing question is if the swale develops itself without being regulated, it will influence partial river regime of the south waterway entrance and the channel stabilization. First, the sand head of Baimao Sand Shoal is scoured continuously and the main channel of the entrance of Baimao Sand Shoal is further broadened; the riverbed will go on developing toward wide-and-shallow type; the channel conditions will be gradually worsened and the downward sediment by lashing is not conducive to the maintenance of downstream channel. In addition, since 1985 when the north bank of Xuliujing was enclosed for cultivation and a manually-controlled constriction was formed, Baimao Sand Shoal reach, although being influenced by the upper reaches, has receded sharply, if huge volume of sediment is drained from the upper reaches, a new side shoal is easy to be formed adjacent to Baimao estuary, thus leading to the change of partial river regime at the entrance of Baimao Sand Shoal reach. Therefore, the stability of Baimao Sand Shoal reach, on one hand, needs to be remained by stabilizing the sand head of Baimao Sand Shoal, adjusting the flow field and arranging the south and north waterway, and enhancing the water power of partial shallow reaches when protecting the sand body; on the other hand, it needs to decrease the excessive drainage of riverbed sediment from the upper reaches with integration of the regulation works of Tongzhou Sand Shoal etc. for creating conditions for the construction of deep-water channel and long-term stabilization (Xu et al. 2011, Yu et al. 2008).

6   CONCLUSION

Baimao Sand Shoal waterway is a key navigation-obstructing waterway first encountered in the extension project of deep-water channel at Yangtze Estuary. This article utilizes the hydrographic and sediment topography data of Baimao Sand Shoal reach measured for years to study the swale evolution characteristics of Baimao Sand Shoal reach as well as the change and influence factors thereof, comprehensively research the impact of riverbed evolution to the channel construction and maintenance. The research shows that, due to the

45

retrogression of the head of Baimao Sand Shoal by lashing and washout of the lower sand body of small Baimao Sand Shoal, the riverbed develops toward the wide-and-shallow type and hydrodynamic force becomes weakened, thereby causing sediment siltation and forming a navigation-obstructing shoal at the transition section of branch. Aiming at the aforesaid riverbed evolution characteristics, a channel regulation idea for fixing shoal and stabilizing channel is proposed. The research results provide key technical support for waterway engineering design and decision making etc. To further improve the stability of channel condition of the south waterway entrance of Baimao Sand Shoal, the research on treatment of small Baimao Sand Shoal is suggested to be carried out in the next stage.

ACKNOWLEDGEMENT

This work was supported by the Key Technology Special Fund of Ministry of Transportation and Communication (Grant No. 201132874660) and the National Natural Science Fund for Young Scholars (Grant No. 51309158).

REFERENCES

Cao, M.X., Xia, Y.F. & Ma, Q.N. 2000. Riverbed Evolution Analysis of Fujiang Sand Shoal Water-way. *Yangtze River* 31(12): 23–27.

Xia, Y.F., Wu, D.W., Zhang, S.Z., et al. 2000. *Riverbed Evolution Analysis on the Waterways of Three Sand Shoals in the Lower Reaches of Yangtze River*. Nanjing: Nanjing Hydraulic Research Institute.

Xia, Y.F., Cao, M.X. & Chen, X.B. 2001. Waterway Evolution Analysis of Three Sand Shoals (Fujiang Sand Shoal, Tongzhou Sand Shoal and Baimao Sand Shoal) in the Lower Reaches of Yangtze River and Assumptions of Deep-water Channel Regulation. *Sediment Research* (3):57–61.

Xu, H., Wu, D.W., Xia, Y.F., Du, D.J., Zhang, S.Z., Wen, Y.C., et al. 2011. *Riverbed Evolu-tion Analysis on Toingzhou Sand Shoal Reach and Baimao Sand Shoal Reach in the Lower Reaches of Yangtze River*. Nanjing Hydraulic Research Institute.

Yu, W.C., Zhang, Z.L. 2008. Some Issues of River Channel Evolution of the Yangtze Estuary in Recent period. *Yangtze River* 39(8): 86–89.

Hydraulic Engineering III – Xie (Ed)
© 2015 Taylor & Francis Group, London, ISBN 978-1-138-02743-5

Distribution features and transport trend of seabed sediment in Leizhou Bay

Zhiyuan Han

Tianjin Research Institute for Water Transport Engineering, Key Laboratory of Engineering Sediment of Ministry of Communication, Tianjin, China

ABSTRACT: The grain size parameters of seabed sediment are the most important information, which can indicate the distribution and transportation features of seabed sediment. A study on distribution features of grain size parameter in Leizhou Bay is conducted based on 109 samples data, which were collected in July of 2009. At the same time, using the Grain Size Transport Analysis model, transporting trend of seabed sediment had been analyzed. The research results are showed as follows: (1) seabed sediment types in Leizhou Bay are mainly sandy sediment, and only a small amount of clayey and silty sediments distribute at western bay and deep grooves in central bay; (2) seabed sediments in Leizhou Bay distribute with discontinuous features, which indicate that the transport trend of seabed sediments is not obvious, and the seabed sediment is not active, so seabed sedimentary environment and underwater topography can maintain a stable state; (3) seabed sediments in Leizhou Bay are mainly from inner continental shelf in northern South China Sea during post-glacial transgression and the sediment had gradually adapt to the modern hydrological and sedimentary environment.

Seabed sediment grain size parameters in costal zone are important to describe depositional environment and reflect the coupling mechanism of dynamic-deposition-topography action. Therefore, based on a combination of characteristics of seabed sediment grain size parameters, it can help understand sedimentary environment, such as sediments deposition condition, sediment sources, sediment transport trends etc.

Leizhou Bay locates in southwestern Guangdong Province, and at eastern coastline of Leizhou Peninsula (see Fig. 1). Leizhou Bay, with area of approximately 1690 km², is enclosed

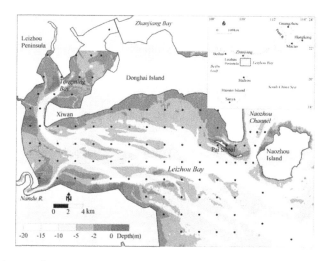

Figure 1. Sketch map of research zone and seabed sampling stations.

by Leizhou Peninsula, Donghai Island and Naozhou Island, and is only open to South China Sea at southwest. There are many shallows and deep grooves staggered in the bay, and underwater terrain is very complex. Some deep grooves can extend to the top of the bay from deep water of open sea. According to long-term economic development plan of Zhanjiang City, Donghai Island will be developed as petrochemical and steel industry, and Leizhou Bay will be developed as deep water harbor and some grooves will be developed for deep water navigating channel. Therefore, it is necessary to study seabed sediment distribution and transportation in Leizhou Bay for understanding its sedimentary and hydrodynamic environment.

Under long-term action of hydrodynamic condition (runoff, tide and waves, etc.) and sand sources, seabed sediments of Leizhou Bay have complex material content and different grain size. As a newly developed port site, sedimentary environment of Leizhou Bay has not specifically studied (Han Zhi-yuan, 2012). Based on mass filed sediment and hydrologic data of Leizhou Bay, this article focus on the sediments deposition condition, sediment sources, sediment transport trends etc., and will provide strong support for the utilization and environmental protection of Leizhou Bay.

1 RESEARCH DATA AND METHODS

In July 2009, 109 seabed sediment samples were sampled using grab sampler in Leizhou Bay, with sampling stations covering the whole bay (sampling station see Fig. 1). All the samples were sent to laboratory for particle size analysis using *Macrotrac S3500* laser particle size analyzer. Seabed sediment were classified and named using Shepard sediment classification method (Shepard F.P, 1954). Particle size parameters including median grain size (d50), sorting coefficient (Qdφ) and skewness (Skφ) were calculated by using Folk and Ward formula (Folk R.L & Ward W.C, 1957). Sediment transport trend was analyzed by two-dimensional particle size based on sediment trend analysis model (GSTA model) proposed by Gao S. and M. Collins (Gao S & M Collins, 1992).

2 RESULTS

2.1 *Distribution of seabed sedimentary types*

Based on field seabed sediment data, seabed sediment of Leizhou Bay contend 6 types, including Coarse Sand (CS), Median Sand (MS), Fine Sand (FS), Sandy Silt (ST), clayey silt (YT) and silty clay (TY) (see Table 1). Seen from Table 1, sandy sediments, including coarse sand, median sand and fine sand, are dominant sediment types in the bay, and account for 67% of total sediment samples; medium particle size of sandy sediments range from 0.16 mm to 2.0 mm. Silty and clayey sediments including sandy silt, silty clay and clayey silt, only account for 33% of all samples, with median grain size ranging from 0.002 mm to 0.04 mm.

Table 1. Statistics of seabed sediment in Leizhou Bay.

Sedimentary type	Number	Percentage	Mean value of grain parameter		Mean value of grain content (%)		
			D50 (mm)	Qdφ	Sand	Silt	Clay
Coarse sand	13	11.9%	0.927	0.48	89.6	0.0	0.0
Median sand	32	29.4%	0.345	0.53	92.8	0.0	0.0
Fine sand	28	25.7%	0.227	0.48	95.4	0.0	0.0
Sandy silt	8	7.3%	0.036	1.69	32.4	46.7	20.9
Clayey silt	12	11.0%	0.014	2.06	22.7	44.5	32.8
Silty clay	16	14.7%	0.005	1.79	14.9	31.8	53.4
Sum/Mean	109	100.0%	0.275	0.95	72.1	13.1	13.0

Distribution sketch of seabed sedimentary type in Leizhou bay is showed in Figure 2. Seen from Figure 2, fine sediments as clayey or silty sediments are only deposited at Tongming Bay, south of Xiwan in western Leizhou Bay, deep grooves in central Leizhou Bay; sandy sediments are widely distributed at shallows and some deep grooves of the whole bay. From west to east, seabed sediment distributes as fine sand—silty clay—coarse sand—silty clay—median sand, and from north to south, sediment distributed as median sand/fine sand—coarse sand—silty clay—median sand/fine sand. Thus, seabed sediments distribute with discontinuous feature.

2.2 Distribution of median grain size

Median grain size (d50) of all seabed sediments ranges from 0.002 mm to 2.0 mm in Leizhou bay. Fine sediments with D50 less than 0.01 mm, mainly distribute at northwestern Tongming Bay, south of Xiwan in western Leizhou Bay, and deep grooves in central Leizhou Bay (Fig. 3). Fine sediments with D50 ranging from 0.01 to 0.05 mm, mainly distribute at southern and eastern Tongming Bay. Sandy sediments with D50 ranging from 0.16 mm to 2 mm, widely distribute at shallows and grooves in the whole bay. Distribution features of D50 are similar to that of sedimentary types. From west to east, d50 distributed as coarse—fine—coarse—

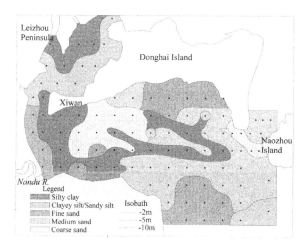

Figure 2.   Sketch of distribution of seabed sedimentary types.

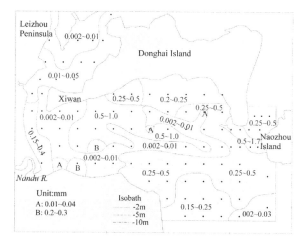

Figure 3.   Sketch of distribution of seabed sediment median particle size (d50).

49

fine—coarse, and from north to south, d50 distributed as coarse—fine—coarse. Thus, d50 of seabed sediments distribute with discontinuous feature.

2.3 *Contours of sorting coefficient of seabed sediment*

Sorting coefficient (Qdφ) of all seabed sediments ranges from 0.17 to 2.7 in Leizhou bay. Very well sorting sediments with Qdφ less than 0.6, mainly distribute at nearshore shallows in northern and western Leizhou Bay and grooves in southeastern Leizhou Bay (see Fig. 4). Well sorting sediments with Qdφ between 0.6 and 1.4, mainly distribute in shallows and grooves at central Leizhou Bay. Medium sorting sediments with Qdφ between 1.4 and 2.2, mainly distribute at Tongming Bay, south of Xiwan in western Leizhou Bay, and deep grooves in central Leizhou Bay. Poorly sorting sediments with Qdφ greater than 2.2, distribute near Xiwan in western Leizhou Bay.

Qdφ of seabed sediment distributed as very well sorting—medium sorting—well sorting—medium sorting—very well sorting from west to east; and Qdφ distributes as very well sorting—medium sorting—very well sorting from north to south. Thus, Qdφ of seabed sediment distributes with discontinuous feature.

2.4 *Contours of sand content of seabed sediment*

Sand content of seabed sediment in Leizhou bay ranges from 5.8% to 100%. Sediments with sand content less than 50%, mainly distribute at Tongming Bay, south of Xiwan in western Leizhou Bay and deep grooves in central Leizhou Bay (see Fig. 5); sediments with sand content greater than 50% widely distribute at shallows and grooves in the whole bay. Sand content of seabed sediment distribute with discontinuous features.

2.5 *Contours of clay content of seabed sediment*

Clay content of seabed sediment in Leizhou bay ranges from 0% to 67.4%. Sediments with clay content great than 30%, mainly distribute at Tongming Bay, south of Xiwan in western Leizhou Bay and deep grooves in central Leizhou Bay (Fig. 6). Sediments with clay content less than 10% widely distribute at shallows and grooves in the whole bay. Clay content of seabed sediment distribute with discontinuous feature.

2.6 *Transport trend of seabed sediment*

Gao S and M Collins (Gao S & M Collins, 1992) had proposed a two-dimensional grain size trend analysis model (GSTA model) based on one-dimensional grain size trend analysis

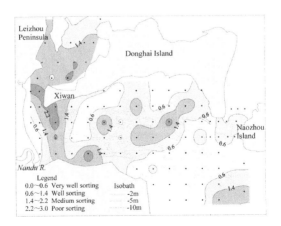

Figure 4. Sketch of distribution of seabed sediment sorting coefficient (Qdφ).

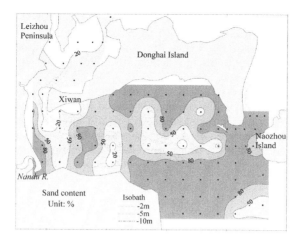

Figure 5.   Sketch of sand content contours of seabed sediment.

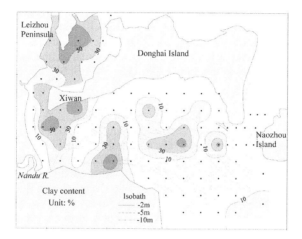

Figure 6.   Sketch of clay content contours of seabed sediment.

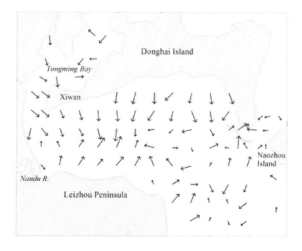

Figure 7.   Transport trend of seabed sediment.

51

mode (McLaren P D, 1985). After the grain size parameters of seabed sediment in Leizhou Bay were input, the vector and direction of transport trend of each sediment sample points were output from GSTA model (see Fig. 7). It is noted that the transport vector arrows indicate the direction of the vector, and vector length only indicates transport trends significantly or not, not the value of transport rate.

Seen from Figure 7, seabed sediment transport features in Leizhou Bay are shown as follows: (1) seabed sediment from Tongming Bay could transport to western Leizhou Bay; (2) seabed sediment from northern and southern Leizhou Bay may transport to the central bay; (3) seabed sediment from open sea can not transport into the central bay; (4) seabed sediment from the western bay can not transport to east and seabed sediment from the eastern bay can not transport to west.

3 DISCUSSION

3.1 *Hydrodynamic conditions*

3.1.1 *Wave*
According to wave data measured from 2002 to 2004 in Naozhou marine station, wave directions of strong wave are SE ~ E, and wave directions of secondary strong wave are NNE ~ ENE; wave directions of prevailing wave are SE and ENE, with annual frequency as 24.2% and 20.9%, respectively. Since Leizhou Bay is only open to southeast, the entire bay has a better cover conditions, and the waves are not big, and annual frequency of $H_{1/10}$ greater than 2.0 m is only 3.9%. So seabed sediment in Leizhou Bay can not be affected significantly under the action of normal wave.

3.1.2 *Tidal current*
Leizhou Bay is a semi-enclosed bay and there are many deep grooves in the bay, so tidal current in the bay is restricted by shoreline shape and underwater topography, and flow reciprocally along the deep grooves (see Fig. 8). During flood tide, the most part of tide prism enters the bay from the gate at south of Naozhou Island, with flow direction from SE to NW; the other part of tide prism enters the bay from Naozhou Channel with flow direction from NE to SW. During ebb tide, the flow direction is the opposite. At southeastern bay, tidal current flows reciprocally with direction from SE to NW, and at central bay, tidal current flows reciprocally with direction from E to W.

Mean current velocity of flood tide in the bay is between 0.22 m/s and 0.58 m/s, and is between 0.27 m/s and 0.58 m/s in ebb tide. Maximum velocity of flood tide in the bay is between 0.62 m/s and 1.05 m/s, and is between 0.82 m/s and 1.28 m/s in ebb tide. Because current

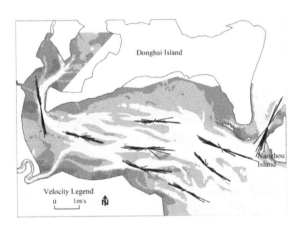

Figure 8.   Spring tidal current ellipses in Leizhou Bay.

velocity is large, and flow direction is almost parallel with the direction of deep grooves, tidal current is the main power to shape the underwater terrain and transport seabed sediment.

3.2 Sedimentary conditions

3.2.1 Suspended sediments

Mean value of Suspended Sediment Concentration (SSC) during flood and ebb tide is between 0.001 kg/m³ and 0.161 kg/m³, with an average value of 0.035 kg/m³. Maximum value of SSC during flood and ebb tide is between 0.008 kg/m³ and 0.339 kg/m³. The SSC does not inflate with velocity during flood and ebb tide. Suspended sediment types are mainly clayey silt and silt, with clay percentage exceeding 20%; median grain size of suspended sediment is between 0.0073 mm and 0.0151 mm, with an average value of 0.0118 mm.

In all, in Leizhou Bay, suspended sediment type are mainly silty sediment, and seabed sediment type are mainly as sandy sediment, so seabed sediment deposited in shallows and deep grooves is difficult to transport under action of current and wave, and seabed sediment is not active.

3.2.2 Sediment sources

1. There are 3 small mountainous rivers flowing into Leizhou Bay, as Nandu River, Tongming River and Lei River. Nandu River is the most long river, with a total length of 88 km and muti-year averaged annual runoff of 866 million m³; Tongming River is 26 km long with muti-year averaged annual runoff of 160 million m³; Lei River is 32 km long, with muti-year averaged annual runoff of 62 million m³. The total value of sediment load of the 3 rivers is not more than 100,000 tons per year. So sediment from runoff is very few.

2. The east shoreline of Donghai Island is straight, so beach sand can transport along the coast from north to south under the action of southeast wave during October and April. The sand may accumulate in the southeast corner of Donghai Island. However, due to current velocity of reciprocal flow in Naozhou Channel is strong, which can block the sand into Leizhou Bay. Therefore, the transporting sand from east shoreline of Donghai Island cannot affect Leizhou Bay significantly.

3. Mean SSC of Leizhou Bay is between 0.007 kg/m³ and 0.134 kg/m³ and the maximum SSC is only between 0.009 kg/m³ and 0.175 kg/m³, so suspended sediment load is very few. Suspended sediment mainly as silty type and seabed sediment mainly as sandy type, show that seabed sediment in shallows and deep grooves of Leizhou Bay is difficult to transport under the action of current and wave, and seabed sediment is not active. So sedimentary environment of Leizhou Bay is stable and underwater terrain can maintain a stable state.

3.3 Source of sandy sediments in Leizhou Bay

Leizhou Peninsula is composed by loose debris of clay, gravel layer which formed in Quaternary of Zhanjiang Group (Q_1) and Beihai Group (Q_2), and could be easily eroded by river including temporary stream (Li Chunchu, 1986a). During Quaternary glacial age sea level was very low, so the coast was easy to develop small scale erosion valley system. When sea level became high during the post-glacial transgression, the valley would be submerge and develop nearshore bay. Leizhou Bay was developed upon such submerged valley after the post-glacial transgression. During the formation of the bay, sea water level rise, then ancient coastal sediment transport landward during shore face shift (Huang Yukun et al. 1982).

When the ancient coastal sediment transported landward, it became the main sediment source of the modern Leizhou Bay seabed sediment (Wang Wenjie, 1985; Li Chunchu, 1986a; Li Chunchu, 1986b). The evidences are showed as follows: (1) all the small river flow into Leizhou Bay take very few sediment load; (2) suspended sediment load transported under action of current and wave are very few; (3) sand ridge in Leizhou Bay with large scale consist of a large amount of sand, so it cannot origin from river or the outer sea, so the sand in Leizhou Bay may come from inner continental shelf at northern South China Sea during post-glacial transgression.

4 CONCLUSION

In this paper, based on analysis of sedimentary features and transport trend of seabed sediment of Leizhou Bay, the main conclusions are showed as follows:

1. Seabed sediment types in Leizhou Bay are mainly sandy sediments, clayey and silty sediments only distribute at western bay and some deep grooves in central bay.
2. Seabed sediment in Leizhou Bay distributes with discontinuous features, which showing that transport trend of seabed sediment is not obvious and seabed sediment is not active, so sedimentary environment and underwater topography can maintain a stable state.
3. Seabed sediments in Leizhou Bay are mainly from inner continental shelf at northern South China Sea during post-glacial transgression and the sediment gradually adapt to the modern hydrological and sedimentary environment.

REFERENCES

Folk R.L, Ward W.C. Brazos River bar: a study in the signification of grain size parameters [J]. Journal of Sedimentary Petrology, 1957, 27: 3–27.

Gao S., M. Collins. Net sediment transport patterns inferred from grain size trends, based upon definition of "transport vectors" [J]. Sedimentary Geology, 1992, 81(1/2): 47–60.

Han Zhi-yuan, Yang Shu-sen. Analysis on deep channel stability in Leizhou Bay [J]. Marine Geology and Quaternary Geology, 2012, (5): 43–48.

Huang Yukun, Xia Fa, Huang Daofan, etc. Holocene sea level changes and modern crystals movement on northern South China Sea coast [J]. Acta Oceanolotica Sinica (in Chinese), 1982, 4(06): 713–723.

Li Chunchu, Luo Xianlin, Zhang Zhenyuan, etc. Formation and evolution of barrier lagoon system of Shuidong Bay in Western Guangdong Province [J]. Chinese Science Bulletin, 1986, (20): 1597–1582.

Li Chunchu. Geomorphological Features of Embayed Coast in South China [J]. Acta geographica sinica, 1986, 41(4): 311–320.

Mclaren P., D. Bowles. The effects of sediment transport on grain-size distributions [J]. Journal of Sedimentary Petrology, 1985, 55: 457–470.

Shepard F.P. Nomenclature based on sand-silt-clay ratios [J]. Journal of Sedimentary Petrology, 1954, 24(3):151–158.

Wang Wenjie. The effect of sediment loads discharged by the rivers in South China on the coast and shelf [J]. Journal of Sediment Research, 1986, (04):27–36.

Hydraulic Engineering III – Xie (Ed)
© *2015 Taylor & Francis Group, London, ISBN 978-1-138-02743-5*

Application of a 3d mathematical model for closure of a cofferdam in Caofeidian Harbour

Bing Yan
Tianjin Research Institute for Water Transport Engineering, M.O.T., China

ABSTRACT: The tidal current is one of the most important factors for closure of a cofferdam in port and coastal engineering. A 3d mathematical model was established to help estimate the hydrodynamic environment in the process of closure of a cofferdam in Caofeidian Harbour. It is shown from simulated results that the model is able to reflect the vertical changes of currents around the project reasonable. The characteristic velocities in final closure from the prediction results were analyzed. Some suggestions were also proposed for the closure of the cofferdam.

1 INTRODUCTION

Caofeidian Harbour is located on the northwest coast of Bohai sea and near Tangshan Province. It is a deep-water port in east china coast. After years of development, first, second and third Phases of Ore Terminal, crude oil terminal, LNG terminal, No. 1 basin, No. 2 basin, breakwaters and channels have been established. In the sea around the harbour, the characteristic of tide belongs to irregular semidiurnal tide and the average tidal range is 1.4 m (Yang et al., 2005). According to survey in field (Zhang, 2011), the mean current velocities of the three locations (#1~#3 in Fig. 1) around the project are 0.52 m/s during spring tide and 0.22 m/s during neap tide. The directions of currents are regular generally, which indicates the currents are rectilinear currents. The flood current is slightly stronger than the ebb one. For instance, the mean velocities of flood and ebb current are 0.54 m/s and 0.49 m/s respectively during spring tide. Recently, a new reclamation project was arranged between

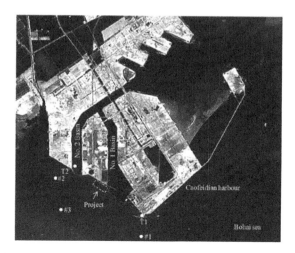

Figure 1.   The position of the project (2012).

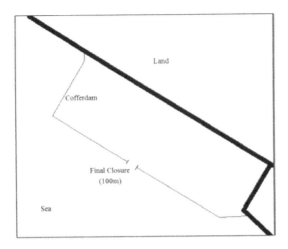

Figure 2. The layout of the cofferdam.

No. 1 basin and No. 2 basin (dashed box in Fig. 1). So a cofferdam needs to be constructed firstly. The closure of a cofferdam is very important for the whole reclamation. Here a 3d mathematical model was established to help estimate the hydrodynamic conditions in the process of closure.

2 NUMERICAL MODEL

2.1 Tide current model

The tidal current fields at the project sea area are simulated by the three-dimensional numerical model (MIKE3 FM) which developed by Danish Hydraulics Research Institute. The model is based on the solution of three-dimensional incompressible Reynolds Navier-Stokes equations, subject to the assumption of Boussinesq and hydrostatic pressure.

$$\frac{\partial u}{\partial x} + \frac{\partial v}{\partial y} + \frac{\partial w}{\partial z} = S \tag{1}$$

$$\frac{\partial u}{\partial t} + \frac{\partial u^2}{\partial x} + \frac{\partial vu}{\partial y} + \frac{\partial wu}{\partial z} = fv - g\frac{\partial \eta}{\partial x} - \frac{1}{\rho_0}\frac{\partial P_a}{\partial x}$$
$$- \frac{g}{\rho_0}\int_z^\eta \frac{\partial \rho}{\partial x}dz - \frac{1}{\rho_0 h}\left(\frac{\partial s_{xx}}{\partial x} + \frac{\partial s_{xx}}{\partial y}\right) + F_u + \frac{\partial}{\partial z}\left(v_t\frac{\partial u}{\partial z}\right) + u_s S \tag{2}$$

$$\frac{\partial v}{\partial t} + \frac{\partial v^2}{\partial y} + \frac{\partial uv}{\partial x} + \frac{\partial wv}{\partial z} = -fu - g\frac{\partial \eta}{\partial y} - \frac{1}{\rho_0}\frac{\partial P_a}{\partial y}$$
$$- \frac{g}{\rho_0}\int_z^\eta \frac{\partial \rho}{\partial y}dz - \frac{1}{\rho_0 h}\left(\frac{\partial s_{yx}}{\partial x} + \frac{\partial s_{yy}}{\partial y}\right) + F_v + \frac{\partial}{\partial z}\left(v_t\frac{\partial v}{\partial z}\right) + u_s S \tag{3}$$

In the above formulas, t is the time; x, y and z are the Cartesian co-ordinates; η is the surface elevation; d is the still water depth; $h = \eta + d$ is the total water depth; u, v and w are the velocity components in the x, y and z direction; f is the Coriolis parameter; g is the gravitational acceleration; ρ is the density of water; s_{xx}, s_{xy}, s_{yx} and s_{yy} are components of the radiation stress tensor; v_t is the vertical turbulent viscosity; P_a is the atmospheric pressure; ρ_0 is the reference density of water; S is the magnitude of the discharge due to point sources and

(u_s, v_s) is the velocity by which the water is discharged into the ambient water; F_u and F_v are the horizontal stress terms.

2.2   Model settings

The three levels nested grids method is adopted here (Fig. 3). The large model includes the Bohai sea mostly. The middle model covers the Bohai Gulf. The small model contains the sea area near the Caofeidian Harbour. The tidal prediction tool in MIKE was used to produce the tidal boundaries of the large model, which provides tide levels for middle model. The small model gets the tide levels in boundary from middle model. The minimum grid step of the small model is 5 m.

2.3   Verification

The models have been calibrated with field observations and the results are given here. Tide and current data from three temporary observation stations are used to verified and validate the present model. By comparison, the numerical calculation results are in good agreement with the measured data (see Fig. 4 and Fig. 5).

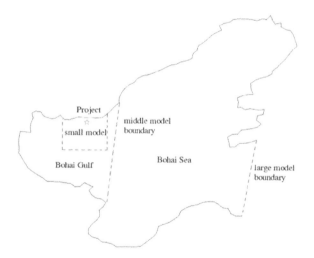

Figure 3.   Three levels nested grids.

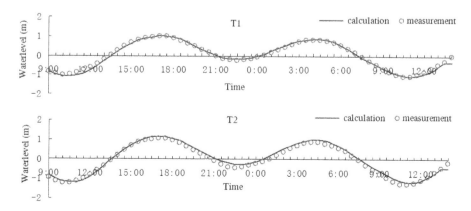

Figure 4.   Comparison of computed and measured tide levels.

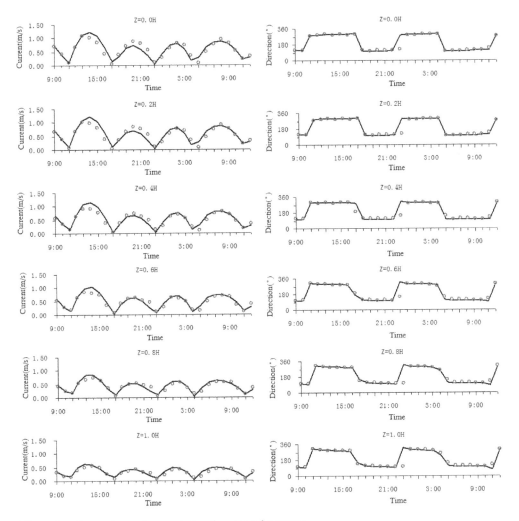

Figure 5.   Comparison of computed and measured currents.

3   HYDRODYNAMIC ENVIRONMENT IN THE PROCESS OF CLOSURE

The final closure is located on the middle of the cofferdam (Fig. 2), whose width is 100 m. The protection layer on bottom has been finished, the top elevation of which is −4 m (Fig. 6). The key work of the closure is to fill the 100 m wide gap from −4 m to 1.5 m. There are two situations. One is that the fill is achieved (Fig. 6-(1)); another is that the fill is only half complete (Fig. 6-(2)). The currents in the two situations are simulated by the verified model above.

Figure 7 and 8 show the currents around the final closure in case 1 and case 2 respectively. It is known that current in the cofferdam is weaker than outside. The current in the closure is strongest. The outside current in flood tide flows from southeast to northwest along the cofferdam; one in ebb tide reverses. The directions of currents in closure are perpendicular to that outside in case 1. The case 2 is different. Because of the half fill, the inflow and outflow mainly flow in the unfilled section. So the current in the filled section in case 2 is weaker than that in case 1.

We set 11 characteristic points (t1~t11) in the closure section (Fig. 9). The maximum and time-averaged velocities of the characteristic points are shown in Figure 10. The velocities in case 1 are larger than that in case 2 because that the discharge area of case 1 is small than

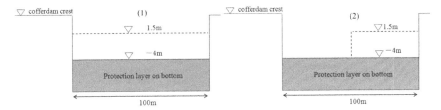

Figure 6.   The two situations in the process of closure ((1) is case 1, (2) is case 2).

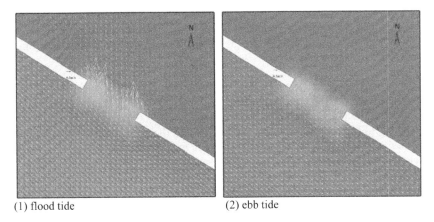

(1) flood tide   (2) ebb tide

Figure 7.   The currents around closure in case 1.

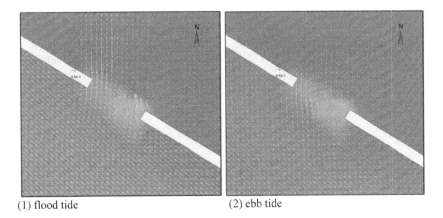

(1) flood tide   (2) ebb tide

Figure 8.   The currents around closure in case 2.

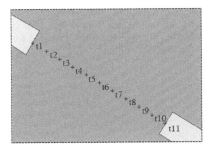

Figure 9.   The layout of characteristic points.

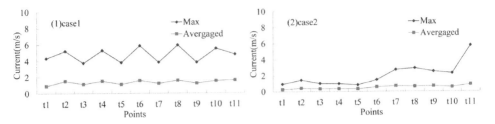

Figure 10.   The statistical velocities in characteristic points.

that of case 2. The time-averaged velocity in case 1 is 1.34 m/s approximately; that in case 2 is 0.52 m/s. The maximum velocity in case 1 is 4.73 m/s approximately; that in case 2 is 2.08 m/s.

4   CONCLUSION AND SUGGESTIONS

1. The model is able to predict the tidal current around the project reasonable. The simulated water level and current basically agreed with the measured data.
2. The current in flood tide outside the cofferdam flows from southeast to northwest along the cofferdam; one in ebb tide reverses. The directions of currents in closure are perpendicular to that outside mainly. The current in the closure section is stronger than that around.
3. The time-averaged velocities of the closure section in case 1 and case 2 are 1.34 m/s and 0.52 m/s respectively. The maximum velocities in case 1 and in case 2 are 4.73 m/s and 2.08 m/s respectively.
4. The tide around the project belongs to the standing wave, in which the maximum currents occur in the middle water levels and the currents are weak in the high and low water levels. So it is advantaged to the closure that the final stage is arranged in the period of the high or low water levels. In addition, it is also advantageous to closure in neap tide.

ACKNOWLEDGMENTS

The research was supported by the National Natural Science Foundation of China under grant no. 51209111.

REFERENCES

MIKE21& MIKE 3 FLOW FM hydrodynamic and transport module Scientific Documentation, MIKE by DHI 2009.

Yang, H., Zhao, H.B., et al. 2005. Analysis of hydrodynamic and sediment transport of Caofeidian sea area. *Journal of Waterway and Harbor*, (3): 130–133. (Chinese).

Zhang, N. 2011. Report of current and sediment transport about the coal terminal in Caofeidian Harbour based on numerical simulation. Tianjin Research Institute for Water Transport Engineering. (Chinese).

Hydraulic Engineering III – Xie (Ed)
© *2015 Taylor & Francis Group, London, ISBN 978-1-138-02743-5*

The land use changes in Qingdao from the last two decades

Qiuying Li
Institute of Geographic Sciences and Natural Resources Research, Chinese Academy of Sciences, Beijing, China
University of Chinese Academy of Sciences, Beijing, China

Chuanglin Fang
Institute of Geographic Sciences and Natural Resources Research, Chinese Academy of Sciences, Beijing, China

ABSTRACT: This paper aims to analyze the land use changes from 1990 to 2008 using remote sensing data taking Qingdao as a case study area. We identify the pattern of land use change using the Geographic Information Systems (GIS) methods. From the analysis of the land use transforming matrix, the most significant change is the cultivated land transiting to construction land. From 1990 to 2008, a total of 123.05 km² of cultivated land has converted to construction land, accounting for 26.63% of the cultivated land in all. Through the analysis of the factors of land use transition, the human dimensions that have influenced the land uses changes are population factor and economy factor.

1 INTRODUCTION

1.1 *Background*

Along with the rapid growth of urban populations comes rapid urban expansion. The total globe urban area quadrupled during the period from 1970 to 2000 (Li et al. 2013; Seto et al. 2005). Though urbanization promotes socioeconomic development and improves quality of life, urban expansion inevitably converts to the natural and semi-natural ecosystems into impervious surfaces and thus has tremendous ecological and environmental consequences, such as forest loss and fragmentation (Miller, 2012; Sutton, 2003), local and regional climate change (Kalnay, et al. 2003), hydrological circle alternation (Simon et al. 2010; Pataki et al. 2009; McMichael, 2000).

With the IGBP and LUCC programs proposed by the United Nation in 1995, the studies have been systematic and extensive. With the fast development of economy, the land use has been changed obviously as tall buildings appeared and cultivated land disappeared. It has shown significant regional variations on spatial patterns. Cultivated land use transition is an important part in the transition because its suitability to live and proximity to city (Li et al. 2013).

The process and mechanism of the transition of cultivated land into non-agricultural land and its relationship with some weakness as labor force change and food security also attract close attention (Zuo et al. 2013).

1.2 *Type area*

Use Qingdao is a core of Shandong peninsula urban agglomeration which is one of the most significant agglomeration in China. It is a coastal city of Huanghai Sea. The study of land use transition in this area is essential and necessary. This study aims to investigate the land uses changes in Qingdao from 1990 to 2008. Specifically, we attempted to address two questions: what are the changes of the land use in Qingdao. Transition during three time periods (1990–2000–2008) was detected based on multi-temporal remote sensing images. We applied variance portioning to examine the importance of the driving human factors.

2 THE STUDY METHODS AND DATA

2.1 *The study area and data*

Qingdao district (between 119°30′–121°00′E and between 35°35′–37°09′N) is located in the south of Shandong peninsula urban agglomeration, with a total area of approximately 1405 km², roughly 37.7% is flat, 21.7% is basin, 25.1% is hill and 15.5% is mountainous. It is a coastal city of Huanghai Sea with the coastline 730.64 km². The mountainous areas are mostly located in the east, with an average elevation of approximately 600–1200 m, while the plains are in the center and south and west, with an elevation ranging from 100 to 400 m (Fig. 1).

2.2 *The land use of Qingdao district*

Landsat TM data in 1990, 2000 and 2008 are used in the study. Before the image analysis, we carry out remote sensing image preprocessing, which includes remote sensing image radiometric correction, geometric correction, image cropping, image enhancement (contrast enhancement, color conversion, spatial filtering, principal component transforming), etc. The main purpose is to eliminate the image distortion, enhance the image quality, and obtain the

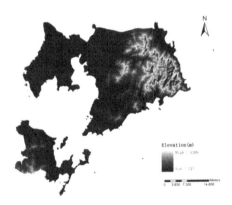

Figure 1. The elevation of Qingdao.

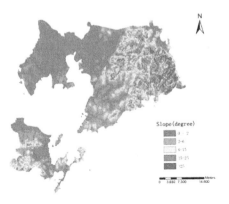

Figure 2. The slope of Qingdao.

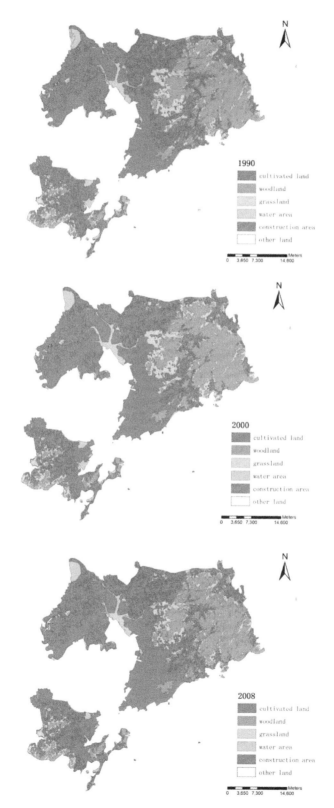

Figure 3.   Land use of Qingdao in 1990, 2000, 2008.

Table 1. Data and its type.

Format	Year	Data	Remarks
Raster	1990	TM	30 m resolution
	2000	ETM+	15 m resolution
	2008	TM	30 m resolution
	2008	DEM	From NASA 30 m resolution
Vector	2008	Shapefile	From National Road Traffic Data

same coordinate system for the images with different temporal and spatial projection in order to conduct the comparison and integration of different date. After being preprocessed in the ArcGIS 9.3, the images are interpreted visually into land use type layers of Qingdao City in the years of 1990, 2000 and 2008 (Fig. 3). Then, according to national land use classification criteria and the purpose of this study, the land use types can be divided into six categories, namely town, cultivated land, orchards, woodland, grassland, water area, other construction land, other land (bare land, beaches, etc.) Elevation (115–1209 m, five categories) and slope (0°–48°, five categories) are generated by DEM data in Arcgis 9.3. (Table 1).

3 RESULTS AND ANALYSIS

3.1 *The analysis of land use transition*

3.1.1 *The analysis of land use from 1990 to 2000*
In 1990, the cultivated land is 462.08 km^2, woodland is 279.67 km^2, the grassland is 149.06 km^2, the water land is 55.49 km^2, construction land is 394.47 km^2, other land is 0.66 km^2. From 1990 to 2000, the cultivated land reduced to 413.54 km^2 with a transition to cultivated land about 47.44 km^2 and a transition to grassland about 1.66 km^2. The water land area increase of 1.11 km^2 with a transition to cultivated land. The construction land is rising 49.09 km^2 of 12.44%.

3.1.2 *The analysis of land use from 2000 to 2008*
In 2008, the cultivated land is 339.03 km^2 from 413.54 km^2 in 2000 with a decrease of 72.01 km^2. There is little changes of woodland and grassland in the total number. The water area drops about 2.58 km^2. The construction land is increasing from 443.57 km^2 to 538.92 km^2 with a transition of 72.01 km^2 from cultivated land, 1.52 km^2 from woodland, 15.79 km^2 from grassland and 6.24 km^2 from water area.

3.1.3 *The analysis of land use from 1990 to 2008*
From the analysis of the land use matrix, the most significant change is the cultivated land transiting to construction land. From 1990 to 2008, a total of 123.05 km^2 of cultivated land has converted to construction land, accounting for 26.63% of the total cultivated land. And 47.44 km^2 has been converted from 1990 to 2000 with 4.85 km^2 per year, 74.51 km^2 has been converted from 2000 to 2008 with 9.13 km^2 per year. The reduced rate in the second stage is 1.92 times than the first stage.

3.2 *The analysis of influencing factors from human dimensions aspect*

The main factors that cause the expansion of urban area increasing were the population, GDP and the Developing District of Industry. Population density was the one of the most vital factors. The factor was relatively more important for population compared to other factors. In 2008, the urban population raised 2.76 million from 1.31 million in 1990. GDP was 1.80 billion in 1990, then it increased rapidly to 44.36 billion in 2008. The industry developing required lots of land as basic condition. Economic development relied on the construction of

Table 2. The land use transition from 1990 to 2000.

	Cultivated land	Wood land	Grass land	Water area	Construction	Other	Total
Cultivated land	413.54	0.00	0.00	0.00	0.00	0.00	413.54
Wood land	0.00	279.67	0.00	0.00	0.00	0.00	279.67
Grass land	0.00	0.00	147.40	0.00	0.00	0.00	147.40
Water area	1.11	0.00	0.00	55.49	0.00	0.06	56.66
Construction	47.44	0.00	1.66	0.00	394.47	0.00	443.56
Other	0.00	0.00	0.00	0.00	0.00	0.60	0.60
Total	462.08	279.67	149.06	55.49	394.47	0.66	1341.43

Table 3. The land use transition from 2000 to 2008.

	Cultivated land	Wood land	Grass land	Water area	Construction	Other	Total
Cultivated land	339.03	0.00	0.00	0.00	0.00	0.00	339.03
Wood land	0.00	278.15	0.00	0.00	0.00	0.00	278.15
Grass land	0.00	0.00	129.04	0.00	0.00	0.00	129.04
Water area	1.11	0.00	2.36	52.03	0.21	0.00	55.71
Construction	72.01	1.52	15.79	6.24	443.36	0.00	538.92
Other	1.40	0.00	0.21	0.02	0.00	0.61	2.24
Total	0.00	0.00	0.00	0.00	0.00	0.00	0.00

infrastructure, including transportation, telecommunications and utility services, all of which required abundant industrial land provisions. Tsingtao Economy & Technology Development Area was the one of the first Development Area of national grade established in 1984. The core of the zone was planned to 20.02 km².

4 CONCLUSIONS

Landsat TM data in 1990, 2000 and 2008 were used in the study. Then, according to national land use classification criteria and the purpose of this study, the land use types can be divided into six categories, namely town, cultivated land, orchards, woodland, grassland, water area, other construction land, other land. From the analysis of the land use matrix, the most significant change is the cultivated land transiting to construction land. From 1990 to 2008, a total of 123.05 km² of cultivated land has converted to construction land, accounting for 26.63% of the total cultivated land. The main factors that cause the expansion of urban area increasing were the population, GDP and the Developing District of Industry.

ACKNOWLEDGMENTS

This research was financially supported by the major research plan of the Ministry of land and resources of public welfare scientific research (No. 201411014-2).

REFERENCES

Kalnay, E. and M. Cai, "Impact of urbanization and land-use change on climate," Nature, 423(6939): pp. 528–531, 2003.
Li, J.X., et al., "Spatiotemporal pattern of urbanization in Shanghai, China between 1989 and 2005," Landscape Ecology, 28(8): pp. 1545–1565, 2013.

Li, X.M., W.Q. Zhou, and Z.Y. Ouyang, "Forty years of urban expansion in Beijing: What is the relative importance of physical, socioeconomic, and neighborhood factors?" Applied Geography, 38: pp. 1–10, 2013.

McMichael, A.J., "The urban environment and health in a world of increasing globalization: issues for developing countries," Bulletin of the World Health Organization, 78(9): pp. 1117–1126, 2000.

Miller, M.D., "The impacts of Atlanta's urban sprawl on forest cover and fragmentation," Applied Geography, 34: pp. 171–179, 2012.

Pataki, D.E., et al., "An integrated approach to improving fossil fuel emissions scenarios with urban ecosystem studies," Ecological Complexity, 6(1): pp. 1–14, 2009.

Seto, K.C. and M. Fragkias, "Quantifying spatiotemporal patterns of urban land-use change in four cities of China with time series landscape metrics," Landscape Ecology, 20(7): pp. 871–888, 2005.

Simon, D. and H. Leck, "Urbanizing the global environmental change and human security agendas," Climate and Development, 2(3): pp. 263–275, 2010.

Sutton, P.C., "A scale-adjusted measure of "Urban sprawl" using nighttime satellite imagery," Remote Sensing of Environment, 86(3): pp. 353–369, 2003.

Zuo, L.J., et al., "Spatial Exploration of Multiple Cropping Efficiency in China Based on Time Series Remote Sensing Data and Econometric Model," Journal of Integrative Agriculture, 12(5): pp. 903–913, 2013.

Hydraulic Engineering III – Xie (Ed)
© 2015 Taylor & Francis Group, London, ISBN 978-1-138-02743-5

Spectrum analysis of measured wave in Andaman Sea, Myanmar

Hui Xiao
Tianjin Research Institute for Water Transport Engineering, Key Laboratory of Engineering Sediment, Ministry of Transport, Tianjin, China

ABSTRACT: Based on the measured wave data within one year in Andaman Sea, Myanmar, the wave frequency spectrum and the directional spectrum are calculate by fast Fourier transform method. The frequency spectrum results show that single peak spectrum is given priority to this sea area, the annual average frequency spectrum peak period for 8 s, the wave spectrum energy is biggest in summer, followed by autumn, spring and winter; the directional spectrum results show that the wave energy concentrate angle is between 200 °~250 °, the main wave direction is southwest.

1 INTRODUCTION

In ocean engineering applications, the appearance characteristics of ocean wave (wave height, period, wave direction, wave steepness, wave crest, wave trough, wave type, and so on) are generally take as the research basis, but with the continuous development of engineering research methods and measure methods, to understand the internal structure of the waves is becoming more and more important. Research and analysis of the wave spectrum, can effectively understand the internal structure and determine the corresponding appearance characteristics of the waves, which has great significance in ocean engineering applications.

In this paper, based on the measured wave data within one year at a fixed point in Andaman sea, Myanmar, the wave frequency spectrum and the directional spectrum are calculate by Fast Fourier Transform (FFT) method, the wave spectrum characteristics of this engineering sea area are obtained, it can provide scientific basis for engineering construction.

2 MEASURE DATA

Wave measure period is from March 7, 2012 to March 6, 2013 (Yang, S.S & Han, X.J 2013), measure site (marked in red) as shown in Figure 1.

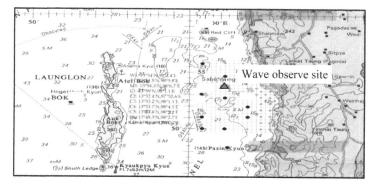

Figure 1. Wave observe site.

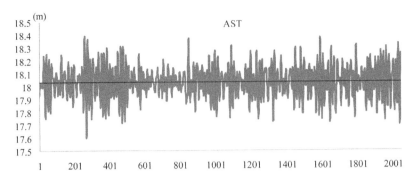

Figure 2. Measured wave process.

The wave measured automatically by the acoustic Doppler current Profiler ("AWAC" wave Dragon) (Liu, J & Lu, Z 2009). Measured data contains pressure, unit velocity along three oblique beams and wave height along a vertical beam (AST), the sampling frequency is 2 Hz, and each pulse emission sampling number is 2048, wave record every hour, each time record 18 minutes (no less than 100 waves).

Figure 2 shows the measured wave process (AST) for a certain period.

3 CALCULATE METHOD

3.1 Frequency spectrum calculate

Frequency spectrum reflects the wave energy distribution relative to the different frequency of each wave component, and the wave spectrum analysis is transforming the wave process from time domain to frequency domain (Sun, L et al. 2008, Wang, Y.Y et al. 2011, Yu. Y.X 2000).

Set the continuous wave surface record at a fixed point for $x(t)$, it can be transformed the time domain $x(t)$ to frequency domain $X(\omega)$ by Fourier transform, and using Parseval theorem can obtain the mean wave energy as

$$\overline{P_x} = \lim_{T \to \infty} \frac{1}{2T} \int_{-T}^{T} \{x(t)\}^2 dt = \lim_{T \to \infty} \frac{1}{4\pi T} \int_{-\infty}^{\infty} |X(\omega)|^2 d\omega \tag{1}$$

For discrete values, there is

$$X(\omega) = \sum_{n=1}^{N} x_n e^{i\omega n \Delta t} \Delta t, \quad T = N\Delta t \tag{2}$$

Therefore, the one-sided spectral density function can be written as

$$S(\omega) = \frac{1}{2\pi N \Delta t} \left| \sum_{n=1}^{N} x_n e^{i\omega n \Delta t} \Delta t \right|^2$$

$$= \frac{\Delta t}{2\pi N} \left| \sum_{n=1}^{N} x_n e^{i\omega n \Delta t} \right|^2 \qquad |\omega| < \pi / \Delta t \tag{3}$$

When $\Delta t = 1$, One-sided spectral value is

$$\hat{S}(\omega) = \frac{1}{2\pi N} \left| \sum_{n=1}^{N} x_n e^{i\omega n} \right|^2 \qquad |\omega| < \pi \tag{4}$$

While $\omega_r = 2\pi r/N$ $r = 0, 1, 2, ..., N/2$, then

$$\hat{S}\left(\frac{2\pi r}{N}\right) = \frac{1}{2\pi N}\left|\sum_{n=1}^{N} x_n e^{i\frac{2\pi r}{N}(n-1)} e^{i\frac{2\pi r}{N}}\right|^2$$

$$= \frac{1}{2\pi N}\left|\sum_{n=1}^{N} x_n e^{i\frac{2\pi r}{N}(n-1)}\right|^2 \tag{5}$$

Set $x_n = \eta_{n-1}, k = n-1$, then

$$\hat{S}\left(\frac{2\pi r}{N}\right) = \frac{1}{2\pi N}\left|\sum_{n=1}^{N} \eta_{n-1} e^{i\frac{2\pi r}{N}(n-1)} e^{i\frac{2\pi r}{N}}\right|^2$$

$$= \frac{1}{2\pi N}\left|\sum_{n=0}^{N-1} \eta_k e^{i\frac{2\pi r}{N}k}\right|^2 \tag{6}$$

Set

$$A_r = \sum_{n=0}^{N-1} \eta_k e^{i\frac{2\pi r}{N}k} \tag{7}$$

Then

$$\hat{S}\left(\frac{2\pi r}{N}\right) = \frac{1}{2\pi N}\left|A_r\right|^2 \quad r = 0, 1, 2, ..., N/2 \tag{8}$$

Therefore, the problem of frequency spectrum calculation is how to solve the coefficient A_r by Equation 7. Separate the odd item and even item of A_r can get

$$A_r = \sum_{k=0}^{N/2-1} \eta_{2k} e^{i\frac{2\pi r}{N}2k} + e^{i\frac{2\pi r}{N}} \sum_{k=0}^{N/2-1} \eta_{2k+1} e^{i\frac{2\pi r}{N}2k}$$

$$= B_r + W^r C_r \tag{9}$$

By discrete Fourier transform of Equation 9, N coefficient A_r can be calculated by the samples $[\eta_k]_N$. First divided $[\eta_k]_N$ into samples $[\eta_{2k}]_{N/2}$ and $[\eta_{2k+1}]_{N/2}$ to compute N coefficient B_r and N coefficient C_r, then split the samples sequence in two, then two to four and always split down, until each sequence has only one entry and then can get A_r. While the samples order will change, the last rearrange order is the "reverse binary" of the original order.

3.2 *Directional spectrum calculate*

Extension the solution of two-dimensional wave equation to a three dimensional wave spectrum expression can get the directional spectrum; it can be decomposed into a frequency spectrum and an orientation distribution function, such as

$$S(f,\theta) = S(f)G(\theta,f) \tag{10}$$

$$\int_{-\pi}^{\pi} G(\theta,f)d\theta = 1 \tag{11}$$

The orientation distribution function can be expressed as

$$G(\theta,f) = \frac{1}{\pi}\left[\frac{1}{2} + \sum_n (a_n \cos\theta + b_n \sin\theta)\right] \tag{12}$$

$$\begin{cases} a_1 = \dfrac{S_{xu}}{\left|S_{xx}\left(S_{uu}+S_{vv}\right)\right|^{1/2}} & a_2 = \dfrac{S_{uu}-S_{vv}}{S_{uu}+S_{vv}} \\[4mm] b_1 = \dfrac{S_{xv}}{\left|S_{xx}\left(S_{uu}+S_{vv}\right)\right|^{1/2}} & b_2 = \dfrac{2S_{uv}}{S_{uu}+S_{vv}} \end{cases} \tag{13}$$

where, S_{xx}, S_{uu}, S_{vv}—auto spectrum; S_{xu}, S_{xv}, S_{uv}—cross spectrum; x—wave process; u,v—horizontal and vertical velocity in the horizontal wave surface direction.

4 RESULTS AND ANALYSIS

4.1 *Frequency spectrum*

Figure 3 shows the average wave frequency spectrum of each quarter (Which March to May for the spring, June to August for the summer, September to November for the autumn, the rest for the winter) and the whole year in Andaman sea area, it shows that:

1. The wave frequency spectrum gives priority to unimodal characteristics; it suggests that the measured sea area is affected by the superimposition of wind wave and swell.
2. The wave energy is relatively large in summer, the peak value of frequency spectrum is 1.00 m²/Hz, peak-period corresponds to 8 s, and it means the wave has the characteristic of big wave height and short wave period.
3. The frequency spectrum peak value in spring, autumn and winter are all below 0.02 m²/Hz, corresponding spectrum peak period between 11 ~ 13 s, and it means the wave has the characteristic of small wave height and long wave period.
4. Annual average frequency spectrum is a single peak spectrum; the main peak (maximum energy) corresponds period is about 8 s.

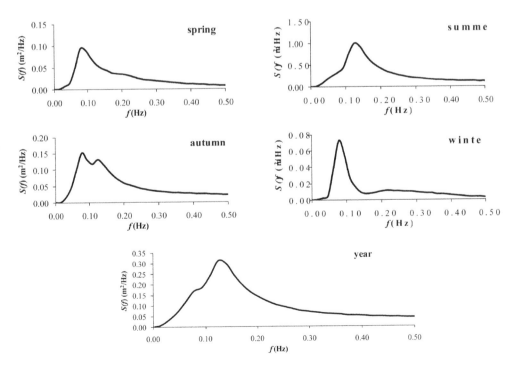

Figure 3. Average frequency spectrum of each quarter and year.

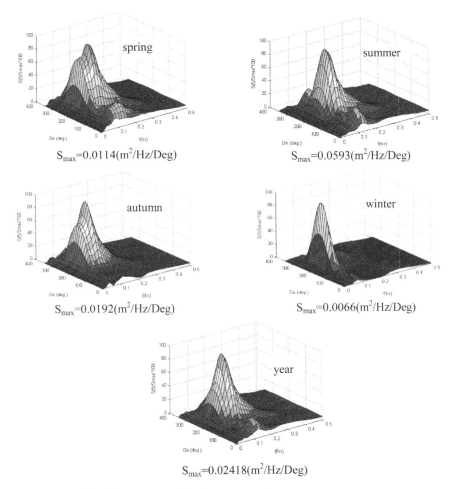

$S_{max}=0.0114(m^2/Hz/Deg)$ $S_{max}=0.0593(m^2/Hz/Deg)$

$S_{max}=0.0192(m^2/Hz/Deg)$ $S_{max}=0.0066(m^2/Hz/Deg)$

$S_{max}=0.02418(m^2/Hz/Deg)$

Figure 4.   Average direction spectrum of each quarter and year.

4.2 *Directional spectrum*

Figure 4 shows the average directional spectrum of each quarter and the whole year. As can be seen from the Figure 4, the mean wave energy is concentrate in the period 11 s~13 s in spring, autumn, and winter, and concentrate in 7 s~8 s in summer, this result is consistent with the frequency spectrum result; Each season and annual average wave energy concentrate angle are between 200 ° ~ 250 °, that means the main wave direction is southwest in this sea area.

5   CONCLUSION

According to the measured wave data within one year in Andaman Sea, Myanmar, the quarterly and the annual average frequency spectrum and average directional spectrum are calculate by fast Fourier transform method, the results show that the sea is given priority to unimodal spectrum, the wave energy is largest in summer, followed by autumn, spring and winter, the wave energy concentrate angle is between 200 ° ~250 °, the main wave direction is southwest. The wave height is smaller but period is long in autumn, spring and winter, which also need pay attention in project construction.

REFERENCES

Liu, J. Lu, Z. 2009. Application of AWAC in Hydrographic in Huanghua Port & Comparative Analysis for AWAC and Ocean Current Instrument, *Port engineering technology* 46(z1):94–96.

Sun, L. Wang, S.L. Lv, H.M. 2008. Measurement and Spectrum Analysis of the High-frequency Waves, *Marine science bulletin* 27(6):9–14.

Wang, Y.Y. Xu, J.W. He, L. 2011. Computational analysis of wave spectral estimation, *Journal of Harbin Engineering University* 32 (10):1283–1289.

Yang, S.S. Han, X.J. 2013. *The 5 million ton refinery project wharf engineering in Myanmar, Guangdong Zhenrong Energy co., LTD. measured data of wave, tide, wind statistics and analysis report*, Tianjin: Tianjin Survey and Design Institute of water transport engineering.

Yu, Y.X. 2000. *Random wave and its applications to engineering*, Dalian: Dalian University of Technology Press.

Hydraulic Engineering III – Xie (Ed)
© 2015 Taylor & Francis Group, London, ISBN 978-1-138-02743-5

Tidal current modeling of the yacht marinas of Double Happiness Island in Xiamen

Chun Chen & Yifeng Zhang
Tianjin Research Institute for Water Transport Engineering, Key Laboratory of Engineering Sediment, Ministry of Transport, Tianjin, China

ABSTRACT: The tidal current problems related to the yacht marinas of Double Happiness Island in Xiamen was studied. Natural conditions of project area were collected and analyzed. The hydrodynamic movement characteristics were mastered. 2D mathematical models with irregular triangular grid of tidal current were set up. The study on tidal current before and after project were carried out. Effects to the hydrodynamic of the around sea area were analyze. This study provides the basis and technical support to the planning and design.

1 INTRODUCTION

The yacht economy status with the development of marine economic activities and the rapid rise In recent years. At present, the Yangtze River Delta, Pearl River Delta, Hainan Island, Zhou Shan islands, Tianjin, Fujian and other regions of the large marinas and yacht base facilities are in the planning or construction. In this paper, by using numerical simulation method to research Xiamen Happiness Island yacht wharf engineering hydrodynamic conditions. Through collecting and analyzing the waters natural conditions near the project, master of engineering waters hydrodynamic characteristics, using a two-dimensional mathematical model of tidal flow to the program after the implementation of the tidal current field of research work has been carried out. The research results can provide basis and support for the design.

2 PROJECT OVERVIEW

Happiness Island planning in Xiamen Bay Dapan shallow, the west is the Nantaiwu beach, engineering geographical position as shown in Figure 1.

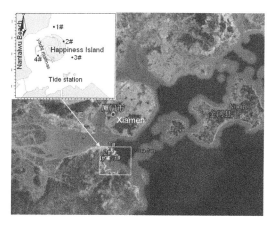

Figure 1. Sketch of the project position.

The yacht wharf area is arranged at the island on the west side of Pisces. the waters of an area of about 200000 m². gate width is 60 m, design elevation of berth port is −7.0 m (referred to the 85 national height datum), channel width is 55 m, bottom elevation of −6.0 m. The yacht dock West Port breakwater is considered as a ramp type pier.

3 THE PROJECT AREA HYDRODYNAMIC CHARACTERISTICS

3.1 *Tide*

According to the 2005 November the tidal observation (Zhen, 2005) statistics, tidal parameter F = 0.36, the average tidal range is 3.85 m, the maximum tidal range 5.51 m. Its Consistent with the tidal data of Xiamen ocean stations.

3.2 *Wave*

During 1995 ~ 1996 the establishment of the short wave station wave observation in Dapan shallow (He, 2004), through analysis, the sea wave direction is ESE, the frequency was 34.26%, secondary wave direction is SE, frequency 14.67%. Strong wave direction is SE, the maximum wave height during the period of observation was 2.7 m, appeared in August 6, 1996, corresponding to the $H_{1/10}$ is 2.5 m.

3.3 *Current*

Tidal current in Xiamen port belongs to the semidiurnal tide, reciprocating flow. According to the analysis of hydrological data observed in Dapan shallow sea tide during 2005 November.
　Dapan shallow tidal power is weak, the average velocity at flood tides is 0.38 ~ 0.46 m/s, the average velocity at ebb tides is 0.32 ~ 0.34 m/s, the maximum flow at flood tides is 0.80 m/s, the maximum flow at ebb tides is 0.74 m/s.

4 2-D TIDAL FLOW MATHEMATICAL MODEL

4.1 *Mathematical model establishment*

1. Calculation software
   The calculation software is used Mike21 series of FM module. The software developed by the Danish hydraulic, using unstructured triangular mesh.
2. The computational domain and grid
   The total area of the model simulation is 6000 km². Consists of 25295 grid node computing the number of mold, the minimum size of less than 10 m, can ensure sufficient grid resolution. Computational grid is shown in Figure 3. Floating pier structure within the marina is not considered in the calculation.
3. Boundary conditions
   The mathematical model of the offshore boundary provides (Li, 2007) by Chinatide software.

4.2 *Model verification*

Tidal level, flow rate verification specific verification sees (Chen, 2012). The model results in conformity with the provisions of the Ministry of transport "Technical code for simulation of tidal current and sediment of Estuary and coast".

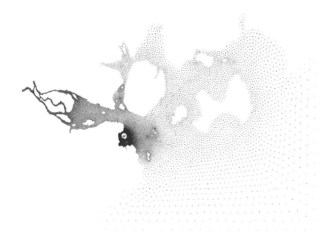

Figure 2. Computation grid.

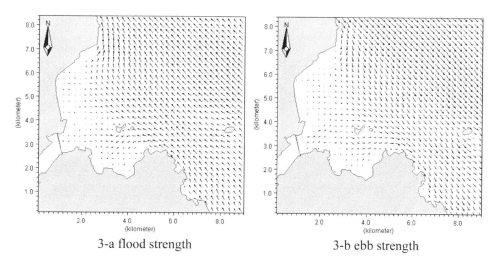

3-a flood strength   3-b ebb strength

Figure 3. Flow field at flood and ebb strength of tide before project.

5 NUMERICAL SIMULATION AND ANALYSIS

5.1 The flow field before and after the project

Figures 3–5 is the flow field at flood and ebb strength of tide before and after project. According to the results of numerical simulation, the project has the following characteristics of tidal current:

1. The trend of a reciprocating motion along the coast line, the flood tide point to NW, ebb tide point to SE; velocity distribution is lower trend from open to coastal.
2. After the project, the construction does not change the trend movement of large range, but only Dapan shallow because of affected by the project.
3. In the eastside of Double happiness Island flowing more smoothly, south breakwater outside Marina the maximum velocity exceeds 0.8 m/s. the maximum velocity in the yacht dock port entrance is about 0.22 m/s; the current outside entrance average velocity at 0.2 ~0.4 m/s; velocity in yacht harbor is low tide, mean velocity below 0.2 m/s.

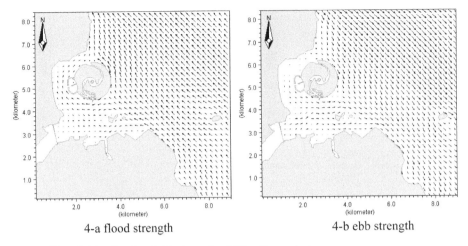

4-a flood strength　　　　　4-b ebb strength

Figure 4.　Flow field at flood and ebb strength of tide after project.

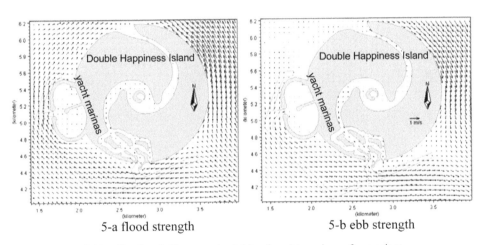

5-a flood strength　　　　　5-b ebb strength

Figure 5.　Flow field at flood and ebb strength of tide of yacht marinas after project.

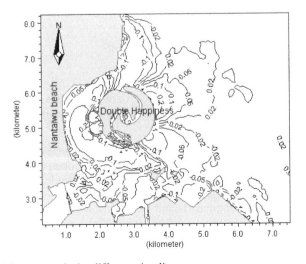

Figure 6.　Total tidal average velocity difference isoclines.

5.2 The impact on the surrounding water tidal power

Figure 6 is a contour map the average tidal velocity variation, compared with the pre engineering, flow velocity in the East west sides of Island increased. The maximum growth rate of about 0.3 m/s; velocity decreases in south and north sides of island, the maximum reduction of about 0.5 m/s.

5.3 The maximum channel cross flow

The cross flow in yacht wharf harbor entrance channel generally within 0.2 m/s; turning section of channel cross flow velocity is within 0.5 m/s; the channel line due to the ebb and flow of main to smaller and channel angle, maximum flow within 0.2 m/s.

6 CONCLUSION

This paper focuses on the research of Xiamen Double Happiness land yacht wharf engineering hydrodynamic, the flow field after project was simulated by means of a 2-D tidal flow mathematical model research, the result shows:

1. The trend of a reciprocating motion along the coast line, the flood tide point to NW, ebb tide point to SE; velocity distribution is lower trend from open to coastal. In the bay mouth outside, the average velocity is about 0.4 m/s, the average velocity inside bay is less than 0.2 m/s.
2. After project the flow is smoothly, the yacht harbor breakwater outside the local flow velocity more than 0.8 m/s. The yacht port door maximum velocity is about 0.22 m/s, the velocity in yacht port is low, the average velocity below 0.2 m/s.
3. The cross flow in yacht wharf harbor entrance channel generally within 0.2 m/s; turning section of channel cross flow velocity is within 0.5 m/s; the channel line due to the ebb and flow of main to smaller and channel angle, maximum flow within 0.2 m/s.

REFERENCES

Chen, Z.X. 1994. Chinese bays in eighth volumes (Southern Fujian Bay). Beijing: Ocean Press.
Chen, C & Cai, Y. 2012. Solution Mingxiao. Pisces Island project of wave current sediment mathematical Marina Model Test Research Report. Tianjin: Tianjin Scientific Research Institute of water transport engineering.
He W.Q. On the Design of Marin. Port and Waterway Engineering, 2004(3):61–64.
Li, M.G. & Zheng, J. 2007, Introduction to Chinatide software for tide prediction in China seas, Journal of Waterway and Harbor, 28(1):65–68.
Zheng, B.X & Shu, F.F & He, J. Zhangzhou Development Zone Investment Promotion Bureau of artificial island project (sea garden project) sediment test and calculation of waves. Journal of Xiamen: the third Oceanography Research Institute, 2005.

Hydraulic Engineering III – Xie (Ed)

Prediction of flow over thin-plate rectangular weir

Yan Li

Tianjin Research Institute for Water Transport Engineering, Key Laboratory of Engineering Sediment of Ministry of Transport, Tianjin, China

ABSTRACT: The thin-plate rectangular weir is a measuring water device commonly used in laboratories; its measure accuracy of flow depends on the precision of weir flow calculated formula, weir manufacture and installation. This paper discusses four possible overflow shapes and its influence on the flux measure precision of the weir flow, summarizes the technical requirements of thin-plate rectangular weir production, and propose a calculation formula of weir flow which theoretically deduced by the weir width, weir head and flume water depth, the weir flow calculation result of this formula is very close to of Rehbock empirical formula which popular applied at present, it can be used for engineering application.

1 INTRODUCTION

The measuring weir composed by weir-body, flume, point gauge and other facilities, in hydraulic engineering and river model test, it is commonly used to measure the flow discharge. Based on the relatively stable relationship of the weir head and the flow rate, the flow rate can be calculated by empirical formula. In laboratory the rectangular weir and triangular weir are commonly used, its precision depends on accuracy of the weir flow experience formula and the weir installation. In this paper, a calculation formula of weir flow which theoretically deduced and technology requirements of the thin-plate equal-width rectangle measuring weir are discussed, and put it forward for proper use.

2 OVERFLOW SHAPE OF THE THIN-PLATE RECTANGULAR WEIR

The width of the thin-plate rectangular weir is the same as the flume, Weir plates for sharp edges on the top and contact with water in a line (see Fig. 1), the overflow is freedom without

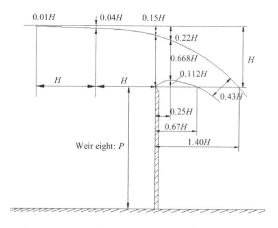

Figure 1. Standard water tongue shape of thin-plate rectangular weir.

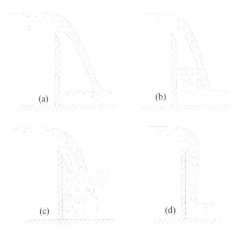

(a)   (b)

(c)   (d)

Figure 2.   Four shape of water tongue.

suppress on both sides, but the top and bottom water surface are shrink by gravity, the over-flow shape like a "tongue", often called "water tongue". For the upper and lower weir flow shape, Bazin and Rehbock has made a detailed experiment, and obtain the tongue shape of free flow without side shrink and submerge as shown in Figure 1. The surface of water tongue on the weir crest sharp edges reduce 0.15 H (H is the weir head) in vertical; horizontal distance from the weir crest of 0.25 H, the upper surface reduce 0.22 H, the lower water surface rise to the maximum height of 0.112 H; horizontal distance from the weir crest of 0.67 H, the lower water surface drop to the weir crest horizontal line, of 1.4 H, the upper surface also dropped to the weir crest horizontal line; when the center of water tongue through the weir crest horizontal line, the tongue thick is 0.43 H (Wu, C.G. 1982).

In fact, a weir flow may have the following four conditions (see Fig. 2): (1) Free flow: when the flow over a standard measuring weir, as the air unblocked under the water tongue, the water flow shape will like a parabola curve (Fig. 2a); (2) Pressure water: when the air pressure under the water tongue less than atmospheric pressure, the weir downstream water rise to a certain height under the water tongue and the upstream water surface of tongue gradually incline to weir walls by the action of atmospheric pressure (Fig. 2b); (3) wet flow: no air exist under the water tongue, filled with turbulence stagnant water (Fig 2c); (4) affix flow: no air, no backwater, the water tongue fall down close to the weir face (Fig. 2d). The flow rate of kind (2) ~ (4) are bigger than that of the kind (1).

The obstruction of the air under the weir flow determines the flow trajectory and flow regime, so to make a thin-plate rectangular measuring weir, ventilation hole must set on the weir wall between the weir plate and the water tongue under the weir, to make sure the weir flow is free-dom, and to guarantee the flow rate calculate accuracy by empirical formula. The diameter of the ventilation hole can be calculated according to the following formula (Sun, J.S. et al. 2003):

$$\phi = 0.11 H_{max} B^{0.5} \qquad (1)$$

where: ϕ—diameter of ventilation hole (m); H_{max}—the biggest water head on the weir (m); B—weir width (m).

3   BASIC FORMULAS FOR CALCULATION OF THIN-PLATE RECTANGULAR MEASURING WEIR FLOW

Set the thin-plate rectangular weir width for B, the weir head for H, the weir high for P, the approach average velocity of whole section in flume for V_0, then the approach velocity head for $h_v = V_0^2/2g$. As the mean velocity above the weir crest is greater than the mean velocity of the whole section in flume, the weir flow velocity head for $aV_0^2/2g$ (α is greater than 1 for a non-uniform velocity coefficient). Divided the cross section into several thin layer, the height for dh, the area for Bdh, the head above the thin layer for h, then the total effective water head for $h + ah_v$, and the ideal flow rate of every thin layer is

$$dQ_h = Bdh\sqrt{2g(h+ah_v)} \tag{2}$$

Integral it from 0 to H can get the total ideal flow Q_0 as follows:

$$Q_0 = B\sqrt{2g}\int_0^H (h+ah_v)^{\frac{1}{2}}dh = \frac{2}{3}B\sqrt{2g}\left[(H+ah_v)^{\frac{3}{2}} - (ah_v)^{\frac{3}{2}}\right] \tag{3}$$

Considering the water beam shrinkage and head loss for local resistance and other factors, the actual flow is less than the ideal flow; it need to correction by multiply a flow coefficient C_e, therefore the actual flow is:

$$Q = C_e\frac{2}{3}B\sqrt{2g}\left[(H+ah_v)^{\frac{3}{2}} - (ah_v)^{\frac{3}{2}}\right] \tag{4}$$

So the approach velocity head can be express as a function of flow rate Q, and the velocity head h_v is relate to the ratio of water head on weir H and water depth of flume d.

Set $C_e\, 2/3\,\sqrt{2g} = C$, multiply and divide by $H^{3/2}$ in the formula right, then:

$$Q = CBH^{\frac{3}{2}}\left[\left(1+\frac{ah_v}{H}\right)^{\frac{3}{2}} - \left(\frac{ah_v}{H}\right)^{\frac{3}{2}}\right] \tag{5}$$

Quote binomial formula to spread items which in the brackets, and omit the smaller items, then:

$$Q = CBH^{\frac{3}{2}}\left[1+\frac{3}{2}\frac{ah_v}{H}\right] \tag{6}$$

The water depth of flume for d, the cross section area for $A = Bd$, the approach mean velocity for $V_0 = Q/A$, consider that h_v is far less than H, temporarily assuming $Q = CBH^{3/2}$, then:

$$h_v = \frac{V_0^2}{2g} = \frac{Q^2}{2gA^2} = \frac{\left(CBH^{\frac{3}{2}}\right)^2}{2g(Bd)^2} = \frac{C^2H}{2g}\left(\frac{H}{d}\right)^2 \tag{7}$$

Then equation (6) can be written as:

$$Q = C\cdot\left[1+\frac{3a}{2}\cdot\frac{C^2}{2g}\left(\frac{H}{d}\right)^2\right]BH^{\frac{3}{2}} \tag{8}$$

$$\text{Or } Q = 2.95C_e\left[1+0.667C_e^2a\left(\frac{H}{d}\right)^2\right]BH^{\frac{3}{2}} \tag{9}$$

It shown that the total flow can be calculate by the formula of weir width B, weir head H and water depth of flume d, and the value of Q is relate to H/d, but the flow coefficient C_e and the non-uniform velocity coefficient α must be obtained by experiment.

4 EXPERIMENT FORMULA OF THIN-PLATE RECTANGULAR WEIR FLOW

In weir flow calculation formula, the flow coefficient C_e is relate to the weir flow shrinkage and head loss for local resistance; non-uniform velocity coefficient α is relate to the inhomogeneous degree of approach velocity in flume cross section and the ratio of weir head H and weir height P. Therefore, the value or expression of C_e and α are not identical by test in specific conditions and certain circumstances at home and abroad.

1. After many experiments, Fraucis get the flow coefficient C_e for 0.623, and not consider the approach velocity, that is $\alpha h_v \approx 0$, then obtain the Fraucis weir flow formula as:

$$Q = 1.839 BH^{\frac{3}{2}} \tag{10}$$

2. In domestic relevant technical specification (Sun, J.S. et al. 2003, Wang, Y.L. et al. 1998), the weir flow can be calculate by Rehbock formula as:

$$Q = \left(1.782 + 0.24\frac{H_e}{P}\right)BH_e^{\frac{3}{2}} \tag{11}$$

In which: $H_e = H + 0.0011\,\text{m}$.

5 COMPARISON OF WEIR FLOW CALCULATE RESULT

In this paper, get the flow coefficient C_e for 0.623 and non-uniform velocity coefficient α for 1.1, then the weir flow formula (9) change as:

$$Q = \left[1.839 + 0.524\left(\frac{H}{P+H}\right)^2\right]BH^{\frac{3}{2}} \tag{12}$$

Assume that the weir width B for 0.6 m, weir height P for 0.6 m, calculate the weir flow of different weir head H by equation (1) ~ equation (12), the result comparison shows in Table 1.

As shown from Table 1, in Fraucis formula, the approach velocity is not considered, which will cause obviously calculated flow errors, and the flow value is very close as calculate by the formula of Rehbock and this article. In fact, if the sectional velocity distribution of weir head under different H/P can be accurate measured, the value α can be determined, and the flow rate can be precise calculated by Equation (9), when for preliminary calculation can also use Equation (12).

6 TECHNICAL REQUIREMENTS OF THIN-WALL RECTANGULAR WEIR MADE

1. The Weir height and Weir width should be determined by the maximum and minimum model flow. When design the weir width, the weir crest head should be guaranteed no

Table 1. Flow value comparison by different formula (unit: *l/s*).

Weir head (m)	Formula		
	Fraucis (10)	Rehbock (11)	This article (12)
0.1	34.892	35.337	35.095
0.2	98.691	100.773	100.449
0.3	181.307	188.594	187.036

less than 3 cm under the minimum model flow, otherwise the overflow water tongue will instability affected by surface tension; when design the weir height, the weir height should be guaranteed greater than 2 times maximum weir crest head under the maximum model flow, otherwise the weir approach velocity impact is too large (Zhou, M.D. et al. 1995).
2. The total flume length should be not less than 19 times the maximum weir head, Weir plate shall be installed not less than 1.6 times the maximum weir head away from the water tank that the overflow water tongue will not spread immediately after weir.
3. The flume must be fixed-width, flume wall should be perpendicular and smooth, weir plate should be vertical orthogonal to the flume, weir crest sharp edges thickness should be no greater than 1 mm, the difference between the water level at the lower tail and the weir crest elevation should be greater than 7 cm, to ensure the flow free.
4. On the flume wall under the water tongue should set ventilation hole to guarantee air smooth, and the diameter of ventilation hole can be calculated by equation (1).
5. The water surface should be smooth for measuring; the measure position can be located at 6 times the maximum weir head in the upstream of weir plate. Wave dissipation gate in the flume should be located at not less than 12 times the maximum weir head in the upstream of Weir plate, in order to get enough smooth length of water surface.
6. The weir should be verified after construction, calibration methods and requirements should according to the "Verification method of common instrument in hydraulic and river model (SL/T 233-1999)" published by the Ministry of Water Resources of the People's Republic of China.

7 CONCLUSIONS

The measuring weir is a measuring water device commonly used in laboratories; its measuring accuracy is associated with the attainment of certain technical requirements of its manufacture and installation. Currently the weir flow calculation formulas are more the empirical formula obtained through experiments under certain conditions, and the Rehbock formula for thin-plate equal-width rectangular weir is relatively simple and precision that always used in domestic hydraulic and physical model test. This paper theoretically deduced a weir flow formula by weir width, weir head and flume water depth, the weir flow calculation result of this formula is very close to of Rehbock empirical formula, it can be used for engineering application.

REFERENCES

Sun, J.S. Hu, Y.A. & Zheng, B.Y. et al. 2003. *Technical Regulation of Modelling for Hydraulics of Navigation Structures (JTJ/T 235-2003)*. China: Ministry of Transport of the People's Republic of China.

Wang, Y.L. Liu, J.M. & Tang, C.B. et al. 1998. *Technical Regulation of Modelling for Flow and Sediment in Inland Waterway and Harbour. (JTJ/T 232-98)*. China: Ministry of Transport of the People's Republic of China.

Wu, C.G. 1982. Hydraulics (Second Edition), Beijing: Higher Education Press.

Zhou, M.D. Zhang, S.M. & Zhou, Y.R. 1995. *Test regulation for normal hydraulic model (SL 155-95)*. China: Ministry of Water Resources of the People's Republic of China.

Hydraulic Engineering III – Xie (Ed)
© 2015 Taylor & Francis Group, London, ISBN 978-1-138-02743-5

Parameters inversion of mortar in concrete considering the hardening process based on meso-mechanics

Yongrong Qiu
Department of Hydraulic Engineering, Tsinghua University, Beijing, China
State Key Laboratory of Simulation and Regulation of Water Cycle in River, Basin, China
Institute of Water Resources and Hydropower Research, Beijing, China

Guoxin Zhang
State Key Laboratory of Simulation and Regulation of Water Cycle in River, Basin, China
Institute of Water Resources and Hydropower Research, Beijing, China

Xiaonan Yin
CGGC International Ltd., Beijing, China

ABSTRACT: It is very important to determine the thermal and mechanical parameters of mortar and concrete in mesoscopic simulation. In this paper, on the basis of the Mori-Tanaka formula of mesoscopic mechanics and the concrete is treated as a two-phase composite material constituted by aggregates and mortar, the inversion of coefficient of thermal expansion, autogenous shrinkage, elastic modulus and creep were studied. This paper proposed some inversion formulas regarding these four mechanical parameters of mortar in concrete. The accuracy of these formulas was verified by FEM numerical test and demonstrated by some examples.

1 INTRODUCTION

From the perspective of material mesostructure, meso-mechanics explores mechanical mechanism of composite materials, and reveals connections between material macro-mechanical behavior and mesostructure. Meso-mechanics is a key method for studying the composite material. As we know, concrete is a typical composite material constituted by coarse aggregate, sand, cement, fly ash and some other admixtures. Because of the hydration reaction of cement and fly ash after mixed with water, the mortar matrix of concrete gradually hardens and bonds together with aggregate. This makes concrete has a unique property: some thermal and mechanical parameters such as hydration heat, elasticity modulus, strength, ultimate tensile strain, autogenous shrinkage are time-depended functions because they will vary according to the age of concrete. As the thermal and mechanical properties of aggregate do not vary with age, the macro-scopical thermal and mechanical characteristics variation with age are mainly caused by cement mortar.

Until now, much research work has been done on the prediction of composite material coefficient of thermal expansion and elastic modulus by forefathers, and many prediction methods have been developed such as the sparse method (Guanlin Shen, et al. 2006), the Self-Consistent Method (Hill R.A. 1965), the Mori-Tanaka method (Mori T, Tanaka K. 1973) and so on. However, none of these formulas take into account the parameters variation with concrete age, and there is little research on the autogenous shrinkage and creep. In the mesoscopic simulation of concrete, thermal and mechanical parameters of mortar and aggregate (coefficient of thermal expansion, autogenous shrinkage, elasticity modulus, creep, strength) are important input parameters. In fact, there is abundant of test data on concrete, but much less data on mortar while it is one of the important components. Also parameter inversion is an essential method to obtain the data, but there are few studies on this so far.

Suppose that concrete is a two-phase composite material constituted by aggregate and mortar, and considering no influence of interface, we will give four inversion formulas of mortar's coefficient of thermal expansion, autogenous shrinkage, elasticity modulus and creep based on Mori-Tanaka method and verifies the veracity of the formulas by numerical load test. Among them, inversion formulas of autogenous shrinkage, elasticity modulus and creep need to consider concrete age and provide parameters for curves.

2 FUNDAMENTAL FORMULAS OF INVERSION

2.1 Mori-Tanaka formula (Mori T, Tanaka K. 1973)

Ignoring the interface's effect, concrete can be considered as a two-phase composite material constituted by aggregate and mortar. If the bulk modulus and shear modulus of mortar and aggregates are given, equivalent bulk modulus K and shear modulus G of concrete can be estimated according to Mori-Tanaka method.

$$K = K_0 + \frac{K_1 - K_0}{1 + 9 f_0 K_p (K_1 - K_0)} f_1, \tag{1}$$

$$G = G_0 + \frac{G_1 - G_0}{1 + 4 f_0 G_p (G_1 - G_0)} f_1, \tag{2}$$

where K_p and G_p are determined by:

$$K_p = \frac{1}{3(4G_0 + 3K_0)}, \tag{3}$$

$$G_p = \frac{3(2G_0 + K_0)}{10 G_0 (4G_0 + 3K_0)}. \tag{4}$$

In Eqs. (1, 2, 3, 4), f_0 and f_1 are volume ratios of mortar and aggregate; K_0, K_0, K_1, G_1 are bulk modulus and shear modulus of mortar and aggregate; The relation formulas between them and elasticity modulus E and Poisson's ratio μ are

$$K = \frac{E}{3(1 - 2\mu)}, \quad G = \frac{E}{2(1 + \mu)}, \tag{5}$$

$$E = \frac{9KG}{3K + G}, \quad \mu = \frac{3K - 2G}{6K + 2G}. \tag{6}$$

2.2 Elasticity modulus

As a hardening material, elasticity modulus of concrete increases as age grows, therefore, the elasticity modulus of concrete and mortar are function related with time. Elasticity modulus of concrete and mortar are usually expressed by Eq. 7 (Bofang Zhu. 1999). If the elasticity modulus of concrete and mortar are given, and the effect of interface is not considered, Eq. 1 and 2 can be utilized to determine the concrete equivalent elastic modulus. For this calculation, concrete age should be discretized firstly, and then the equivalent elastic modulus for each increment of time can be obtained.

$$E(t) = E_u \left(1 - e^{-\alpha t^\beta}\right), \tag{7}$$

where E_u is ultimate modulus, α and β are some fitting parameters.

Conversely, if both of the elasticity modulus of aggregate and concrete are given, the mortar elasticity modulus can be obtained by the inverse function of Mori-Tanaka formula. Because it is difficult to obtain the explicit expression of inverse function for mortar bulk modulus and shear modulus, iterative computation becomes a good choice. Eqs. 8 and 9, which are the iterative formulas to obtain mortar bulk modulus and shear modulus respectively, are transformed by the simplification and deduction of Eqs. 1~4:

$$K_0 = \frac{1}{1-m_K}(K - m_K K_1),$$

(8)

$$G_0 = \frac{1}{1-m_G}(G - m_G G_1).$$

(9)

where m_K and m_G are the function of K_0 and G_0 respectively. Specific formulas are as follow:

$$m_K = f_1 / \left[1 + 9 f_0 K_p \left(K_1 - K_0\right)\right],$$

(10)

$$m_G = f_1 / \left[1 + 4 f_0 G_p \left(G_1 - G_0\right)\right].$$

(11)

2.3 Creep

In fact, concrete is not an ideal elastic material. Under constant stress, strain will increase as the time increases, which is known as creep effect. Creep is not only connected with load time τ, but also connected with concrete's age t. The creep degree is usually be expressed as a variable by the formula as follow (Bofang Zhu.1999):

$$C(t,\tau) = \left(A_1 + A_2/\tau^{\alpha_1}\right)\left(1 - e^{-k_1(t-\tau)}\right) + \left(B_1 + B_2/\tau^{\alpha_2}\right)\left(1 - e^{-k_2(t-\tau)}\right)$$
$$+ De^{-k_3(t-\tau)}\left(1 - e^{-k_3(t-\tau)}\right),$$

(12)

where A_1, A_2, B_1, B_2, D, k_1, k_2, k_3, α_1 and α_2 are some fitting parameters.

If the elasticity modulus and creep degree are given, and the creep effect of the aggregate is neglected, equivalent creep of concrete can be calculated from Eq. 13:

$$K(\tau) C(t,\tau) = f_0 K_0(\tau) C_0(t,\tau),$$

(13)

where $K(\tau)$ is the equivalent bulk modulus of concrete, $C(t, \tau)$ is equivalent creep degree of concrete; $K_0(\tau)$ is bulk modulus of mortar; $C_0(t, \tau)$ is creep degree of mortar; f_0 is volume ratio of mortar. Thus, the creep of mortar can be obtained conveniently by Eq. 13.

$$C_0(t,\tau) = K(\tau)C(t,\tau)/\left[f_0 K_0(\tau)\right].$$

(14)

2.4 Coefficient of thermal expansion

Concrete coefficient of thermal expansion affects the thermal stress significantly. Thus, for the calculation of concrete thermal stress, coefficient of thermal expansion is a key parameter. In the reference (Guanlin Shen, et al. 2006), a method to determine the equivalent coefficient of thermal expansion was proposed. Suppose that coefficients of thermal expansion of mortar and aggregate are known, the formula is

$$a = a_0 + \left(\frac{1}{K} - \frac{1}{K_0}\right) / \left(\frac{1}{K_1} - \frac{1}{K_0}\right)(a_1 - a_0),$$

(15)

where a is the coefficient of thermal expansion of concrete; a_0 and a_1 are the coefficients of thermal expansion of mortar and aggregate; K_0 and K_1 are the bulk modulus of mortar and aggregate respectively; K is the equivalent bulk modulus of concrete, which can be determined by Eq. 1.

If concrete and aggregate coefficient of thermal expansion is given, mortar's coefficient of thermal expansion can be calculated by Eq. 16 which is transformed from Eq. 15:

$$a_0 = \frac{1}{1-m}(a - ma_1),\tag{16}$$

where m was obtained by Eq. 17:

$$m = \left(\frac{1}{K} - \frac{1}{K_0}\right)\Big/\left(\frac{1}{K_1} - \frac{1}{K_0}\right).\tag{17}$$

2.5 Autogenous shrinkage

Under the condition of constant temperature and humidity, volume deformation caused by the hydration of cementitious material is called autogenous shrinkage, the range of which is almost $(20\sim100) \times 10^{-6}$, it is an vital parameter. For hardening material like concrete, its autogenous shrinkage increases as the time grows, If the autogenous shrinkage of mortar is given, the changing curve of concrete's equivalent autogenous shrinkage is available by referring to the method of coefficient of thermal expansion.

Suppose that mortar autogenous shrinkage is $s_0(t)$ and age t can be divided into several increment of time $\Delta t_1 = t_1 - t_0$, $\Delta t_2 = t_2 - t_1$, ..., $\Delta t_i = t_i - t_{i-1}$, then mortar autogenous shrinkage increment for any time interval can be denoted as $\Delta s_0(t_i) = s_0(t_i) - s_0(t_{i-1})$. By referring to the method of coefficient of thermal expansion, the concrete equivalent autogenous shrinkage for this period can be calculated by:

$$\Delta s = \left[1 - \left(\frac{1}{K} - \frac{1}{K_0}\right)\Big/\left(\frac{1}{K_1} - \frac{1}{K_0}\right)\right]\Delta s_0,\tag{18}$$

where K_0, K_1 and K is same in Eq. 15, but K_1 and K will change at every time interval.

After all the autogenous shrinkage increments for each period are determined, concrete total autogenous shrinkage is available by accumulation calculation as follow:

$$s(t) = \sum_{k=1}^{i} \Delta s(t_k).\tag{19}$$

If concrete autogenous shrinkage is given, mortar autogenous shrinkage can also be determined by Eq. 20, which is transformed from Eq. 18.

$$\Delta s_0 = \Delta s\Big/\left[1 - \left(\frac{1}{K} - \frac{1}{K_0}\right)\Big/\left(\frac{1}{K_1} - \frac{1}{K_0}\right)\right].\tag{20}$$

3 NUMERICAL LOAD TEST METHOD

Numerical test is an important tool for exploring concrete behavior and performance. Based on material meso-structural characters and combining random distribution theory and computational mechanics, the overall process analysis of concrete progressive failure can be achieved by the FEM method. Due to no need of the real experiment, numerical test has taken the place of traditional test gradually, and it becomes a significant research method

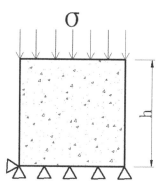

Figure 1.   Numerical load test model.

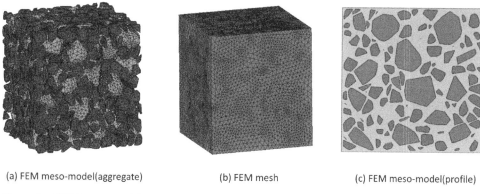

(a) FEM meso-model(aggregate)   (b) FEM mesh   (c) FEM meso-model(profile)

Figure 2.   FEM meso-model.

for concrete. In this paper, a numerical load test can also be used to verify the inversion calculation formulas. Substituting inversion values of mortar parameters into numerical test, results can be calculated by 3D-FEM. By comparing the numerical result and target value, accuracy of the inversion calculation can be verified.

In this paper, a numerical test specimen in size of $150 \times 150 \times 150$ mm and secondary aggregate grading was generated by random aggregate packing model (Huaifa Ma, et al. 2008). The number of small size particles (diameter 5~20 mm) is 781 and that of medium size particles (diameter 20~40 mm) is 103. The number of aggregate is confirmed by the Fuller curve. The volume of aggregate accounts for 46% of the total volume. The FEM model of the test specimen is shown in Figure 2, including 77035 nodes and 435761 elements.

In this paper, all finite element simulation analysis was conducted within **SAPTIS** software (Guoxin Zhang, et al. 2013), which was self-developed by China **IWHR** and it can be utilized to simulate various problems on the temperature, flow, deformation, stress fields of concrete structures. This software adopts efficient compression data storage mode, and has several equation solvers such as the Conjugate Gradient Method, the Gauss elimination method, the parallel direct method. It can solve million DOF problem on general computer, and also solve more large structure problem if it is run on high-performance computing clusters. This software has been developed for over 20 years, and was widely used in the analysis of temperature, stress field, deformation, safety evaluation of the Three Gorges dam, Ertan dam, Xiaowan dam, Xiluodu dam, JinPing dam, Laxiwa dam and so on.

The various values of FEM simulation results, which were used to compare inversion target values, were obtained by the ways as follows:

1. Elasticity modulus

   In the finite element analysis model, normal constraints were applied at the bottom and the vertical load was applied at the top as shown in the Figure 1. Similar to the load test in laboratory, $\sigma \sim \varepsilon$ curve is available by numerical load test. ε is

   $$\varepsilon(t) = \Delta h(t)/h, \tag{21}$$

   where Δh is the average displacement of the top node.

   The slope of the $\sigma \sim \varepsilon$ curve is elasticity modulus of concrete.

   $$E(t) = \left[\sigma(t_i) - \sigma(t_{i-1})\right]/\left[\varepsilon(t_i) - \varepsilon(t_{i-1})\right]. \tag{22}$$

2. Creep

   The modal of numerical calculation is same with elasticity modulus. Constant face load σ was applied at the age 3d, 7d, 14d, 28d and 90d. After measuring the average displacement of the top node $\Delta h(t, \tau)$, strain and Creep can be calculated by formula:

   $$\varepsilon(t, \tau) = \Delta h(t, \tau)/h, \quad C(t, \tau) = \varepsilon(t, \tau)/\varepsilon(\tau), \tag{23}$$

   where $\varepsilon(\tau)$ is the instantaneous elastic strain, the value of which is $\varepsilon(\tau, \tau)$ when $t=\tau$.

3. Coefficient of thermal expansion

   With the thermal load ΔT applied on the model, the average displacement Δh was measured, the average strain was calculated by Eq. 21, and the concrete coefficient of thermal expansion calculated by dividing the temperature increment as follow:

   $$a = \varepsilon/\Delta T. \tag{24}$$

4. Autogenous shrinkage

   Suppose that the autogenous shrinkage of aggregate is 0, given mortar autogenous shrinkage curve, measure Δh at each age, concrete's autogenous shrinkage $\varepsilon(\tau)$ can be determined by Eq. 21.

4 EXAMPLES

Example 1: Given concrete's elasticity modulus $E(t) = 35.0[1-\exp(-0.3t^{0.5})]$ (Gpa); Poisson's ratio is 0.2; Aggregate's elasticity modulus is 50.0Gpa; Poisson's ratio is 0.2; volume ratio is 0.46. Determine the mortar's elasticity modulus by inversion calculation.

According to Eq. 8 and Eq. 9, mortar elasticity modulus can be calculated (see Table 1 and Fig. 3) by subsection inversion method. After fitting the result data, mortar's elasticity modulus is $E(t) = 26.16[1-\exp(-0.15t^{0.60})]$ (Gpa).

Result of numerical load test (Table 1 and Fig. 3) shows that the FEM result of concrete elasticity modulus which is calculated by putting the mortar's elasticity modulus from inversion calculation into numerical load test is almost the same with target elasticity modulus.

Table 1. Inversion result of mortar's elasticity modulus (unit (GPa)).

Age (d)	3	7	14	28	90
Concrete's elasticity modulus (E1)	14.18	19.17	23.61	27.84	32.97
Mortar's elasticity modulus (E2)	6.84	10.24	13.85	17.90	23.65
Concrete's FEM elasticity modulus (E3)	14.30	19.58	23.84	27.79	32.75
Relative error (E1–E3)/E1(%)	−0.85%	−2.11%	−0.96%	0.19%	0.66%

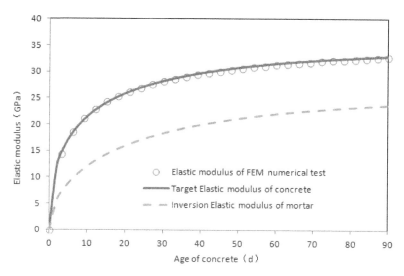

Figure 3.   The inversion result of mortar's elasticity modulus and FEM test result.

Table 2.   FEM result of creep* (Unit: 10–12/Pa).

Time of loading	3	7	14	28	90
3	25.90 (24.79)	33.70 (31.99)	38.20 (36.08)	43.60 (41.08)	53.10 (49.70)
7	12.90 (14.53)	17.50 (17.83)	20.80 (20.81)	25.00 (24.79)	32.10 (31.43)
14	7.82 (8.03)	10.80 (10.99)	13.20 (13.41)	16.40 (16.58)	21.70 (21.91)
28	4.95 (5.24)	6.79 (7.13)	8.28 (8.63)	10.30 (10.65)	13.80 (14.37)
90	2.74 (3.00)	3.53 (3.80)	3.98 (4.21)	4.64 (4.80)	6.33 (6.65)

* The values in the table are FEM creep result and concrete's target creep, concrete's target creep in bracket.

The error between them is slight, so it verifies that the inversion result of mortar elasticity modulus has high accuracy.

Example 2: Given concrete's elasticity modulus $E(t) = 35.0[1-\exp(-0.3t^{0.5})]$ (Gpa); Poisson's ratio is 0.2; Creep is

$$C(t,\tau) = \left(2.55 + 74.1/\ \tau^{0.97}\right)\left(1 - e^{-0.55(t-\tau)}\right) + \left(6.64 + 24.2/\ \tau^{1.01}\right)\left(1 - e^{-0.0052(t-\tau)}\right)$$
$$+ 18.6e^{-k_3\tau}\left(1 - e^{-0.039(t-\tau)}\right)\left(10^{-12} \times \mathrm{Pa}^{-1}\right);$$

Aggregate elasticity modulus is 50.0 Gpa; Poisson's ratio is 0.2; Creep is 0; volume ratio is 0.46. Determine the mortar creep by inversion calculation. Inversion result and numerical result are list in Table 2 and Figure 4.

Example 3: Given concrete's elasticity modulus is 26.0 Gpa and coefficient of thermal expansion is $10.0 \times 10^{-6}/°C$; aggregate's elasticity modulus is 50 GPa and coefficient of thermal expansion is $7.5 \times 10^{-6}/°C$. Mortar's coefficient of thermal expansion is calculated to be $12.8 \times 10^{-6}/°C$ by Eq. 16. Putting the mortar's coefficient of thermal expansion into FEM numerical model, the concrete's coefficient of thermal expansion is calculated to be $10.03 \times 10^{-6}/°C$. It shows that the FEM result is very close to the target result.

Example 4: Given mortar elasticity modulus $E(t) = 26.16[1-\exp(-0.15t^{0.60})]$ Gpa; Aggregate elasticity modulus is 50.0 Gpa; concrete's autogenous shrinkage is shown in Table 3 concrete row. Concrete's autogenous shrinkage is shown in Table 3 concrete row. Determine the mortar autogenous shrinkage. The inversion result is seen in Table 3 mortar row. By comparing the FEM result and target result, inversion result is approved to be accurate.

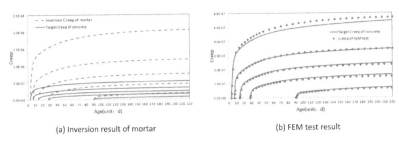

| (a) Inversion result of mortar | (b) FEM test result |

Figure 4. Comparing between inversion result and FEM result of concrete's creep (10^{-4}/Pa).

Table 3. Comparison between inversion result and FEM result of autogenous shrinkage (Unit: 10^{-6}).

Age (d)	3	7	28	90	180	365
Concrete target value	0.20	−2.40	−15.70	−26.90	−31.30	−32.50
Mortar inversion value	0.52	−5.94	−37.13	−61.85	−71.30	−73.86
FEM test value	0.11	−1.94	−14.50	−26.20	−30.80	−32.00
Error	0.09	−0.46	−1.20	−0.70	−0.50	−0.50

5 CONCLUSIONS

Concrete is treated as a two-phase composite material constituted by aggregate and mortar. This paper provides the inversion formulas of several vital thermal and mechanical parameters of mortar by Mori-Tanaka theory in meso-mechanics. With these formulas, the mortar coefficient of thermal expansion, autogenous shrinkage curve, elasticity modulus curve and creep curve can be determined conveniently. However, this paper takes no consideration of the influence of interface between aggregate and mortar. Thus further studies are needed to be done to show the effect of this factor.

ACKNOWLEDGEMENTS

This work was financially supported by the *National Basic Research Program of China (973 Program 2013CB036406 and 2013CB032904)*, the Twelfth-five Science and Technology Support Project *(2013BAB06B02)*, *IWHR and the Basin Water Cycle Simulation and Regulation of State Key Laboratory Special Research Foundation.*

REFERENCES

Bofang Zhu. 1999. Thermal stress and temperature control of mass concrete [M]. China Electric Power Press.
Guanlin Shen, GengKai Hu. 2006. Mechanics of composite materials [M]. Tsinghua University Press.
Guoxin Zhang. 2013. Development and application of SAPTIS-a software of multi-field simulation and nonlinear analysis of complex structures (part I) [J]. Water Resources and Hydropower Engineering, 44(1): 31–35.
Hill R.A. 1965. Self-consistent mechanics of composite materials [J]. Journal of the Mechanics and Physics of Solids, 13(4): 213–222.
Huaifa Ma, HouQun Chen. 2008. The dynamic damage mechanism research and meso-mechanics analysis method of full-graded aggregate concrete [M]. China Water Power Press.
Mori T, Tanaka K. 1973. Average stress in matrix and average elastic energy of materials with misfitting inclusions[J]. Acta metallurgica, 21(5): 571–574.
Qiujin Zhou, Guoxin Zhang. 2013. Development and application of SAPTIS-a software of multi-field simulation and nonlinear analysis of complex structures (part II) [J]. Water Resources and Hydropower Engineering, 44(9): 39–43.

Hydraulic Engineering III – Xie (Ed)
© 2015 Taylor & Francis Group, London, ISBN 978-1-138-02743-5

Hydrological responses of the upper reaches of Yangtze River to climate change

D.L. Xu
Bureau of Hydrology, Changjiang Water Resources Commission, Wuhan, China

Z.Y. Wu & Y. Yang
College of Hydrology and Water Resource, Hohai University, Nanjing, China

Z.Y. Hu
Business School of Hohai University, Nanjing, China

ABSTRACT: The upper reaches of the Yangtze River are located on the Tibetan Plateau, where climate change imposes serious effects on hydrological process. Based on the data collected from 21 Coupled Model Intercomparison Project phase 5 global climate models combined with the Variable Infiltration Capacity model, this paper predicts temporal and spatial hydrological changes in the upper reaches of the Yangtze River under the Representative Concentration Pathway scenarios RCP2.6, RCP4.5, and RCP8.5 within the next 30 years (2011–2040). The results show that compared with a base period of 1970–1999, multi-year temperature will increase by 1.2–1.5°C, annual precipitation rate will increase by 3.1–3.6%, and annual runoff will decrease by 1.2–3.5% over the next 30 years under the three scenarios. The significant increase in temperature and a decrease in autumn and winter precipitation could be main reasons for the decrease in runoff. In the studied Panzhihua section, the likelihood of flood is estimated to increase within the next 30 years, whereas in the Zhimenda section, the likelihood of flood is predicted to decrease. However, water resources may decline, but by a small amount, in both sections through the next 30 years.

1 INTRODUCTION

The Yangtze River is the largest river in China and has an important strategic position. Upper reaches of the Yangtze River located on the Tibetan Plateau are very sensitive to climate change (Yang et al. 2012). In the past 40 years, an aridification trend characterized by increases in temperature, decreases in precipitation, and increases in evaporation has emerged and seriously impacted hydrological process, leading to a series of ecological and environmental problems (Chen et al. 2007). Recent studies on the impact of climate change on the Yangtze River drainage basin areas have been based mainly on climate change data from the Coupled Model Intercomparison Project phase 3 (IPCC CMIP3) (Ju et al. 2011; Tao et al. 2013). To further promote research on climate change, CMIP5 published a new series of greenhouse gas emission scenarios, namely, the Representative Concentration Pathways (RCPs) (Vuuren et al. 2011). Application of these latest scenario model data to project future climate change has become a hot research topic (Guo et al. 2009; Yao et al. 2012; Alkama et al. 2013; Jiang et al. 2013). Studies have shown that the CMIP5 model is superior to the CMIP3 model for simulating average temperature in China (Guo et al. 2009). In this paper, we used temperature and precipitation data derived from the CMIP5 global climate model under the RCP2.6 (low emissions), RCP4.5 (medium emissions) and RCP8.5 (high emissions) scenarios in combination with the Variable Infiltration Capacity (VIC) hydrological model to project spatio-temporal changes in the hydrological process of the upper reaches of the

Yangtze River in the next 30 years (2011–2040). This study should serve as a reliable reference for studies on the effects of climatic changes on the regional hydrological processes.

2 DATA AND METHODS

2.1 *Study area*

The drainage area of the upper reaches of the Yangtze River upstream from Panzhihua is located in southwestern China (Fig. 1). It has an altitude range of more than 5000 m, catchment area of 260,000 km², average multi-year temperature of –0.2°C, average annual rainfall of 536.1 mm, and average annual runoff of 57 billion m³.

2.2 *Data*

The climate scenario data were collected from 21 CMIP5 global climate models (http://www-pcmdi.llnl.gov/) in the China Regional Climate Change Projections Dataset Version3.0 (http://www.climatechange-data.cn/en/) provided by the China National Climate Center. We studied the monthly average temperature and precipitation in the base period (1970–1999) and in the future 30 years (2011–2040) under the RCP2.6, RCP4.5, and RCP8.5 scenarios. Observed meteorological data of daily temperature and precipitation during the period from 1970 to 2007 were collected from 49 weather stations in the upper reaches of the Yangtze River.

2.3 *Downscaling and testing of the model data*

We used a bilinear interpolation method to interpolate data collected from various models onto 0.25° × 0.25° grid points. Then, we used an equidistant quantile matching method to correct monthly average data. Finally, we applied the Delta method to generate daily data for the base period and future scenarios. By these means, we obtained daily temperature and precipitation data for 518 0.25° × 0.25° grids of the upper reaches of the Yangtze River in the base period and in the future period under the three scenarios and then input these data into the VIC model. Detailed introductions to the equidistant quantile matching-based

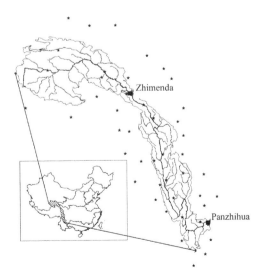

Figure 1.   Map of rivers and weather stations in the drainage area of the Yangtze River upstream from Panzhihua.

correction method and the Delta method are described in the previous studies (Zhao and Xu 2007; Yu et al. 2008; Li et al. 2010; Xiao et al. 2013).

Table 1 compares the downscaling results of the CMIP5 multi-model temperature and precipitation data with the measured data during the base period. From Table 1, we see that the downscaling regional precipitation and temperature averages in the CMIP5 multi-model resemble the measured values. The simulated precipitation was greater than the measured value, with an approximately 5% margin of error. The simulated temperature was lower than the measured value, with a margin of error of approximately −0.24°C. Seasonally, compared with the corresponding measured values, the simulated precipitation was greater in the spring, summer, and autumn, and smaller in winter; the simulated temperature was smaller in all seasons, but these had very small margins of error. Therefore, the downscaling method used in this study is suitable for study of these drainage areas of the Yangtze River.

2.4 VIC model

The VIC model is a large-scale hydrological model developed by Washington University and Princeton University (Liang et al. 1994). It is a hydrological model that considers the grid distribution of water and energy balance, and it has been widely used in regional hydrological process simulations (Wu et al 2007; Guo et al. 2009; Lu et al. 2013).

The VIC was used to study 518 $0.25° \times 0.25°$ grids of the river drainage areas, where gridded meteorological data (daily precipitation, highest and lowest temperature) were obtained from the interpolation of data from 49 weather stations. Vegetation parameters were set according to global data on the amount of vegetation/1 km of land, and soil parameters were set according to global data on the amount of soil/10 km.

The measured flow rates of the Zhimenda and Panzhihua sections from 1970 to 2007 were used to calibrate the model, and two parameters, the relative error (Er) and Nash-Sutcliffe index (NS) (Nash and Sutcliffe 1970), were used to evaluate the model. Model calibration and evaluation results are shown in Table 2. In Table 2, NS values during the calibration and validation periods all exceeded 0.7, and even reach 0.94, whereas the margins of relative error are between −5% and 5%. Therefore, the VIC model is applicable for the study areas.

Table 1. Downscaling results of annual/seasonal and regional average precipitation and temperature data in the base period of CMIP5 multi-model ensemble (1970–1999).

Observation/CMIP5 model estimations	Multi-year average preci. (mm d⁻¹)					Multi-year average temp. (°C)				
	Ann.	Spr.	Sum.	Aut.	Win.	Ann.	Spr.	Sum.	Aut.	Win.
Observation	1.45	0.77	3.54	1.32	0.15	−0.22	0.08	8.11	0.25	−9.51
Pre-downscaling	3.08	2.66	5.87	2.86	0.93	−2.81	−2.97	7.03	−1.67	−13.62
Post-downscaling	1.53	0.78	3.85	1.37	0.11	−0.47	−0.13	8.06	−0.03	−9.94

Table 2. Calibration and validation results of the VIC model.

Simulation period	Period	Zhimenda			Panzhihua		
		Er (%)	NS_day	NS_mon	Er (%)	NS_day	NS_mon
Calibration period	1970–1990	−0.8	0.8	0.9	0.36	0.89	0.94
Validation period	1991–2007	−3.24	0.72	0.87	−2.31	0.89	0.93

3 RESULTS AND DISCUSSION

3.1 *Changes in temperature and precipitation*

Figure 2 is a histogram of the multi-model average of temperature changes under the three scenarios from 2011 to 2040 compared with those of the base period (1970 to 1999). From this figure, we can see that the average temperatures (2011 to 2040) under all three scenarios of the upper reaches of the Yangtze River are all greater than those of the base period, with increases of 1.2–1.5°C. The RCP8.5 scenario has the greatest increase, and the RCP2.6 scenario has the smallest increase. Seasonally, temperature increases are greatest in the winter and spring, with a maximum increase of 1.6–1.7°C.

Figure 3 is a histogram of the multi-model average precipitation changes under the three scenarios from 2011 to 2040 compared with those of the base period. This figure demonstrates that the multi-model average precipitations (2011 to 2040) under all three scenarios of the upper reaches of the Yangtze River are all greater than those of the base period, by 3.1–3.7%. The increase is greatest under the RCP2.6 scenario and smallest under RCP4.5 scenario. From the perspective of monthly changes, precipitation decreases in January, October, November, and December, and the greatest decrease occurs in December under the RCP8.5 scenario (−8.5%). Precipitation increases in all other months and scenarios. The increase in spring is most remarkable, with the largest increase (in May) reaching 5.4–7.2%.

3.2 *Changes in runoff*

3.2.1 *Interannual changes in runoff*

Figure 4 shows the multi-model average values of estimated annual runoff changes in the Yangtze River drainage area upstream from Zhimenda and Panzhihua under the RCP2.6,

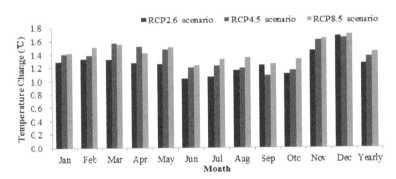

Figure 2.   Change of multi-annual and monthly mean temperature during 2011–2040 relative to the base period.

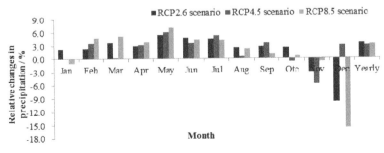

Figure 3.   Change of multi-annual and monthly mean precipitation from 2011 to 2040 relative to the base period.

96

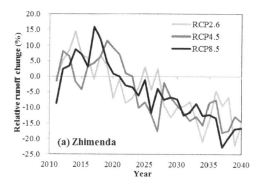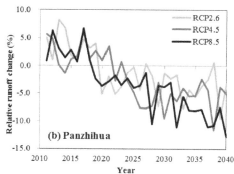

Figure 4.   Runoff changes of the drainage areas upstream from Zhimenda and Panzhihua from 2011 to 2040 compared with those of the base period.

RCP4.5, and RCP8.5 scenarios during the period of 2011 to 2040 relative to those of the base period.

From Figure 4(a), we can see that, under the different scenarios and in the drainage area upstream from Zhimenda, the estimated annual runoff change trend resembles that during the base period: the runoff increases in the first 10 years, begins to decline in approximately 2021, and reaches a reduction of 15–20% in 2040. Figure 4(b) shows that the change trends under all scenarios for the drainage area upstream from Panzhihua generally resemble those of the drainage area upstream from Zhimenda: the runoff in both areas shifts from an increase to a decrease around 2020. One difference, however, is that the runoff in the area upstream from Panzhihua experiences a smaller decrease, decreasing by only –5 to –10% by around 2040, than that in the area upstream from Zhimenda. Overall, the drainage areas upstream from Zhimenda and Panzhihua both experience a runoff change of an initial increase followed by a decrease in the period of 2011 to 2040; however, the change is greater in the drainage area upstream from Zhimenda than in the drainage area upstream from Panzhihua.

3.2.2   *Annual runoff changes*
Figure 5 is a histogram of the multi-model averages of the annual and monthly mean changes of runoff in the drainage areas upstream from Zhimenda and Panzhihua obtained by comparison of the simulated values for the period 2011 to 2040 with those of the base period. Figure 5(a) shows that for the drainage area upstream from Zhimenda the 2011–2040 multi-model annual average runoffs under all three scenarios change by –5.0 to 4.5% compared with those of the base period; the RCP8.5 scenario has the largest decrease, and the RCP2.6 scenario has the smallest decrease. For the monthly changes, runoff increases in April, May, August, and October in some scenarios; it increases by 3.0% in October in the RCP2.6 scenario. Other months experience decreases in all three scenarios, especially in the winter, with the greatest decrease (in November) exceeding –10%. Figure 5(b) shows that under the RCP2.6, RCP4.5, and RCP8.5 scenarios the multi-model annual average runoff in the drainage area upstream from Panzhihua during the period from 2011 to 2040 decreases by –1.2, –2.4, and –3.5% compared with those of the base period, respectively. RCP8.5 has the largest decrease, and RCP2.6 has the smallest. For the monthly changes, relative to those of the base period, runoff increases in April, June, July, August, and September in some scenarios, with August having the largest increase (4.0%) under the RCP2.6 scenario. The other months experience decreases in all three scenarios. The greatest decreases, exceeding –8.0%, occur in May and November.

3.2.3   *Changes in annual maximum flow*
Based on the Panzhihua and Zhimenda flow results from the VIC simulation model, we calculated the maximum flows and annual runoff during the base period and the period from

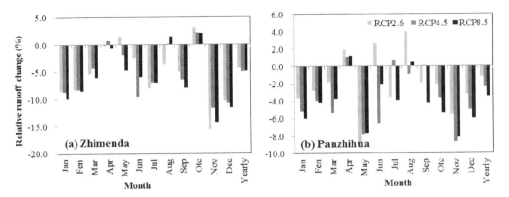

Figure 5. Relative changes of yearly and monthly runoff in the drainage areas upstream from Zhimenda and Panzhihua during the period from 2011 to 2040 compared to those during the base period.

Table 3. Relative magnitudes of future changes in maximum flows and annual runoff in the Panzhihua and Zhimenda sections under the three scenarios at different frequencies relative to those during the base period (%).

| Stations | Climate scenarios | Flood flow | | Annual runoff | | |
		20-year flow	10-year flow	High-flow year	Normal year	Low-flow year
Zhimenda	RCP2.6	−6.2	−7.4	−2.3	−4.3	−5.9
	RCP4.5	−4.1	−4.6	−2.2	−4.8	−6
	RCP8.5	−4.1	−6.8	−1.1	−4.9	−6.7
Panzhihua	RCP2.6	8.8	7	0.1	−1.1	−1.9
	RCP4.5	9.8	8.1	−1.1	−2.2	−3
	RCP8.5	3.9	1.2	−1.2	−3.6	−5.1

2011 to 2040 under the three scenarios, with a 30-year-long study period for each. Through a fitting process, we obtained P-III frequency curves to estimate the frequency changes of extreme floods and the abundance or shortage of surface water resources in the future. Tables 3 shows the relative magnitudes of the changes in the maximum flows and annual runoff in the Panzhihua and Zhimenda sections, respectively, under the three scenarios at different frequencies compared with those of the base period.

Table 3 shows that under the RCP2.6, RCP4.5, and RCP8.5 scenarios, the 20-year flood-flow (P = 5%) in the Panzhihua section during the period from 2011 to 2040 may increase by 8.8, 9.8, and 3.9%, respectively, and that the 10-year floodflow (P = 10%) may increase by 7.0, 8.1, and 1.2%, respectively. Therefore, under the three scenarios, there may be a higher chance that flooding would occur at both the 10- and 20-year periods in the Panzhihua section; however, the change in flood magnitude is not large. Conversely, under the three scenarios, the 20-year floodflow in Zhimenda during the period from 2011 to 2040 may decrease by 6.2, 4.1, and 4.1%, respectively, and the 10-year floodflow may decrease by 7.4, 4.6, and 1.2%, respectively. Therefore, under the three scenarios, there may be a lower chance of flooding at both the 10- and 20-year periods in the Zhimenda section, but the change of flood magnitude is also small.

Under the RCP2.6, RCP4.5, and RCP8.5 scenarios, the annual runoff in water-abundant years (10%) in the Zhimenda section within the next 30 years may decrease by 2.3, 2.2, and 1.1%, respectively. The annual runoff in normal years (50%) may decrease by 4.3, 4.8, and 4.9%, and the annual runoff in low-flow years (75%) may decrease by 5.9, 6.0, and 6.7%, respectively. Therefore, under the three scenarios, annual runoff in the Zhimenda section could decrease in high-, normal-, and low-flow years, but the magnitude of the decreases are relatively small. Similarly, in the Panzhihua section under the three scenarios, annual runoff could decrease under all three flow conditions, but with a reduction of less than −5.1%, smaller than that of the Zhimenda section.

4 CONCLUSIONS

1. Under RCP2.6, RCP4.5, and RCP8.5 scenarios, the multi-year average temperature and monthly average temperature in the upper reaches of the Yangtze River during the period from 2011 to 2040 are estimated to be 1–2°C higher than that during the base period (1970–1999). In particular, winter and spring will experience relatively large increases. Average annual precipitation may increase by 3 to 4%, but precipitation could decrease in autumn and winter, especially in December when it could reach a maximum decrease of −8.5%.
2. Under the three scenarios, the multi-year average runoff in 2011–2040 is estimated to decrease by 1.2–3.5% compared to that of the base years. Except for a small number of months in which the monthly runoff increases, the majority of months will experience runoff decreases, with the greatest decreases occurring in autumn and winter. The decrease in runoff is closely related to the significant future temperature increases. The decrease in precipitation in autumn will also lead to fluctuations in runoff levels in the season.
3. The frequency of flooding in 2011–2040 is predicted to increase in the Panzhihua section and decrease in the Zhimenda section, but flood magnitude will not change significantly. Both areas are predicted to experience a decrease in annual runoff during high-, normal-, and low-flow years within the period of 2011 to 2040, but the magnitude of decreases should not be remarkable.

ACKNOWLEDGEMENTS

This work is supported by the Foundation for the Author of National Excellent Doctoral Dissertation of PR China (Grant no. 201161), the Qing-Lan Project and Program for New Century Excellent Talents in University (Grant no. NCET-12-0842).

REFERENCES

Alkama, R., Marchand, L., Ribes, A. et al. 2013, Detection of global runoff changes: results from observation and CMIP5 experiments. *Hydrology and Earth System Sciences* 17(8): 2967–2979.
Chen F., Yingfang M.A., & Liu X. 2007. Study on climate change and its causes in the source area of Yangtze River. *Journal of Qinghai Meteorology* (2): 11–15. (in Chinese).
Guo S.L., Guo J., Zhang J. et al. 2009. VIC distributed hydrological model to predict climate change impact in the Hanjiang Basin. *Science in China: Technological Sciences* 52(11): 3234–3239.
Guo, Y., Dong, W., Ren, F. et al. 2013. Assessment of CMIP5 simulations for China annual average surface temperature and its comparison with CMIP3 simulations. *Progressus Inquisitiones DE Mutatione Climatis* 9(3): 181–186. (in Chinese).
Jiang, X., Maloney, E.D., Li, J.L.F. et al. 2013. Simulations of the eastern North Pacific intraseasonal variability in CMIP5 GCMs. *Journal of Climate* 26(11): 3489–3510.
Ju, Q., Hao, Z., Yu, Z. et al. 2011. Runoff prediction in the Yangtze River Basin based on IPCC AR4 climate change scenarios. *Advances in Water Science* 22(4): 462–469. (in Chinese).
Li, H.B., Sheffield, J. & Wood, E.F. 2010. Bias correction of monthly precipitation and temperature fields from Intergovernmental Panel on Climate Change AR4 models using equidistant quantile matching. *Journal of Geophysical Research* 155, D10101. doi: 10.1029/2009JD012882.
Liang X., Lettenmaier, D.P. Wood, E.F. et al. 1994. A simple hydrologically based model of land surface water and energy fluxes for general circulation models. *Journal of Geophysical Research: Atmospheres* 99(14): 415–428.

Lu, G.H., Xiao, H., Wu, Z.Y. et al. 2013. Assessing the impacts of future climate change on hydrology in Huang-Huai-Hai Region in China using the PRECIS and VIC models. *Journal of Hydrologic Engineering* 18(9): 1077–1087.

Nash, J.E. & Sutcliffe, J.V. 1970. River flow forecasting through conceptual models: Part I: A discussion of principles. *Journal of Hydrology* 10(3): 282–290.

Tao, H., Huang, J., Zhai, J. et al. 2013. Simulation and projection of climate changes under the RCP4.5 scenario in the Yangtze River Basin based on CCLM. *Progressus Inquisitiones DE Mutatione Climatis* 9(4): 246–251. (in Chinese).

Vuuren, D., Edmonds, J., Kaimuma, M. et al. 2011. The representative concentration pathways: an overview. *Climatic Change* 109(1/2): 5–31.

Wu, Z.Y., Lu, G.H., Wen, L. et al. 2007. Thirty-five year (1970–2005) simulation of daily soil moisture using the variable infiltration capacity model over China. *Atmosphere-Ocean* 45(1): 37–45.

Xiao H., Lu, G., Wu, Z. et al. 2013. Flood response to climate change in the Pearl River Basin for the next three decades [J]. *Journal of Hydraulic Engineering* 44(12): 1409–1419. (in Chinese).

Yang, Y., Lu, G., Wu, Z. et al. 2012. Variation characteristics analysis of hydrological cycle factors in upper reaches of Jinshajiang Basin. *Water Resources and Power* 30(3): 8–10. (in Chinese).

Yao, Y., Luo, Y. & Huang, J. 2012. Evaluation and projection of temperature extremes over China based on 8 modeling data from CMIP5. *Progressus Inquisitiones DE Mutatione Climatis* 8(4): 250–256. (in Chinese).

Yu, F., Zhang, G. & Liu, Y. 2008. Analysis on effects of global climate change on water resource in the Yellow River Basin. *Journal of China Hydrology* 28(5): 52–56. (in Chinese).

Zhao, F. & Xu, Z. 2007. Comparative analysis on downscaled climate scenarios for headwater catchment of Yellow River using SDS and Delta methods. *Acta Meteorologica Sinica* 65(4): 653–662. (in Chinese).

Hydraulic Engineering III – Xie (Ed)

Research on the evolution of drainage networks in a simulated loss watershed based on fractal dimension

Youfu Dong & Yucheng Wang

College of Geomatics Engineering, Nanjing Tech University, Nanjing, Jiangsu, P.R. China

ABSTRACT: To describe the evolution of drainage networks in a simulated loess watershed, high resolution DEMs in six different periods are obtained by artificial experiments. Then, drainage networks are extracted and their fractal dimension values are figured out and analyzed. The experiments show that the fractal dimension values of the drainage networks increase in early stages and decrease in later stages with the development of the drainage networks. In contrast, the hypsometric integral values of the simulated watershed keep reducing and the entropy values of the watershed keep rising from the 1st to 6th period all the time. It reveals that the drainage networks experienced from simple to complicated and balanced in the different periods. Meanwhile, there are some differences in the evolution process of drainage networks and the watershed which would deserve to be discussed in the future.

1 INTRODUCTION

The geometric pattern of the stream network of a drainage basin can be viewed a "fractal" with a fractional dimension (Mandelbrot, 1982). The investigation of stream networks as one sort of fractal has been of great interest in geomorphology in recent years (Andrle and Abrahams, 1989; La Barbera and Rosso, 1990; Rosso et al., 1991; Fan and Hu, 2012; Jiang and Chen, 2013; Yuan et al., 2013). Although some progress toward the goal of understanding hydrological processes at the basin scale has been made in estimating fractal dimension of stream networks in a given time, the fractal structure and properties of drainage networks in a watershed during different periods have been poorly understood because of the dynamic data being not available in a short term. As we know, the need for a comprehensive quantitative theory of channel networks "reflecting the constraints of space filling and available potential energy as well as climatic, hydrologic, and geologic controls which are in dynamical equilibrium with channel network forms" (Mesa and Gupta, 1987). Therefore the measurement and quantitative expression of the drainage networks in different periods plays a major role in river basin hydrology. The aim of this paper is to make a study on the evolution of drainage networks in a simulated loess watershed by analyzing their fractal dimension values and relational factors.

2 MATERIALS AND METHODS

2.1 Experimental data

The experimental watershed was simulated in the State Key Laboratory of Soil Erosion and Dryland Farming on Loess Plateau, Institute of Soil and Water Conservation, Chinese Academy of Science, Yangling, Shannxi province. It lasted for ten months. The basis for the experiment is to create a lab-based suitable and manmade test site where the loess material and surface configuration could correctly represent the true bess surface through a series of statistic analysis and pre-experiment.

The loess is tamped after 5 cm soil being filled each time. While completing the first soil filling, 4 times pre-raining are carried out in a specially designed rainfall laboratory, so as to keep the validation of soil filling. A carefully work was done in order to make the bess surface and the gully networks similar to the true loess surface. Then, more soil will be filled with similar method till a satisfied test site, which completes all the necessary preparation for the experiment and waiting for the normal raining experiment, was setup. The rain strength, duration and term in the experiment will be adjusted as much similar as the true rainfall of loess region. Close-range photogrammetry is applied at the stage of normal rainfall experiment. The interval between the neighboring two shoots is approximately a week and 2 to 5 times rainfalls were applied during the same time. 9 shoots are conducted in normal rainfall experiment (6 shoots are selected to study in this paper). And its procedure is constitutive of control-survey and close-range photogrammetry. DEMs are generated by JX4 digital photogrammetry. The grid cell size of DEM is 1 cm, scale is 1:20 and elevation RMES (Root mean-square error) is not greater than 2 mm.

2.2 Methods and procedures

The drainage networks in the selected six periods in the watershed are acquired firstly. The main extraction steps are as follows: (1) the sinks in the initial DEMs are filled to get proper flow direction, (2) the flow direction for each grid is determined, (3) the flow accumulation for each grid is calculated, (4) the drainage networks are gotten by setting the accumulation threshold value as 200, (5) Based on Strahler's ordering scheme, the grid and vector drainage networks belonging to different orders are extracted (Fig. 1).

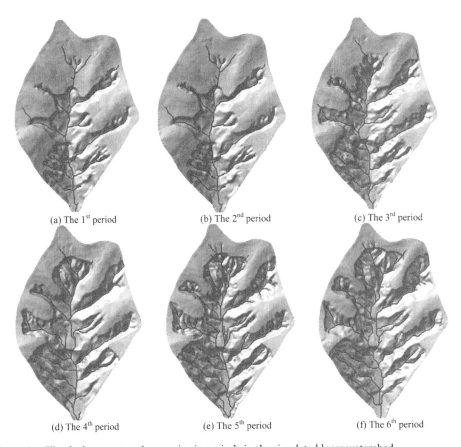

(a) The 1st period   (b) The 2nd period   (c) The 3rd period

(d) The 4th period   (e) The 5th period   (f) The 6th period

Figure 1. The drainage networks map in six periods in the simulated loess watershed.

There are two common indexes for calculating the fractal dimension value of drainage networks, which one is based on basic definition of fractal theory and another is based on Horton Law. For the former index which called box dimension in the following text, its main algorithm is described as follows: if the object lines are overlapped with different square boxes of various side length r, the number of the square boxes $N(r)$ will be changed. That is, there are different box number $N1(r)$, $N2(r)$, $N3(r)$ ..., corresponding to the various box length $r1$, $r2$, $r3$ .... As a result, one line can be fitted according to a series of points coordinated by $(\ln r, \ln N(r))$ based on least square method which can be expressed as equation (1).

$$\ln N(r) = A - D\ln r \qquad (1)$$

where r denotes box length; $N(r)$ denotes the box number; D denotes the box dimension value; A denotes undetermined constant.

For the latter index which called Horton dimension in the following text, its calculation formula given by La Barbera and Rosso based on Horton law is expressed as equation (2).

$$D = \max(1, \log R_B/\log R_L) \qquad (2)$$

where R_B is the bifurcation ratio; R_L is the stream length ratio; D is the Horton dimension value.

The value of R_B is gotten by linear fitting with each drainage order value and the corresponding number of the drainage networks summarized from the attribute table of the channels in each period. The value of R_L is acquired by linear fitting with each drainage order value and the corresponding average length of the drainage networks calculated from the attribute table of the channels in each period in the simulated watershed.

3   RESULTS AND ANALYSIS

3.1   *The difference of box dimension*

According to the algorithm of box dimension, log-log graphs of the six periods are obtained (Fig. 2) by fitting with box length r varying from 20 to 1000 and the corresponding box

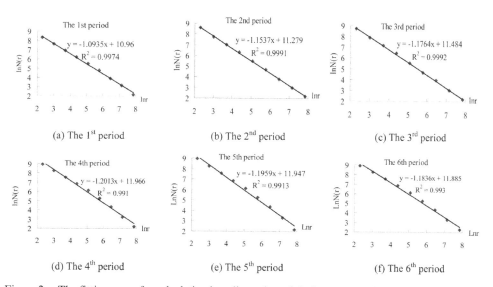

(a) The 1st period   (b) The 2nd period   (c) The 3rd period

(d) The 4th period   (e) The 5th period   (f) The 6th period

Figure 2.   The fitting maps for calculating box dimension of drainage networks in the simulated loess watershed.

number N(r). Then the box dimension values of the drainage networks can be gotten from the rates of the curves (Table 1).

As can be seen in the Table 1 and Figure 2, the values of box dimension increase from 1.0935 in 1st period to 1.2013 in 4th period, and then decrease to 1.1836 in 6th period. At the same time, correlation coefficient r^2 in all periods are over 0.99. It shows that the drainage networks became more and more complicated and unbalanced in their early development stage and they tended towards stability with the evolution of the watershed.

3.2 The difference of Horton dimension

By means of summarizing the number and the length of drainage networks belonging to each order, the Horton dimension values of the drainage networks in each periods can be calculated when R_B and R_L are acquired (Table 2).

Table 2 shows that the values of Horton dimension are beginning to rise at the early stage and reach the max. 1.83 in the 4th period and then diminish to 1.74 in the 6th period. Meanwhile, the values of R_B almost keep the same tendency as the values of Horton dimension and the values of R_L decrease from the 1st to 5th period continuously.

According to Horton's research that the R_B values of natural drainage networks vary from 3 to 5 and their R_L values vary from 1.5 to 3.5, the values of R_B and R_L in the simulated watershed are in accordance with the conclusion. As we know, the value of Horton dimension is determined by the ratio of R_B and R_L, the values of Horton dimension would increase when R_B rises and R_L reduces with the development of the watershed in the early stages. While the watershed came to balanced, the values of Horton dimension would decrease and keep regular as the drainage networks get simple and dynamic.

3.3 The comparison with HI and H(S)

Hypsometric Integral (HI) is a terrain analysis factor with apparent physical and geomorphologic meanings which could reflect the landform erosion stage and evolution process (Strahler A N, 1952). As a macroscopic parameter and method in terrain analysis, the applications of HI could reveal the quantitative characteristic of landform evolution in catchment scale. Base on the HI value, Landform evolution is divided into three stages which are young stage (HI > 0.6), mature stage (HI > 0.35 and HI < 0.6) and old stage (HI < 0.35). Entropy is another useful index in the research on the development and evolution of landform (Ai, 1987). Entropy is expressed as H(s) because it can be calculated by the value of the Strahler's integral S. In the same way, landform evolution is classified into young stage (H(S) < 0.11), mature stage (H(S) > 0.11 and H(S) < 0.4) and old stage (H(S) > 0.4) on the basis of the value of H(S). Table 3 is the results of HI and H(S) of the simulated watershed in all six periods.

Table 1.   The box dimension values of drainage networks in the simulated watershed.

Period	The 1st	The 2nd	The 3rd	The 4th	The 5th	The 6th
Box dimension	1.0935	1.1537	1.1764	1.2013	1.1959	1.1836
Coefficient r^2	0.9974	0.9991	0.9992	0.9910	0.9913	0.9930

Table 2.   The Horton dimension values of drainage networks in the simulated watershed.

Period	The 1st	The 2nd	The 3rd	The 4th	The 5th	The 6th
R_B	4.56	4.66	4.78	4.98	4.64	4.68
R_L	2.68	2.57	2.48	2.41	2.38	2.42
Horton dimension	1.53	1.63	1.72	1.83	1.77	1.74

Table 3. The Hypsometric Integral (HI) and entropy H(S) of the simulated watershed.

Period	The 1st	The 2nd	The 3rd	The 4th	The 5th	The 6th
HI	0.6043	0.5887	0.5847	0.5659	0.5594	0.5481
H(S)	0.1080	0.1185	0.1214	0.1352	0.1403	0.1494

It can be seen from the above table that the 1st period of the simulated watershed landform is of young stage and other periods belong to mature stage based on the HI or H(S) value of the watershed. Meanwhile, the HI values of the simulated watershed in all six periods keep increasing and the H(S) values keep decreasing all the way. From what has been analyzed above, the values of box dimension or Horton dimension increase at the early stages and decrease at the later stages. Consequently, there are some differences between the evolution of the drainage networks and the development of the simulated watershed perhaps because the latter is influenced by more complicated factors.

4 CONCLUSIONS AND DISCUSSIONS

Using a simulated loess watershed as study area and high resolution DEMs in six periods as experimental data, the drainage networks are extracted and their box dimension and Horton dimension are calculated and analyzed. The results show that the fractal dimension values of the drainage networks increase in early stages because of drainage networks being more and more complicated and unbalanced and then the values decrease in later stages with the evolution of the drainage networks.

Compared with the box dimension and Horton dimension of drainage networks in six periods, the HI values of the simulated watershed keep reducing and the H(S) values of the watershed keep rising from the 1st to 6th period all the time. It reflects that there are some differences in the evolution process of drainage networks and the watershed which would deserve to be discussed in the future.

ACKNOWLEDGMENT

Thanks for financial support by National Natural Science Foundation of China (No. 41101360).

REFERENCES

Ai N.S. 1987. Comentropy in erosional drainage system. Journal of Soil and Water Conservation, 1(2): 1–8.
Andfie, R., and A.D. Abrahams. 1989. Fractal techniques and the surface roughness of talus slopes, Earth Surf. Processes Landforms, 14, 197–209.
Fan L.F., Hu R.L., Zhang X.Y., et al. 2012. Calculation of 3D Box Dimension of river System Based on GIS and DEM. Geography and Geo-Information Science, 28(6), 28–30.
Jiang T and Chen D.L. 2013. Fractal Geometry Analysis of River System Based on DEM: An Example of Taoyuan County, Changde City. Chinese Agricultural Science Bulletin, 29(2), 166–171.
La Barbera, P., and R. Rosso. 1990. Reply, Water Resour. Res., 26(9), 2245–2248.
Mandelbrot B B. 1982. The Fractal Geometry of Nature. Freeman, San Francisco:11–13.
Mesa, O.J., and V.K. Gupta. 1987. On the main channel length-area relationship for channel networks, Water Resour. Res., 23(11), 2119–2122.
Rosso, R., B. Bacchi, and P. La Barbera. 1991. Fractal relation of mainstream length to catchment area infiver networks, Water Resour. Res., 27(3), 381–387.
Strahler A N. 1952. Hypsometric (area- altitude) analysis of erosional topography. Geological Society of America Bulletin, 63: 1117–1142.
Yuan X.P., Liu S.F., Tian G.Z., et al. 2013. Analysis of the fractal dimension in the Golmud River basin based on DEM. Remote Sensing for Land and Resources, 25 (1): 111–116.

Hydraulic Engineering III – Xie (Ed)
© 2015 Taylor & Francis Group, London, ISBN 978-1-138-02743-5

Construction of campus 3D scene based on 3dsMax and Unity3D

Lin Lin, Zhihui Tian & Shan Zhao
School of Water Conservancy and Environment, Zhengzhou University, Zhengzhou, China

ABSTRACT: Faced with the low fidelity and low efficiency of scene operation by using the traditional visualization technology to construct 3D scene, This paper proposes a method to build a three-dimensional campus scene, we take Zhengzhou University as an example, based on the Unity3d platform, combine JavaScript with c# language, and make use of the 3dsMax software to construct the 3D scene. Then based on the LOD model, we come up with the 3D scene optimization method.

1   INTRODUCTION

With the development of geographic information system, visualization techniques and virtual reality techniques, three-dimensional model reconstruction techniques are widely used in the constructions of smart city. Digital campus with three-dimensional scene can more intuitively show the style of a campus, which is of important meaning to the planning and development of the campus. Virtual scenes built via traditional visualization techniques have disadvantages like low-fidelity, low interactivity and so on. In this article, Unity3D platform is used to build the three-dimensional scene of Zhengzhou University, with the support of 3dsMax modelling software and interacting programming language JavaScript and C# used in combination. In order to improve the fidelity and interactivity, detailed models and maps of high quality are adopted to accomplish the construction of the model in the scene, and for higher efficiency, LOD model is used to optimize the three-dimensional scene.

2   CONSTRUCTION OF 3D CAMPUS MODELS

Model building is an important part of the whole three-dimensional scene and its quality directly affect the fidelity of three-dimensional scene. The three-dimensional scene includes many models, Such as terrain, buildings, bridges, roads, rivers, vegetation and so on (Ruokun, 2010, Ying, 2013). This paper takes Zhengzhou University as the research object and mainly involved two kinds of models that is terrain and feature model.

2.1   *Construction of three-dimensional terrain models*

First of all, terrain modeling need to from GIS data include planning and engineering drawings, topographic maps, digital remote sensing image, geographic coordinate data and so on. There are many methods of three-dimensional modeling, but the most commonly used is DEM model and it's construction method as follows:

1. Use of the existing terrain elevation data (contour and elevation point), and save it as. SHP format map layer and layer respectively.
2. Import contour and elevation point data into ArcGIS, and generate DEM data as TIN format, then by the TIN generate elevation data as Grid format, use Photoshop converting data format, set appropriate resolution and size of DEM, Transfer the data as RAW format.

3. Import DEM data into Unity3D, and generate three-dimensional terrain model based on DEM of Zhengzhou University.

2.2 Construction of three-dimensional building models

In the process of building the three-dimensional scene model, different kinds of modeling methods are required according to the accuracy of the model. This paper is based on AutoCAD and 3DSMAX modeling method. The modeling process of building is shown in Figure 2.

2.2.1 Building classification

Building is the main object in the three-dimensional scene, the campus buildings can be divided into primary and secondary structures. The Fineness of model building, modeling efficiency and modeling inputs are important issues plagued the people. We can classify the building model in two types based on campus scene effect and the benefit of comprehensive analysis. One is detailed model, focusing on the building, administration building, library and lab building, etc; the other is simple model, mainly aimed at the canteen, dormitory, etc.

Figure 1.   Three-dimensional terrain model of Zhengzhou University.

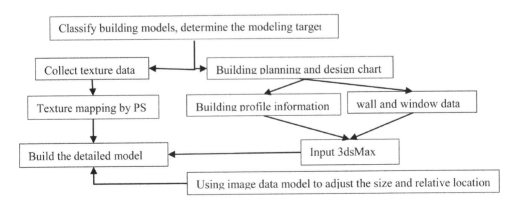

Figure 2.   The modeling process of building.

108

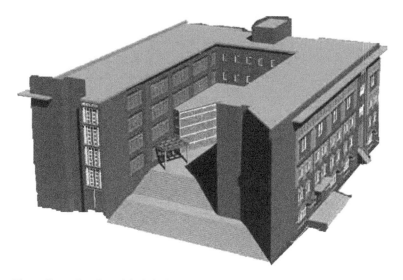

Figure 3.   Three-dimensional model of the institute of water conservancy and environment.

2.2.2   *Building data acquisition*

Architectural entity data is mainly composed of geometric data and high data. Building height data can be roughly estimated in accordance with the building level number. The texture of building can be divided into flat texture and facade texture. The flat texture of buildings can be obtained by matching orthogonal projection image of aviation image processing with the ground texture. The facade texture of buildings are formed by a digital camera on the ground, and then we can use Photoshop to make the façade texture.

2.2.3   *Building models building*

Based on the institute of water conservancy and environment of the Zhengzhou University, we studied the three-dimensional reconstruction of buildings. For most regular buildings, their modeling process can be divided into three types: the construction of the roof, the construction of the wall and the texture map of the buildings.

1. Construction of the roof: Pitched roof building and two-sided gable roof building are obtained based on solid geometry models by Boolean operators (differential operation).
2. Construction of the wall: We get doors and walls with the use of a polygon or entity Boolean (Difference), in addition, we should pay attention to the latch modeling and handle modeling as well as the modeling of the window sash and the glass. Wall unit is filled by a number of doors and windows, in the modeling process we use arrays, mirroring, cloning and other algorithms to improve the efficiency and accuracy of modeling to achieve accurate modeling. Metope is showed by surface extrusion, extrusion and rectangular polygon,
3. Building Texture Editor: The texture image of the building is assigned to the corresponding parts of the building by using 3DS Max material editor bitmap tools in the form of maps. The building model is like Figure 3.

3   THE THREE-DIMENSIONAL SCENE OF ZHENGZHOU UNIVERSITY

The work of the construction of three-dimensional scene is mainly to make each individual object integrated in the model of Unity3D (integration of terrain) while adding relevant environmental, physical and other elements to form a complete three-dimensional virtual reality simulation scenario. With the management function of Unity3D model, we integrate and build three-dimensional simulation scene with the object model and environmental factors

Figure 4. Three-dimensional scene of Zhengzhou University.

around Zhengzhou University, and then the scene is rendering by using three-dimensional render, and finally, we carry out visualization of three-dimensional scene.

Zhengzhou University of three-dimensional landscape is constructed as follows:

1. The DEM generated data into Unity3D terrain model, and it is shown in Figure 1.
2. Using three-dimensional modeling tool—3dsMax, to construct the feature model, using Photoshop software to process texture data and create a simulation model of the object feature. The building model is shown in Figure 3.
3. Terrain and Feature models are managed by Unity3D, the three-dimensional campus is also constructed, and the rendering and display of campus is achieved as well. The three-dimensional scene of Zhengzhou University is shown in Figure 4.

4 SCENE OPTIMIZATION

In the process of building three-dimensional scene of Zhengzhou university campus, terrain and feature object is mainly by model building, while the model is exquisite, rendering efficiency is low; on the contrary, the high fidelity is lower when rendering efficiency is high. View of this situation, this paper proposes two optimization scheme, after the research of optimization model by Unity3D and the methods of scene optimization: for building model, reducing the complexity of the model, for the efficiency of model rendering, cutting down the amount of model real-time rendering in the scene by using LOD (Yan & Qing, 2006).

4.1 Model optimization

Model optimization is aimed at buildings of campus, construction model building on the basis of building detailed expression degree of build model, buildings can be divided into detailed model and simple model levels according to the Zhengzhou University field situation. The modeling requirements of models in two level and its data control quantity are as follows:

1. Detailed model
   Detailed model is applied to the teaching buildings, laboratory buildings, office buildings, integrated management building and so on. The texture is the same with the actual construction in detailed modeling, the detail structure model is expressed by model; the rows and columns of texture for clipping is best n times of 2, the size of texture is not less than 16×16 and not exceeding 1024×1024, the number of faces is around 200.

2. Simple model

A Simple model is suitable for roads, residential buildings, and other ancillary features. Building structures which use the model to express the detailed part and the texture express other parts, the texture is the same with the actual construction in modeling; the number of model faces is not more than 50.

4.2 *Rendering optimization*

From the beginning of the scene rendering, the real-time rendering optimization reduces the number of 3D scene for rendering model, improves the efficiency of rendering and enhances real-time scene. At present, the mainly methods of real-time rendering optimization have the covered excluding technology, LOD technology, multi-scale expression technology and so on, combined with the Zhengzhou University field Situation, LOD model was used to optimize the rendering of terrain and buildings.

LOD model namely the level of detail model (levels of details model) uses a variety of precision to show (Frank, 2004), the three-dimensional objects, and select different precision of the model based on the location of the observation point change to better realize real-time dynamic display of three-dimensional scenes.

5 CONCLUSION AND FUTURE WORK

This paper generates three-dimensional terrain model by Unity3D and constructs campus three-dimensional scene models by 3dsMax. In practice, Unity3D is simple, construction rapid, high fidelity, strong interactive advantages in three dimensional visualization. Unity3D will be used in more areas and industry fields, giving full play to its advantages is the content of future research.

ACKNOWLEDGEMENTS

We would like to thank all the reviewers for their recommendations that improved the comprehensiveness and clarity of our paper. This research was supported by the Key Technology and Application Demonstration of Spatial and Temporal Information Cloud Platform (No. 201412003).

REFERENCES

Frank Losasso, Hugues Hoppe, "Geometry Clipmaps: Terrain Rendering Using Nested Regular Grids", ACMSIGGRAPH 2004, pp. 769–776, 2004.

Ruokun Chen, Qiushuo Wang, Jinyuan Jia. Interactive virtual campus roanming system based on Unity3D [A]. The 7th Yangtze forum on science and technology [C]. The international Digital Technology Museum, 2010:188–196.

Yan Zhou, Qing Zhu, Duo Huang. Research on LOD model of the building in 3D city [J]. Science of Surveying and mapping, 2006, 31(5):74–77.

Ying Yu, Qingquan Zhang. Research on the method making city ground scene by using 3DMAX [J]. Surveying and spatial information. 2013, 36(8):166–168.

Hydraulic Engineering III – Xie (Ed)
© 2015 Taylor & Francis Group, London, ISBN 978-1-138-02743-5

Evaluation on harmoniousness index system of water resources allocation

Houhua Zhu
China Institute of Water Resources and Hydropower Research, Beijing, China
The Center of Water Resource Management of MWR, Beijing, China

Xianfeng Huang
College of Water Conservancy and Hydropower Engineering, Hohai University, Nanjing, China

Yamei Wang
River and Reservoir Area Administration Bureau of Shanxi Province, China

ABSTRACT: With the water shortages and water competition issues becoming increasingly prominent and global water crisis continuing to emerge, the rational allocation of water resources has gradually become a concern in the field of water resources. The purpose of water resources allocation is to achieve the coordination between social economy and bio-environment, as well as the harmony between water and people through optimizing the utilization of water resources, and to enhance the ability of the sustainable use of water resources. Currently, there are many studies on methods of water resources allocation and allocation model, but studies on the aftereffect evaluation of water resources allocation are very rare, therefore a complete aftereffect evaluation system of water resources allocation including theory, method and index has not been formed. In this paper, the content of the aftereffect evaluation of water resources allocation in terms of multi-system such as society, economy, bio-environment and water management is discussed based on the theory of harmony, and the relevant factors of harmoniousness evaluation are analyzed, and then an initial harmoniousness evaluation index system of water resources allocation is proposed. Finally, the method for calculating harmoniousness degree of water resources allocation is discussed using the analytic hierarchy process.

1 INTRODUCTION

The purpose of the evaluation of water resources allocation is to strengthen water resources management, improve water use efficiency, and enhance the ability of the sustainable use of water resources. Although there are many studies on methods of water resources allocation and allocation model, and these theories and methods are basically mature, studies on the aftereffect evaluation methods and index system of water resources allocation are very rare, resulting in the lack of new theories and methods in the application of water resources allocation evaluation. Meanwhile, methods on the quantitative screening and weight determination of indexes are not mature enough to form a consensus in the aftereffect evaluation theory, method and index system of water resources allocation. Based on the theory of harmony, an aftereffect evaluation method of water resources allocation based on the concept of harmoniousness is discussed, and an initial harmoniousness evaluation index system of water resources allocation is proposed.

2 HARMONIOUSNESS OF WATER RESOURCE SYSTEM

Harmoniousness is a description whether the system has conditions and environment that can fully unleash the initiativeness and creativity of members of the system and subsystems, as well as the overall coordination of the activities of members. Specific performance in these two areas is the degree of matching between system structure, organization and management, the internal environment within themselves and each other, and the degree of adapting between internal and external system (Xi, 2002). Water resources system is an open large-scale system, which has extensive contacts with systems of society, economy and bio-environment: 1) economic system influences the degree of development of water resources, and determines the degree of the intensive use of water resources; 2) social system determines the authorization relations of water resources, and influences the degree of development and utilization of water resources (Ji, 2005). The essence of the coordination and sustainable development between water resources system and other systems is to achieve an orderly exchange between systems, i.e., the relationship between systems is exchange of materials and energy (Tong, 2004).

Specifically, the harmoniousness of constitution mainly consists of elements adapt to system functions, such as precipitation, surface water and groundwater resources quantity, socio-economic and bio-environmental water demand and other factors, which reflect the harmoniousness of the elements between the natural assets and human activities demand. The harmoniousness of organization and management is to achieve organic combination and coordinated operation of system elements emphasized by means of organization and management, specifically including elements like water demand management, regional water resources allocation management and water conservation policies. The harmoniousness of environment is composed of the internal and external environment. The former includes water use efficiency and fairness between water users in the system, such as reuse rate of industrial wastewater, utilization coefficient of irrigation water, wastewater treatment and reuse, etc. The latter focuses on the relationship between the system and the external environment, such as social and environmental factors, including order and stability on politics and society, stability on social water demand and security on the water supply project. The overall harmoniousness is the comprehensive harmoniousness of the system, consisting of harmoniousness of constitution, organization and management, environment, which could form a harmonious group by match all aspects above with each other.

3 ARCHITECTURE OF HARMONIOUSNESS EVALUATION OF WATER RESOURCES ALLOCATION

Rational allocation of water resources is essential to the harmoniousness of water resources system, and the harmoniousness of water resources allocation is a specific performance of the overall harmoniousness water resources system. The harmoniousness evaluation of water resources allocation needs to establish evaluation architecture first. Based on the theory of harmony and sustainable development and methods of systematic science, a three-level hierarchical structure, i.e., the target layer, criterion layer, and index layer of the harmoniousness evaluation system of water resources allocation could be established. As a total control of the system, the target layer is measured mainly through whether the promotion of coordinated development of society, economy, bio-environment and resource management under fair conditions. The criterion layer includes the harmoniousness of systems of society, economy, bio-environment and water resources management. The index layer is the major evaluation factors which can reflect the characteristics of the criterion.

3.1 *Target layer*

Aiming at the sustainable use, the purpose of rational water resources allocation is to promote the harmonious development of society, economy, ecology and resource under fair conditions, reflecting the overall harmoniousness of the rational allocation of water resources.

3.2 Criterion layer

The harmoniousness target of water resources allocation lies specifically in the harmoniousness of systems of society, economy, bio-environment and water resources. The harmoniousness of these four criterion layers within themselves and each other could affect the overall harmoniousness of water resources allocation.

3.3 Index layer

The index layer is composed of the evaluation factors capable of reflecting the characteristics of water resources allocation system. They are measured by quantification, analysis and comparison of the index, which could obtain the degree of the harmoniousness of water resources allocation more scientifically. The corresponding evaluation indexes for each system of the criterion layers are set up, which is widely ranged thanks to the complexity of these systems. Therefore, in order to seize the key to determine the mode of water resources development and facilitate the calculation of the evaluation, the most representative and key indicators will be selected, and the indicators which have little to do with the research purpose should be abandoned.

From the above we note that $\sin \theta = (x + y)z$ or:

$$K_t = \left(1 - \frac{R^2 \tau}{c_a + \nu \tan \delta}\right)^4 k_1 \tag{1}$$

where c_a = interface adhesion; δ = friction angle at interface; and k_1 = shear stiffness number.

4 HARMONIOUSNESS EVALUATION INDEX SYSTEM OF WATER RESOURCES ALLOCATION

Establishment of harmoniousness evaluation index system is the basis of the study on the rational allocation of water resources. The rational determination of the index will have a

Table 1. The harmoniousness evaluation index system of water resources allocation.

Target layer	Criterion layer	Index layer
Harmoniousness degree of water resources allocation	Social system	Guarantee rate of residential water consumption
		Guarantee rate of industrial water consumption
		Guarantee rate of agricultural water consumption
		Guarantee rate of bio-environmental water consumption
	Economic system	Water consumption per unit of RMB 10,000 gross domestic product
		Water consumption per unit of RMB 10,000 industrial added value
		Utilization coefficient of irrigation water
		Leakage rate of urban water supply network
	Bio-environmental system	Exploitation and utilization rate of water resources
		Water quality compliance rate of water function areas
		Treatment rate of sewage
		Green area per capita
	Water resources system	Investment in water resources management
		Equilibrium degree of regional water shortage rate
		Metering rate of water consumption
		Popularizing rate of water-saving appliance

decisive influence on the subsequent evaluation. The guidelines of selecting evaluation index is the theory of sustainable development, i.e., the selected indexes should reflect the main aspects of sustainable development, including sustainable use of water resources, fairness between water users of various regions and departments, water use efficiency, sustainable development of society and economy, healthy development of the bio-environment, coordination between population, resources, environment and economy (Wang, 2003).

Taking into account the rationality of harmoniousness evaluation indexes and the convenience of the quantification in practice, combined with the related indicators of the water conservancy statistics in China, the harmoniousness evaluation indexes are selected. In the social system, guarantee rate of residential, industrial, agricultural and bio-environmental water consumption can be selected. The harmoniousness evaluation index system of water resources allocation is shown in Table 1.

5 METHOD FOR THE HARMONIOUSNESS EVALUATION OF WATER RESOURCES ALLOCATION

5.1 Harmoniousness degree of water resources allocation

System is developing in the conflicting elements. Therefore, discord of system is absolute while harmony is relative in general. The real system is always in a state between ideal harmony and absolute discord (Shen, 2004). In this paper, $Hx = h(x)$ is employed as the degree of harmoniousness of system state x, referred harmoniousness degree. Harmoniousness degree is a quantitative measure of the system harmoniousness, reflecting the coordination between the force inside and outside of the system and its direction of development. It is an indicator measured psychologically, subjectively, or through scientific analysis and calculation based on a certain phenomenon, which reflects the coordination and consistency of system. It is a measurement of harmoniousness of a state or a program under certain conditions (Wang, 2003).

The harmoniousness degree of water resources allocation is a comprehensive indicator measuring the relations between the function of water resources system and the external environment. It has many uses: 1) evaluates the internal functions of water resources system systematically and reflects the composition of each element, relations between them and the coordination between elements and subsystems; 2) reflects the balance between the water supply and the water demand of society, economy, and bio-environment comprehensively; 3) reflects the coordination of the basic relationship between water resources system and the external environment under a certain period of time, certain regions and certain technical conditions, generally; 4) reflects the reasonable degree of the allocation of water resources specifically (Wang, 2003).

According to the architecture and index system of the harmoniousness evaluation of water resources allocation, the harmoniousness degree of water resources allocation can be expressed as:

$$H = h(h_1(s), h_2(e), h_3(b), h_4(m)) \tag{2}$$

where H represents the harmoniousness degree of water resources allocation; h_1, h_2, h_3, h_4 represents the harmoniousness degree of systems of society, economy, bio-environment and water resources respectively. The harmoniousness degree of each subsystem, h, can be expressed by weights and corresponding indexes:

$$h = \sum \varpi_i C_i \tag{3}$$

where ϖ_i, C_i represents weight and value of evaluation index i.

5.2 Method for the harmoniousness evaluation

The harmoniousness evaluation is based on the theory of harmony, systematic science and sustainable development. It is a comprehensive method using a combination of qualitative analysis and quantitative analysis though computer simulation. The AHP is a commonly used approach in the evaluation and analysis of a large index system. It is a method that can simulates the decision-making thinking process of human to solve the problems of multi-factor complex systems, especially the social system which is difficult to describe quantitatively. So the AHP is a useful tool in the analysis of multi-target, multi-criteria, complex systems (Tan, 1999; Wang, 2003). This method is used to evaluate the harmoniousness of water resources allocation whose process mainly includes five steps, namely, establishment of the hierarchical model, establishment of the judgment matrix, sorting within the subsystems, sorting within the whole system, examining the consistency of each judgment matrix (Tan, 1999).

5.2.1 Calculation of index weight

The weight reflects the importance of each index, which means the greater the weight is, the more important the index is. Generally, weights need to be normalized between 0 and 1, at the same time the sum of the weights is 1 (Shen, 2004). The AHP can be carried out to calculate the weights.

5.2.2 Normalization of index

Indexes are parameters reflecting the harmoniousness of water resources allocation, but the initial properties of indexes might have some issues such as different dimensions and huge difference between the original data that must be normalized before the evaluation (Shen, 2004). According to the positive and negative correlation with the system, indexes can be divided into forward-type indexes and backward-type indexes, which can be normalized using the following formula:

$$P = (P_{eva}-P_{min})/(P_{max}-P_{min}) \text{ (forward-type index)} \tag{4}$$
$$P = (P_{max}-P_{eva})/(P_{max}-P_{min}) \text{ (backward-type index)} \tag{5}$$

where P is the normalized value; P_{eva} is the actual value; P_{max} is the maximum value; P_{min} is the minimum value.

5.2.3 Calculation of harmoniousness degree

The harmoniousness degree of water resources allocation can be calculated based on the weights of indexes and the normalized value. The harmoniousness degree of systems of society, economy, bio-environment and water resources is as follows:

$$[h_1(s), h_2(e), h_3(b), h_4(m)] \tag{6}$$

The harmoniousness degree of water resources allocation is as follows:

$$H = \sum_{i=1}^{4} \omega_i h_i \tag{7}$$

where $H = 1$ represents the whole system is in a completely harmonious state, called ideal state.

6   CONCLUSION

The harmoniousness evaluation of water resources allocation is an expansion and extension of aftereffect evaluation of water resources allocation, which not only can be an important reference for the comparison and selection of water resources allocation program, but also a basis for aftereffect evaluation of rational water resources allocation. In this paper, the harmoniousness evaluation index system of water resources allocation is proposed initially, and the method for the harmoniousness evaluation is discussed. This initial architecture and index system and the method could be applied to the aftereffect evaluation of water resources allocation in practice.

REFERENCES

Ji, X.Y., Cui, G.B. 2005. Study on water resources coordinative management system. Yellow River, 27(5), 42–43. (in Chinese).

Shen, J.Q., Wang, Y. 2004. Studies on resettlement system of Baoying Station Project of the Eastern Route Scheme of SNWDP. China Water Resources, P: 33–35. (in Chinese).

Tan, Y.J., Chen, Y.W., Yi, J.X. 1999. Systems Engineering Principles. Changsha: National Defense University Press. (in Chinese).

Tong, C.S., Huang, Q., Liu, J.P, et al. 2004. The theory of water resources system Jieke and its application prospects. Journal of Xi'an University of Technology, 20(0), 21–26. (in Chinese).

Wang, H., Chang, B.Y., Qin D.Y., et al. 2003. Report on the water resources management information system in Heihe River Basin. (in Chinese).

Wang, J.G. 2003. Regional water resources allocation and harmonious analysis on water resources system. Hohai University doctoral dissertation. (in Chinese).

Xi, Y.M., Shang, Y.F. 2002. Harmonious Management Theory. Beijing: China Renmin University Press. (in Chinese).

Hydraulic Engineering III – Xie (Ed)
© *2015 Taylor & Francis Group, London, ISBN 978-1-138-02743-5*

Mean flow characteristics of wall-jets for hydraulic engineering applications

Maneesha Sebastian & B.S. Pani
Department of Civil Engineering, IIT Bombay, Mumbai, India

ABSTRACT: The mean flow features of two types of wall jets used often in hydraulic engineering are analyzed. They include the results of a plane jet and a three-dimensional wall jet. The required flow field was obtained from CFD simulations, the point source method and some limited experiments. These are compared with the available equations in vogue, which predict the growth of the jet, the decay of the maximum mean velocity and concentration of tracers. The variation of the wall shear stress along the flow is also analyzed. The CFD results for the distribution of the mean velocity and the tracer concentration exhibit 'self-similarity'. However, the predicted growth rate of the jets differs from the available data.

1 INTRODUCTION

Wall jets are bounded by a solid surface, the wall, on one side while the outer region of the flow is in contact with the ambient fluid. Wall jets find many applications such as cooling of surfaces, boundary layer control, building ventilation, energy dissipation etc. Jets emanating from sluices and other hydraulic structures close to the bed of channels need special attention to provide the necessary protective works. Prediction of the velocity profiles in the case of a wall jet is more difficult compared to the free jet. After a short distance downstream of the outlet, all jets tend to behave in a similar fashion. The schematic diagrams of a plane wall-jet and the boundary conditions used in the simulation are shown in Figure 1.

The definition sketch for a three dimensional wall jet is depicted in Figure 2. In both the cases, the length of the potential core is approximately equal to six times the opening height. The total thickness of the jet, i.e. the shear layer, may be divided into two zones—the

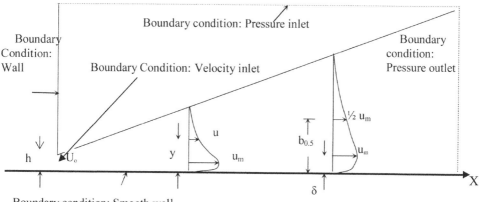

Figure 1. Schematic diagram of plane wall-jet.

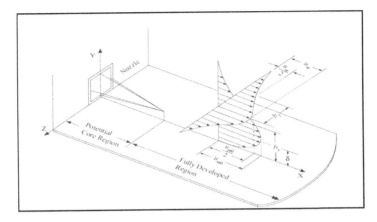

Figure 2.   Definition sketch of a three-dimensional wall jet.

wall region and the outer free shear zone. However, for practical applications it will be more convenient to analyze the entire thickness as a single entity. An empirical expression, which describes the distribution of the mean velocity normal to the wall in the developed zone, was proposed by Verhoff (Rajaratnam, 1976).

$$\frac{u}{u_m} = 1.48\eta^{\frac{1}{7}}[1 - \mathrm{erf}(0.68\eta)] \qquad (1)$$

where $\eta = y/b_{0.5}$ and the length scale $b_{0.5}$ is the value of y at which the velocity $u = 0.5\, u_m$ and $\partial u/\partial y = 0$. The maximum velocity occurs at a distance of $y = \delta$. Equation 1 was derived primarily based on plane wall-jet data. It has been observed that the same relation describes the velocity distribution even for the three dimensional wall-jet. Furthermore, in case of the three-dimensional wall-jet the transverse distribution of the velocity in a plane $z = \delta$ has been found to follow a Gaussian distribution. If the normalized coordinate in the z-direction is denoted as $\eta_z (= z/b_z)$, then the Gaussian distribution can be expressed as

$$\frac{u}{u_m} = e^{-0.693\,\eta_z^2} \qquad (2)$$

In Equation 2, b_z represents the half width of the jet in the $z = \delta$ plane. For a three-dimensional wall-jet, it has been observed experimentally that the half width b_z is significantly larger than b_y. It can be shown that the half widths for the plane and the three dimensional jets must grow linearly with the distance x. This has been corroborated experimentally and the constant of proportionality has been found from the observed data.

The purpose of this investigation is to find the flow features of the aforementioned jets from CFD simulations and establish their efficacy in predicting the mean flow field for which experimental results are available. For the plane and the three-dimensional wall jets, the simple point-source technique was also applied to predict the decay of the maximum velocity and the tracer concentration with the distance x. In the computations of the present study, GAMBIT ver 2.4 was used as the preprocessor and Fluent ver 6.3.26 as the processor and post processor. Based on the earlier experiences, the κ-ε realizable model was considered suitable for the computations. The efflux velocity was selected as 2.0 m/s and the slot height h = 0.01 m. The intensity of turbulence at the efflux section was considered as 1%. For finding the distribution of the tracers, the density of the particles was made equal to that of the ambient fluid.

2 EXPERIMENTS

For validating the simulation results, some experiments were conducted for a plane wall-jet. The size of the efflux section was 0.01 m × 0.3 m and water was used as the medium in a re-circulating flume measuring 7 m long, 0.3 m wide and 0.5 m high. The velocity measurements were conducted using a Pitot tube and an Acoustic Doppler Velocitimeter (ADV). The Reynolds number of flow in the four runs was in the range of 19400 to 24500. The boundary shear stress τ_0 was measured using the Preston tube technique and the outer diameter of the flattened tube was 2.06 mm. A typical velocity distribution at various downstream locations, normal to the wall, is shown in Figure 3. The values predicted by the CFD method for x/h = 10–80 show some deviation from the Verhoff's curve. The present experimental data, not shown in the diagram, too follow the Verhoff's curve very closely.

3 RESULTS AND ANALYSIS

The distribution of the velocities computed at various locations of the plane wall-jet show that the self-similarity is satisfied in the fully developed zone, i.e. x/h ≥ 10. In Figure 3, besides the Verhoff's equation, the Gaussian distribution and the experimental results of Forthmann (from Rajaratnam, 1976) are also shown for comparison. Excepting for $y/b_{0.5} < 0.16$, the Gaussian distribution seems to be a good fit to the actual distribution of the normalized velocity.

The computed distribution of the tracer concentration follows the Gaussian profile as shown in Figure 4. The length scale used for the tracer distribution is b_{yc} which is larger than $b_{0.5}$. It can be expressed as $b_{yc} = \lambda \, b_{0.5}$, where the spread coefficient $\lambda > 1$.

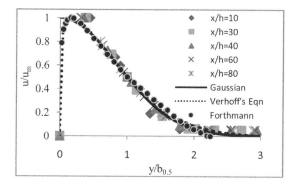

Figure 3.   Similarity of velocity distribution of a plane wall jet.

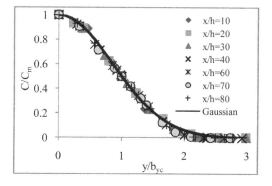

Figure 4.   Computed concentration distribution of a plane wall jet.

The average decay of the maximum velocity u_m of a plane wall jet on a smooth boundary was found empirically as (Rajaratnam, 1976)

$$\frac{u_m}{U_o} = \frac{3.5}{\sqrt{\dfrac{x}{h}}}$$

(3)

Following the recent point-source technique, the wall was treated as a frictionless reflector and the principle of superposition was adopted to arrive at the velocity decay relation

$$\frac{u_m}{U_o} = \left\{ \mathrm{erf} \left(\frac{12.514}{\dfrac{x}{h}} \right) \right\}^{0.5}$$

(4)

The CFD results pertaining to the decay of the maximum velocity in a plane wall-jet are shown in Figure 5, along with the values computed from Equations 3 and 4. From these plots it may be inferred that there is good agreement between the CFD results, the point-source method, the empirical relation proposed earlier and the present observed data. It was also observed that in the developing zone, the predictions from the point-source method agreed better with the observed data. The same problem was solved by using the κ-ω model and the velocity distributions did not agree that well with Equation 1. Hence, the results reported in this article are based mostly on the κ-ε model.

The present measured rate of growth of the half width $b_{0.5}$ for all the experimental runs give an average value of the jet width $b_{0.5} = 0.082x$, whereas the CFD results yield a value of $b_{0.5} = 0.092x$. Interestingly, both the growth rates are significantly larger than the conventional value quoted from experiments (Rajaratnam, 1976). Similarly, the spread coefficient λ from the present CFD results was found as 1.08 compared to the average value of 1.12 observed earlier.

The variation of the skin-friction coefficient, $C_f = \tau_0/(0.5\rho U_o^2)$, for the plane wall jet is shown in Figure 6. There seems to be a fair agreement between the predicted values and the ones given by Myer's expression (Rajaratnam, 1976). The measured τ_0 values of the current four runs exhibit a significant amount of scatter and are not shown here.

For simulating the flow in the three-dimensional wall-jet, a square (10 mm × 10 mm) and a circular outlet of the same area of cross-section were used. The computed velocity profiles for both the shapes were found to be identical which indicates that for three dimensional wall-jets, the shape of the outlet is not important. This agrees with the past experimental results on three-dimensional wall-jets (Pani, 2012). The normalized velocity distributions in the $z = \delta$ plane match with the past experimental observations.

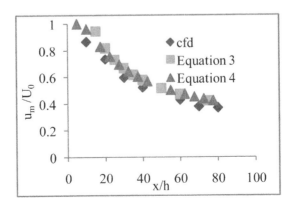

Figure 5.   Decay of the maximum velocity of a plane wall jet.

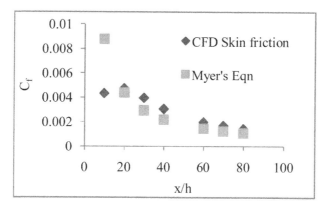

Figure 6. Variation of skin-friction coefficient for a plane wall jet.

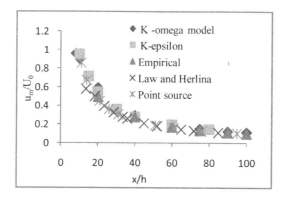

Figure 7. Decay of maximum velocity for 3D wall jet.

By the method of superposition, the decay relation for the maximum velocity of a three dimensional wall jet was derived as (Pani, 2012)

$$\frac{u_{mo}}{U_o} = \frac{0.94}{C}\left(\frac{x}{\sqrt{A}}\right)^{-1} \qquad (5)$$

where the coefficient C was assigned a value of 0.122 from experiments. A slightly different expression was proposed by Law & Herlina (2002). The decay of the maximum velocity u_{mo} predicted by various methods is depicted in Figure 7. The results of all the methods agree well and any one of the methods can be adopted to predict the maximum velocity u_{mo} at a given x-location of a submerged hydraulic outlet.

The predicted thickness and spread of the jet in the y and z directions, respectively, are nearly identical as shown in Figure 8. This is very different from the observations that the spread rate in the z-direction is 4 to 5 times larger than the growth rate normal to the wall.

At any given distance x from the outlet, the boundary shear stress τ_o was predicted for various transverse z locations. Even though the non-dimensional distribution τ_o/τ_{om} followed a Gaussian profile as was found earlier in the experiments, the absolute values of the predicted maximum shear τ_{om} and the length scale b_τ are in variance with experimental observations.

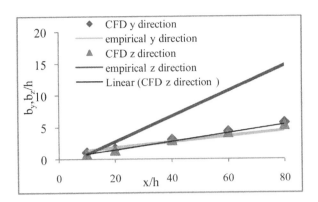

Figure 8. Growth of half-widths b_y and b_z for a 3D wall jet.

4   CONCLUSIONS

Based on comparison of various equations of the plane and the three-dimensional wall-jets with the results of the CFD simulation, the following have been observed.

1. For a plane wall-jet while the velocity distribution in the y-direction can be described by the Verhoff's equation whereas the tracer concentration follows a Gaussian distribution with the maximum value occurring on the wall.
2. Notwithstanding the fact that the predicted b_y values from CFD are always somewhat higher than the experimental values, for applications in hydraulic engineering the flow field of a plane wall jet can be simulated based on the available software.
3. In case of the three-dimensional wall jets, the predicted decay of the maximum velocity u_{mo} agrees quite well with the existing relationships, including the one based on the point-source method.
4. The results from the CFD simulation on the spread of the jet b_y and b_z of three dimensional wall jets are quite different from the experimentally observed values. Discrepancies also exist for the boundary shear distribution on the wall. Hence, the available software may not serve much purpose in designing the protective work below hydraulic outlets located near the channel bottom.

REFERENCES

Law, A.W.K and Herlina. 2002. An experimental study on turbulent circular wall jets. *J. Hyd. Engg.* ASCE. 28(2). 161–174.
Pani, B.S. 2012. *Turbulent jets: A point source method for hydraulic engineering.* Delhi: Cambridge University Press India.
Rajaratnam, N. 1976. *Turbulent jets.* Amsterdam: Elsevier Scientific Pub. Co.

Hydraulic Engineering III – Xie (Ed)
© 2015 Taylor & Francis Group, London, ISBN 978-1-138-02743-5

Numerical modeling and analysis of shoreline change on Beidaihe new district coast

Y.C. Shan, C.P. Kuang, J.Y. Wang & C.F. Hu
Department of Hydraulic Engineering, College of Civil Engineering, Tongji University, Shanghai, China

J. Gu
College of Marine Sciences, Shanghai Ocean University, Shanghai, China

ABSTRACT: As an artificial structure-Purple Grape Island and Beach Nourishment Project is being built in Beidaihe new district, a preliminary modeling study is presented in this paper to evaluate the impact of the project on the shoreline change. The verified numerical model-GENESIS based on one-line theory is used to model the shoreline change caused by the Purple Grape Island Project and beach nourishment project. The result shows that most of the nourished sand remains 10 years after the project and the Purple Grape Island does play a positive role in beach protection at the southwest end of the beach, by blocking the longshore sediment transport. By using the same model parameters, three subsequent projects involving the Purple Grape Island, beach nourishment and submerged breakwaters are discussed. The numerical predictions present that different arrangements of breakwaters with the same total length have different effects on the shoreline change, and each segment of the breakwaters can protect the coastline behind it more than that before it. On the basis of the comparison of protected shoreline length, a reasonable layout of breakwaters is proposed.

1   INTRODUCTION

The erosion of sandy beaches is a worldwide problem and 70% of the sandy beaches in the world are retreating at a rate of 0.5–1.0 m per year (Bird 1985). Beach erosion not only damages the coastal environment, but also processes a threat to the coastal economy, especially tourism. To meet the increasing demand of beach resources, there exists a growing interest in developing effective engineering solutions to protect the beach against further erosion. Among the various ways, beach nourishment is regarded as one of the most popular and effective methods (Dean 2005). To reduce the loss of nourished sand, the nourishments are often combined with artificial structures such as groins and breakwaters (Evans and Ranasinghe 2001).

In the USA, beach nourishment project can date back to the 1920s, and in the 1990s quantities of beach nourishments were put into use (Valverde et al. 1999). In Europe, over the last decades, there has gradually formed the understanding that beach nourishment is an environment friendly method (Hanson et al. 2002). In China, Hong Kong used beach nourishment on Repulse Bay in 1990 (Ji et al. 2006), which benefits the beach tourism evidently. Since then, many large-scale beach nourishment projects have been implemented in China, like the beach nourishment project in Xinghai Bay, Dalian (Song et al. 2005), Anping, Taiwan (Chen et al. 2006) and Xiangshan-Changweijiao, Xiamen (Wang et al. 2009).

As the beaches of Beidaihe are of great importance to the tourism of Qinhuangdao, the Bedidaihe District government has implemented the beach nourishment project since 2008 on West Beach, Middle Beach and East Beach (Kuang et al. 2010). Now an artificial island—Purple Grape Island is being built in Beidaihe New District, and the shoreline change caused by the project is modeled, which guarantees the feasibility of the project. Subsequent projects involving the Purple Grape Island, beach nourishment and submerged breakwaters are also predicted and discussed by using the same model parameters.

2 STUDIED AREA

Beidaihe New District, established in December 2006, is affiliated with one of the first open coastal cities-Qinhuangdao, Hebei Province. The studied area is located in the northeast of Bedidaihe New District, between Yanghe River estuary and Renzaohe River estuary (Fig. 1). The studied shoreline is about 4 km long and the Purple Grape Island is being built at the southwest part of the beach as shown in Figure 1.

Statistical analysis of wave condition in the region from 2012 to 2013 (Fig. 2) shows that the normal wave is swell wave coming from south and southeast.

Figure 1.   The location of studied area.

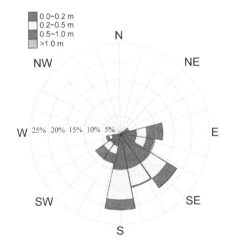

Figure 2.   Rose diagram of wave direction frequency.

3 GENESIS MODEL

GENESIS (GENEralized model for SImulating Shoreline change) model is a numerical model which is widely used in simulation of long-term shoreline change and response of the shoreline to structures constructed in the nearshore. It is developed by the Coastal Engineering Research Center (CERC) (Hanson and Kraus 1989).

GENESIS predicts the sediment transport and calculates the shoreline change. The governing equation for the rate of change of shoreline position is:

$$\left(\frac{dy}{dt}\right)+\left(\frac{1}{D_B+D_C}\right)\left(\frac{\partial Q}{\partial x}-q\right)=0 \tag{1}$$

where y indicates the distance offshore, x indicates the distance alongshore, D_B is the berm height, D_C is the closure depth, q is the offshore sand transport rate, Q is the longshore sand transport rate, which is given by:

$$Q=(H^2 C_g)_b\left[a_1\sin 2\theta_{bs}\left(\frac{\partial H}{\partial x}\right)\right]_b \tag{2}$$

where H is the wave height, C_g is the wave celerity, b is a subscript denoting breaking wave condition, θ_{bs} is the angle of the wave relative to the shoreline, and a_1 and a_2 are dimensionless coefficients given by:

$$a_1=\frac{K_1}{16\left(\frac{\rho_S}{\rho}-1\right)(1-\rho)(1.416)^{5/2}} \tag{3}$$

$$a_2=\frac{K_2}{8\left(\frac{\rho_s}{\rho}-1\right)(1-p)\tan\beta(1.416)^{7/2}} \tag{4}$$

where ρ_s is the density of sand, ρ is the density of water, p is the porosity of sand on the bed (taken to be 0.4), $\tan\beta$ is the average bottom slope between the shoreline and the seaward extent of the surf zone, 1.416 is a factor used for the conversion of significant wave height, and K_1, K_2 are empirical coefficients, treated as a calibration parameter.

4 MODELING SHORELINE CHANGES BY THE PROJECT

4.1 *Model setup and verification*

The modeling area covers the beach from Yanghe River estuary to Renzaohe River estuary. x axis of the computational grid has an azimuth angle of 229°and a length of 6000 m; y axis has a length of 5000 m and the mesh spacing is 50 m. Time step is set to be 3 hr. According to the measurements, berm height is 1 m and the median grain size is 0.2 mm; closure depth is set to be 7 m, on the basis of the previous study of this area; the setting of K_1 is 0.2 and K_2 is 0.2, adjusted to the model.

On basis of qualitative observation, the shoreline is regarded stable in the long term. Hence, using the measured wave data, the shoreline change in 10 years is predicted under the natural situation and compared with the original shoreline in Figure 3. It can be seen that the predicted shoreline position in 10 years is very close to the original shoreline, which also matches with the preliminary observation in the field that the whole shoreline changes little except the two ends.

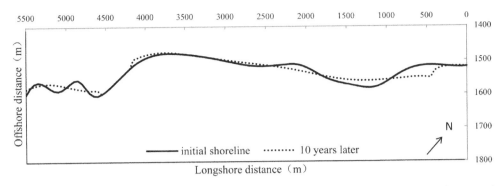

Figure 3. Comparison of predicted shoreline in 10 years and initial shoreline under the natural situation.

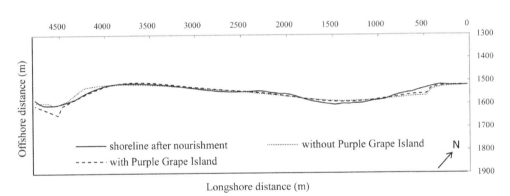

Figure 4. Comparisons of shoreline changes with and without Purple Grape Island.

4.2 Shoreline change predictions under projects

4.2.1 Effect of the Purple Grape Island

The Purple Grape Island Project is being constructed in this area and a beach nourishment project is proposed (Fig. 1). The average width of proposed beach nourishment is 30 m, the berm height is 2 m and the median grain size of filled sand is 0.3 mm. Using measured wave data and same numerical parameters used in the verified model, the effect of Purple Grape Island Project on shoreline change of beach nourishment project are predicted and shown in in Figure 4. It revealed that the beach nourishment increases the beach width and the construction of the Purple Grape Island protects the southwest end of the beach. The 1 km beach near the southwest end is widened to about 18.5 m on the average, owing to the construction of Purple Grape Island. It can be explained that the Purple Grape Island works as a groin and blocks the sediment transport from northeast to southwest, which result in sediment deposition at the southwest end of the beach.

4.2.2 Effect of the submerged breakwaters

The subsequent projects with different arrangements of submerged breakwaters are tested. The submerged breakwaters are designed approximately 400 m offshore, the width of 50 m and the elevation of −0.9 m. The transmission coefficient is set to be 0.7 according to the formula proposed by Seabrook and Hall (1998). The x coordinate of submerged breakwaters is set in: 800 to 1600, 2100 to 2800 and 3300 to 3800 for three breakwaters in scheme 1, from

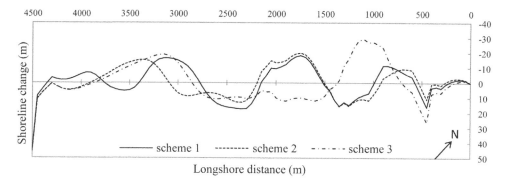

Figure 5. Predicted shoreline changes (relative to the shoreline after pure nourishment) under three breakwater schemes.

700 to 1700 and 2100 to 3100 for two breakwaters in scheme 2 and from 1100 to 3100 for one breakwater in scheme 3, i.e. the total length of three schemes is the same with 2000 m. The simulated shoreline changes in 10 years under the three schemes are shown in Figure 5. It can be seen shoreline salients are formed behind the breakwaters and the locations of salients correspond to the center positions of submerged breakwaters. The protected shoreline lengths are 1500, 1350 and 1450 m by scheme 1, 2 and 3 respectively, which implies that scheme 1 has more advantages than other two schemes, and also forms a more moderately curved shoreline.

5 CONCLUSIONS

By use of a shoreline-change numerical model (GENESIS), a preliminary modeling study on shoreline changes caused by the Purple Grape Island Project and beach nourishment projects in Beidaihe New District, China. The model is verified first, then the long-term shoreline changes by projects are predicted, which shows that the Purple Grape Island plays a positive role in beach protection. The shoreline changes with different arrangements of submerged breakwaters are also predicted and the results show that reasonable distributed breakwaters can protect the shoreline more effective. Hence, the scheme 1 with the Purple Grape Island, beach nourishment and three submerged breakwaters is found to have the best performance.

REFERENCES

Bird, E.C.F. 1985. *Coastline Changes: A Global Review*. Chichester, U.K.: Wiley-Interscience.
Chen, J.Z., Wu, N.J. & Zhu Z.C. 2006. Discussion on topographical change between before and after the artificial beach nourishment in Anping Harbor, *China Technology*, 70: 50–57. (in Chinese).
Dean, R.G. 2005. Beach Nourishment: Benefits, Theory and Case Examples. *NATO Science Series, Environmentally Friendly Coastal Protection.*
Hanson, H. Brampton, A. & Capobianco M. et al. 2002. Beach nourishment projects, practices, and objectives—a European overview, *Coastal Engineering*, 47(2): 87–111.
Hanson, H. & Kraus, N.C. 1989. GENESIS: Generalised model for simulating shoreline change: Report 1. *Technical reference. Tech. Rep. CERC-89-19*, U.S. Army Engineer Waterways Experiment Station, Coastal Engineering Research Centre, Vicksburg, Mississippi.
Ji, X.M., Zhang, Y.Z. & Zhu, D.K. 2006. Research development on artificial beach nourishment, *Marine Geology Letters*, 22 (7): 21–25. (in Chinese).

Kuang, C.P., Pan, Y. & Zhang, Y. et al. 2010. Shoreline change modeling on emergency beach nourishment project on west beach of Beidaihe, China. *China Ocean Engineering*, 24(2): 277–289.

Seabrook, S. & Hall, K. 1998. Wave transmission at submerged rubble-mound breakwaters. In: *Proceedings of the 26th Coastal Engineering Conference (ASCE, Copenhagen, Denmark)*, 2000–2013.

Song, X.Q., Guo, Z.J. & Chen, S.M. 2005. The planning and design of artificial beach in the Xinghai Bay, *China Civil Engineering Journal*, 38(4): 135–140. (in Chinese).

Valverde, H.R. Trembanis, A.C. & Pilkey, O.H. 1999, Summary of beach nourishment episodes on the US East Coast barrier islands. *Journal of Coastal Research*, 15(4): 1100–1118.

Wang, G.L. Cai, F. & Cao, H.M. et al. 2009. Study on the practice and theory of beach replenishment of Xiangshan-Changweijiao Beach in Xiamen, *The Ocean Engineering*, 27(3): 66–75.

Hydraulic Engineering III – Xie (Ed)
© *2015 Taylor & Francis Group, London, ISBN 978-1-138-02743-5*

Technical research on analysis of dam-break emergency evacuation

Xiao-Hang Wang & Jin-Bao Sheng
Nanjing Hydraulic Research Institute, Nanjing, China
Dam Safety Management Center of the Ministry of Water Resources, Nanjing, China
Key Laboratory of Earth-Rock Dam Failure Mechanism and Safety Control Techniques,
Ministry of Water Resources, Nanjing, China

Shu-Hong Song
Hydrology and Water Resources Survey Bureau of Shaanxi Province, Xi'an, China

ABSTRACT: Analysis of dam-break emergency evacuation is one of the important non-engineering measures to avoid or reduce loss of life due to dam-break. It is mainly reflected in the perspective of time and space. Aimed at escape, of the dam-break emergency evacuation is to determine the response-time requirements by analyzing transfer time before dam-break and arrival time of flood. Space analysis primarily includes calculation of flood submerging range and space requirements of dam-break evacuation, depending on submerging range of dam-break and population at risk. Both dam breach simulation and dam-break flood routing simulation are the fundamental work. The distribution forecast of population at risk and calculation of time-consuming of emergency evacuation are based on flood analysis simulation, which are considered as the important parts of analysis of dam-break emergency evacuation. This paper is to analyze and summarize the researches on these four above-mentioned technologies and currently applications.

1 INTRODUCTION

Dam break is considered as the typical public safety event. Once occurred, it may cause catastrophic destroy and impact on downstream life, property, infrastructure, eco-environment and economic and social development. At present, there are altogether 98,002 reservoirs in China, ranking the first place all over the world. From the year of 1954 to 2010, 3515 dams were broken completely, some of which caused heavy casualties. But also some dam-break events had done less damage or even no damage at all. The important reason that causes such difference is to strengthen the analysis of dam-break results and forecast of dam-break

The analysis of dam-break emergency evacuation is mainly reflected in time and space. To escape with the life, the time for dam-break emergency evacuation; therefore, we should determine the time requirements of transfer notification by analyzing the time required for emergency transfer before dam-break and flood arrival time. In consideration of developed communication currently, we can send the notification via telephone, text message of mobile phone (information transmission time can be ignored.) therefore, the time analysis of emergency evacuation before dam-break mainly includes the calculation of flood arrival time and calculation of dam-break emergency evacuation. Whereas, the space analysis is mainly reflected in calculating flood submerging range and space requirements of dam-break evacuation, which is depending on the submerging range of dam-break and population at risk.

This paper has analyzed key technologies of dam-break emergency evacuation, including dam breach simulation, dam-break flood routing analysis, and distribution forecast of population at risk and time estimation of emergency evacuation. Among them, the first two technologies belong to the fundamental tasks for emergency evacuation analysis and the latter ones are the important contents impacting on dam-break evacuation analysis which should be implemented based on the flood analysis simulation.

2 DAM BREACH SIMULATION

The dam breach flow and its changing process have a close relation with the shape of dam breach and its developing process, which can directly influence the downstream flood of dam-break. The flow process of dam-break is the upper boundary condition of downstream routing model of dam-break's flood. Dam breach simulation includes the simulation of dam breach's shape and calculation of dam breach's flow process. According to different dam types, the dam break can be divided into instantaneous dam break and gradual dam break.

Instantaneous dam break (such as concrete dam) is divided into instantaneous and full dam break, horizontal and partial dam break and vertical and partial dam break. The peak flow of dam breach and the whole process of dam break can be calculated by empirical formula (Xie Renzhi, 1993) (Li Lei, Wang Renzhong, Sheng Jinbao, etc, 2006) (Fang Conghui, Fang Kun, 2012).

The earth dam is usually broken gradually; the main reason that causes earth dam break is the destruction of overtopping and permeation. Compared with concrete dam, the earth dam break features various forms and complex dam break path. At present, the research both at home and abroad primarily focuses on the position and shape of dam breach, developing rules of dam breach and the establishment of corresponding numerical model. The research on earth dam break adopts numeral simulation, hydraulic model test and centrifugal model test, etc. Due to the limitation of test conditions, the actual researches are mainly subject to numeral simulation. In 1965, Cristofano (Cristofano E A, 1965) established the first mathematical model used to simulate earth dam break. Afterwards, many scholars and professionals in many countries proposed multiple mathematical models for simulating earth dam break process based on the summarization of field investigation data of dam break cases and testing results of dam-break mechanism model. Foreign dam-break's numerical simulation models mainly include H-W Model (Harris G W, Wagner D A, 1967), Brdam Model (Brown R J, Rogers D C, 1981), P-T Model (Ponce V M, Tsivoglou A J, 1981), Lou & Nogueiga Model (Nogueira V D Q, 1984), BEED Model (Singh V P, Quiroga C A, 1988), Fread Series Mathematical Model (Fread D L, 1988), "Scrap Model (Hanson G J, 2000) (Robinson K M, Hanson G J, 2001) and HR-BREACH Model (Mohamed M A A, 2002). In China, there are mathematical model from China Institute of Water Resources and Hydropower Research (Zhang Baowei, HUANG Jinchi, 2007) and Nanjing Hydraulic Research Institute (Chen Shengshui, 2012).

The dam break process of concrete faced rockfill dam is different from that of ordinary earth dam, the face plate still works to retain water under the supporting of undestroyed downstream dam, along with continuing of downstream dam's scoured, the face plate will be broken off because its suspending length can not bear the self-weight and water load, and at this time, the water head will be increased, water flow at the dam breach will be increased sharply and the dam break process will be accelerated. Then, along with the decreasing water head, flow rate of dam breach will be reduced correspondingly; the face plate works again until it becomes stable in a balanced position. Partial scholars have implemented related researches, derived calculation formula of dam-break flood flow process and proposed broken condition of face plate by taking advantage of material mechanics formula (Li Lei, 1996) (Chen Shengshui, Zhong Qiming, Cao Wei, 2011).

3 DAM-BREAK FLOOD ROUTING SIMULATION

The main task of dam-break flood routing simulation is to calculate the water depth, flow rate, flood arrival time and submerging duration of each downstream position based on flow process route of dam break site and provide basis for the forecast of emergency evacuation requirements.

As the dam-break occurs, the stored water of the reservoir discharges suddenly which causes the rising of downstream water level and dropping of reservoir's water level. The water fluctuation of such kind of unsteady flow in both upstream and downstream riverway is called dam-break wave. It belongs to intermittent wave; some physical parameters of water flow are disconnected in fault surface, whose mathematical model is the Remann Problem of Saint-venant equations. Compared with other kinds of flood, dam-break flood has its own characteristic featuring complex influenced factors. Dam-break flood generally has rapid and tranquil flow as well as bore and rarefaction waves (Wang Lihui, 2006). Besides large water flow and high water level, both vertical flow and lateral dispersion are obvious. When the floodplain routing of flood occurs, due to the complex landform, the phytal zone in the flow field and the flow friction loss should be treated specially and it is required that the numerical calculation should not generate nonphysical oscillation to avoid instable calculation.

Generally speaking, the dam-break flood routing process can be described by two-dimension shallow water hydrodynamics model in conservation form. After several years' research, it is proved that finite volume method with TVD is considered as one of the most successful calculation method for simulating dam-break flood flow (Xia Junqiang, Wang Guangqian, Lin Bin-liang, etc, 2010). At present, this method is widely applied in the numerical simulation of dam-break flood flow process. The actual shallow water flow is generally composed of closed boundary (land boundary) and open boundary. As for the former, it usually adopts slipping boundary conditions, that is to say, in the normal direction of the boundary, the water depth does not change, the flow rate is zero in normal direction but is not zero in tangent direction. As for open boundary, the water level process, water flow process or water level flow is given in general. As for open boundary conditions with free outflow, the normal gradient of each variant in the boundary is set as zero (Zhou J G, Gauson D M, Mingham C G, et, 2004) (Liao C B, Wu M S, Liang S J, 2007).

4 DISTRIBUTION FORECAST OF POPULATION AT RISK

Population at risk refers to all persons directly exposed to certain depth of a flood zone in the influenced area due to dam break. Based on the foreign experience, the flood depth is 0.3 m (Li Lei, Wang Renzhong, Sheng Jinbao, etc, 2006). The distribution of population at risk is the most key factor impacting on the dam-break emergency evacuation requirements. It includes the quantity of population at risk, space distribution and time distribution. At present, most population data come from statistic materials of multi-level governments, which take administrative boundary as the statistic unit. However, such method can not satisfy the research needs with specific purpose. As for the statistics of dam-break population at risk, probably only partial people in certain administrative unit belong to population at risk, therefore, the population at risk in dam-break submerging range should be calculated based on the population of each administrative region provided from statistic departments.

There are many research methods for spatial distribution of population. The main method for population estimation is to apply areal interpolation model. Surface interpolation is divided into surface interpolation without auxiliary data and that with auxiliary data on the basis that whether to use the auxiliary data or not during the process of surface interpolation. It is assumed that the population density is the same or is different from different land-use type (Lv Anmin, Li Chengming, Lin Zongjian, etc, 2005) (Fan Yida, Shi Peijun, etc, 2004) (Jiang Dong, Yang Xiaohuan, Wang Naibin, etc, 2006) (Tian Yongzhong, Chen Shupeng, Yue Tianxiang, etc, 2004). Along with the widely application of GIS and RS technology, the population data gridding has become the hot spot of spatial distribution of population. It is to distribute statistic data of population with the administrative region as the unit into the gridding with specific dimension according to certain mathematical model, so as to realize the transformation of statistic unit from administration unit to gridding. Currently, the population data gridding models are composed of two categories; one is to apply areal interpolation and statistic analysis theory, such as area weight model and core estimation model. The other is to establish the function between influenced factors of spatial distribution of population and population data, including land-use type influencing model, gravity model and multi-source data fusion model.

As for distribution of population at risk of dam-break, if there are enough data, the research should be completed by judging and interpreting high-resolution remote sensing image as well as field verification and obtaining the height and number of floors of the buildings. Field verification is to check and rectify building's use nature, number of floors and households as well as the information collection of household population quantity and travel conditions (Fu Haiyue, Li Manchun, Zhao Jun, 2006).

Time distribution of population is mainly reflected in different time contexts, population at risk is distributed dynamically. The requirements forecast of dam-break evacuation is to analyze population time distribution by combining spatial distribution of population based on different dam-break time in order to satisfy the requirements on dam-break emergency decisions. As for enterprises and public institutions and schools, the quantity of population at risk is relatively small if the dam break occurs in holidays and weekends as well as non-working time of workday; if the dam break occurs in the working time of workday, the quantity will be large. As for residential area or scenic area, it is exactly the opposite situation. Besides, the distribution of population at risk is in connection with the season. In some rural areas, migrant workers are away from home unless farming season comes. And the quantity of population at risk in the reservoir located in the upstream or in the scenic area will be increased greatly in peak tourist season (such as Beijing Miyun Reservoir is regarded as the summer resort in summer). If there is small population in dam-break risk area of reservoir, we can obtain distribution situation of population in different time and different seasons by using typical investigation method. The time distribution of population at risk differs significantly in different regions, which is difficult to establish and adopt uniform model. The existing research results are achieved through the typical investigation method of different land use types and reflect time and spatial distribution of population by taking segmented functions (Zhang Lu, Tian Yongzhong, Yang Hai, 2012).

5 CALCULATION OF EMERGENCY EVACUATION TIME

Dam-break emergency evacuation time and flood arrival time have directly influenced the arrangement and planning of emergency decisions. The flood arrival time can be calculated by dam-break flood model. And the dam-break emergency evacuation time can be divided into two parts, namely time generated for evacuation activity and time used for evacuation activity, which can be obtained by field investigation and establishing road resistance function.

Time generated for evacuation activity (t_{EGT}) refers to the time interval from starting to evacuate from affected area to the settlement position after governmental departments send evacuation alarms and people have received such alarms, which is the summation of time needed for alarming people t_w and for people ready for evacuation t_p (see Formula 1 below). In order to calculate the time generated for evacuation activity, time needed for completing each evacuation activity should be estimated firstly, and then add all these estimated time. Because the time needed for completing each evacuation activity is different from different individual, the time generated for evacuation activity can be obtained by field investigation or probability statistics.

$$t_{EGT} = t_w + t_p \tag{1}$$

Time needed for evacuation activity mainly depends on the evacuation road condition and evacuation method. Strictly speaking, any evacuation activity is the combination of horizontal and vertical movement. If the evacuation path relief has small impact on the evacuation distance, it is approximately considered as horizontal evacuation activity. Or else, vertical evacuation activity should be considered.

Based on the evacuation method, the evacuation activity can be divided into by car and on foot. Evacuation on foot is suitable for the area with near journey and poor road condition as well as vertical movement-dominated evacuation, such as to evacuate to the hill or highland. Or else, most evacuations should be implemented by car, and the vertical movement can be ignored.

As for evacuation on foot, it is generally thought that the evacuation speed is not affected by road conditions. Therefore, the best evacuation route is the shortest path of actual geographical distance. The evacuation time can be calculated according to the path and average walking speed. In view of related research (Revi A, Singh AK, 2005), in flat area, the evacuation speed on foot is 5 km/h with the old person 4 km/h and child 2 km/h.

As for the evacuation by car, the calculation of evacuation activity time is based on both road right of each section of roadway and selected evacuation path. The road right of evacuation path means the travel (migration) time of certain section of evacuation path, including the vehicles' travel time in the transfer road and waiting time in the intersection. Road right is expressed with road resistance function, including, the road resistance function proposed by Bureau of Public Road is widely applied now, as shown below:

$$t = t_0 \left[1 + \alpha \left(\frac{V}{C} \right)^{\beta} \right]$$
(2)

whereas, t refers to the travel time in intersection (min), t_0 refers to the travel time when the traffic volume is zero (min), V refers to the traffic volume of motor vehicle (vehicle/h), C refers to the actual traffic capacity of this section (vehicle/h), α and β refer to parameters ($\alpha = 0.15$ and $\beta = 4$ are recommended).

During the process of dam-break evacuation, besides traffic capacity of motor vehicle, non-motor vehicle's traffic load also impacts on the travel time (or travel speed) (Wang Wei, Xu Jian, 1992), therefore, some scholars adopt half-theory and half-experience road resistance function method to calculate the travel time under the heavy mixed traffic situation (Zhang Xingxing, 2011). The major thought is as follows: firstly determine the theoretical model of road resistance function based on the relationship of flow, speed and density. In such theoretical model, only consider the influence of traffic volume of motor vehicle, then correct non-motor vehicle's traffic volume, traffic lane number, traffic lane width and traffic disconnection (intersection) and calculation formula is shown below (Li Chaojie, 2007) (Wan Qing, Li Huiguo, 1995)

$$t(i, j) = L(i, j) / V(i, j)$$
(3)

$$V(i, j) = U_0 / 2 \pm \sqrt{(U_0 / 2)^2 - Q(i, j) \cdot U_0 / K_m}$$
(4)

wherein, $t(i, j)$ refers to the travel time in section $[i, j]$, $L(i, j)$ refers to the length of section $[i, j]$, U_0 refers to the travel speed when the traffic volume is zero, $Q(i, j)$ is the traffic volume in section $[i, j]$, K_m refers to the section jam density. Refer to related literature for the determination method of U_0 (Cheng Cuiyun, 2010).

6 SUMMARY

Technical Research on Analysis of Dam-break Emergency Evacuation involves hydraulics, hydrology, soil mechanics, mechanics of sediment transport, dam engineering, demography, sociology, traffic, geographic information system and risk analysis. The paper mainly includes research on time and spatial analysis of dam-break emergency evacuation in the perspective of dam-break simulation, dam-break flood routing analysis, population distribution forecast and emergency evacuation time. Among them, the first two perspectives are the fundamental tasks of emergency evacuation and the latter two ones are key factors influencing dam-break evacuation analysis. The analysis research on dam-break emergency evacuation is still in the initial stage, it is recommended to give more attention to flood and earthquake evacuation.

ACKNOWLEDGEMENT

Special Project of Scientific Research in Public Welfare Sectors of Ministry of Water Resources "Evaluation Method and Technical Research on Disaster-causing Results of Earth and Rockfill Dam" (201001034), Key Project of Basic Scientific Research Business Expense Special Funds of Nanjing Hydraulic Research Institute "Dynamic Simulation of Composite Situation-based Dam-break Emergency Evacuation" (Y713009), Surface Project of Basic Scientific Research Business Expense Special Funds of Nanjing Hydraulic Research Institute "Risk Map Making of Emergency Help for Reservoir" (Y712004), Special Project of Scientific Research in Public Welfare Sectors of Ministry of Water Resources "Key Technologies of Small Reservoir's Safety and Management" (201101028).

REFERENCES

Brown R.J, Rogers D.C.1981. BRDAM users manual. Denver: Water and Power Resources Service, U.S. Department of the Interior.

Chen Shengshui, Zhong Qiming, Cao Wei. 2011. Centrifugal model test and numerical simulation of the breaching process of clay core dams due to overtoppin. Advances in water science 22(5): 674–679.

Chen Shengshui. 2012. Dam-break Mechanism and Process Simulation of Earth and Rockfill Dam. Beijing: China Waterpower Press.

Cheng Cuiyun. 2010. Study on Evaluation of Emergency Planning for Reservoir. Nanjing: Nanjing University.

Cristofano E.A. 1965. Method of computing erosion rate for failure of earth-fill dams. Denver: Bureau of Reclamation.

Fan Yida, Shi Peijun, etc. 2004. A Method of Data Gridding from Administration Cell to Gridding Cell. Scientia Geographica Sinica 24(11):105–108.

Fang Conghui, Fang Kun. 2012. Derivation and verification of a new generalized formula set for calculating maximum instantaneous dam breach discharge. Advances in water science 23(5): 721–727.

Fread D.L. 1988. Breach: An erosion model for earthen dam failure. Silver Spring: National Weather Service (NWS) Report, MA NOAA.

Fu Haiyue, Li Manchun, Zhao Jun. 2006. Summary of GRID transformation models of population data. Human Geography 6:21(3).

Hanson G J. 2000. Preliminary results of earthen embankment breach tests. ASAE Annual International Meeting. Milwaukee.

Harris G.W, Wagner D.A. 1967. Outflow from breached earth dams. Salt Lake City: Department of Civil Engineering, University of Utah.

Jiang Dong, Yang Xiaohuan, Wang Naibin, etc. 2002. Study on Spatial distribution of population based on remote sesing and GIS. Hydroscience Development 17(5):734–738.

Li Chaojie. 2007. Research and Application of Key Technology of Flood Disaster Evacuation Migration Decision Supporting System. Beijing: Capital Normal University.

Li Lei, Wang Renzhong, Sheng Jinbao, etc. 2006. Dam Risk Evaluation and Risk Management. Beijing: China Waterpower Press.

Li Lei. 1996. Preliminary Simulation Analysis on Dam-break Process of Gouhou Faced Rockfill Dam. Memoirs of Research on Mechanism of Fracture and Dam-break Process of Gouhou Faced Gravel Dam. Nanjing,: 68–74.

Liao C.B,Wu M.S, Liang S J. 2007. Numercial simulation of a dam break for an actual river terrain environment. Hydrological Processe 21:447–460.

Lv Anmin, Li Chengming, Lin Zongjian, etc. 2005. Population data abstraction based on GIS. Journal of Tsinghua University (Science and Technology) 45(2):1189–1192.

Mohamed M.A.A. 2002 Embankment breach formation and modeling methods. England: Open University.

Nogueira V.D.Q. A mathematical model of progressive earth dam failure. 1984. Fort Collins: Colorado State University.

Ponce V.M, Tsivoglou A J. 1981. Modeling gradual dam breaches. Journal of Hydraulic Division (ASCE) 107(7): 829–838.

Revi A, Singh AK (2007) Cyclone and storm surge, pedestrian evacuation and emergency response in India. Pedestrian and Evacuation Dynamics 2005: 119–130.

Robinson K.M, Hanson G.J. 2001. Headcut erosion research. Proc. 7th Federal Interagency Sedimentation Conf. Reno.

Singh V.P, Quiroga C.A. 1988. Dimensionless analytical solutions for dam breach erosion. Journal of Hydraulic Research 26(2): 179–197.

Tian Yongzhong, Chen Shupeng, Yue Tianxiang, etc. 2004. Simulation of Chinese Population Density Based on Land Use. Advance In Earth Sciences 59(2): 283–292.

Wan Qing, Li Huiguo. 1995. Dynamic Simulation of Victim Evacuation Process in Flood Storage Area (I)—Technology and Method Study. Journal of Geographical Sciences, Supplement: 62–68.

Wang Lihui. 2006. Study on Dam-Break Flow Numerical simulation and Dam-Break Risk Analysis. Nanjing Hydraulic Research Institute.

Wang Wei, Xu Jian. 1992. Theory and Method of Urban Traffic Planning. Beijing: China Communications Press.

Xia Junqiang, Wang Guangqian, Lin Bin-Liang, etc. 2010. Two-dimensional modelling of dam-break floods over actual terrain with complex geometries using a finite volume method. Advances in water science 21(3):289–298.

Xie Renzhi. Dam-break Hydraulics. Jinan: Shandong Science & Technology Press, 1993.

Zhang Baowei, Huang Jinchi. 2007. Research on Overtopping Dam Breach Model of Earth Dam. China Flood and Drought Management(3): 28–31.

Zhang Lu, Tian Yongzhong, Yang Hai. High-Time Resolution Daily Dynamic Simulation for Urban Population [J], Journal of Chongqing Technology and Business University: Natural Science Edition, 2012.4:29(4).

Zhang Xingxing. 2011 Research on Behavior and Route of Evacuation in Flood Disater. Tianjin: Tianjin University.

Zhou J.G, Gauson D.M, Mingham C.G, et al. 2004. Numerical prediction of dam-break flows in general geometries with complex bed topography. ASCE Journal of Hydraulic Engineering 130(4): 332–340.

Hydraulic Engineering III – Xie (Ed)
© 2015 Taylor & Francis Group, London, ISBN 978-1-138-02743-5

The significance and modes of the construction of rural and urban historical and cultural landscape areas of Chenzhou

Yan Chen
Art and Design College, Xiangnan University, Chenzhou, Hunan, China

ABSTRACT: This paper takes urban landscape and the surrounding rural landscape as a whole study object, which is based on the present situation of rural and urban landscape of Chenzhou, and then it will carding the relations of rural-urban historical and cultural landscape systematically. Finally, bring up a historical and cultural landscape zone construction mode which including three methods listed below: the protection of a unique natural substrate, the shaping of three culture landscape chains, and the construction of four urban cultural landscape areas.

1 INTRODUCTION

The term urban landscape first appeared in the journal of *Architectural Review* in January 1944. In the whole world, there is plenty of material space planning study based on cultural landscape. In this area, "Ideal City", "Garden City Theory" developed up by Ebenezer Howard, and Chandigarh and Brasilia constructed from architectonics perspective, they all reflected distinctive landscape feature. Rural landscape is the general expression of economy, culture, society, nature and the like. Rural landscape contains the elements of human's earliest farming culture, therefore it possess strong culture attribute. The present studies mostly regard urban and rural cultural landscape as independent research object. Besides the admitting the difference between the landscape characters, the researcher shall not ignore the continuity of urban-rural landscape substrate as well as the Link and integration effect of urban to the surrounding rural areas. This paper will employ the urban-rural landscape as the breakthrough point, and the present situation of rural and urban landscape of Chenzhou as basis to explore the construction modes of urban-rural landscape areas.

2 THE ANALYSIS OF THE BASIC SITUATION OF CHENZHOU

Chenzhou is Located in the southeast of Hunan province, the cross place of Nanling Mountains and Luoxiao Mountains, the shunting place of the Yangtze river system and the pearl river system. Its geographic coordinates is 112°13′E–114°14′E, 24°53′N–26°50′N. It position is south on Shaoguan, Guangdong, west on Yongzhou, Hunan, north on Hengyang, Hunan, and Zhuhou, it may be called "the wonderful place to appreciate the beauty of Heng and Yue mountains and the commanding of the Five Ridges", since ancient times, this place is called the throat of the central plains to the south China coast, a place of great military importance as well as the gathering place of literati.

Chenzhou governs 11 county (including town and district), Beihu district, Suxian district, Guiyang district, Yizhang district, Yongxing county, Jiahe county, Linwu county, Rucheng city, Guidong county, Anren county, and Zixing county, there are 164 township (including 11 ethnic Yao), 86 towns and seven sub district offices. It has rich county and town landscape resources; meanwhile, the radiation and link effect from urban landscape to the rural landscape is very obvious.

In recent years, a crucial green city activity Green Chenzhou is on its way, Chenzhou is honored as the national garden city, now it is steeping toward the aim of the national forest city; its urban and rural ecological environment and landscaping are improved significantly. Now it is in a critical period of improving the quality of landscape, and excavating the connotation of landscape culture. So it is quite necessary to research the continuity of urban and rural landscape system and landscape dynamic role.

3 THE SIGNIFICANCES OF HISTORICAL AND CULTURAL LANDSCAPE AREAS CONSTRUCTION

3.1 *The primitive attribute of protecting distribution of urban-rural areas*

As the joint work of man and nature, earthscape is a typical cultural landscape. As a component part of cultural heritage of mankind, it reflects basic and typical cultural characteristics of its location. The construction of urban-rural landscape area is helpful to trace back the historical context of urban-rural development, to discover and make advantage of their specialties, to protect diversity of region culture. Besides the promoting of artistic charm, cohesion as well as the molding of charter, it can also improve the cultivation of urban-rural areas and people's aesthetic judgment.

3.2 *The call for humanistic spirit of urban-rural areas*

As the thriving development of construction of new countryside, the dramatic changing looks of urban and rural areas shows its great progress, but the disclosure of urban-rural problem is still common. Although the humanism-focused post-modernity era comes after Industrial Revolution in design field, the present urban and rural landscape of our country still cannot help people to find the spirit support like "the homeland". It not only shows people's concern on their living environment, but also reflects the call of the society to subjectivity and the wake of humanism spirit. The research of culture landscape system cannot separate from the materiality, technicality and utility of urban-rural development, however, it possesses the criteria of humanism and becomes the bond of man and nature, and thus it is helpful to the wake of humanism spirit.

4 THE CONSTRUCTION MODES OF URBAN-RURAL HISTORICAL CULTURAL LANDSCAPE OF CHENZHOU

There exists a foundation for the construction modes of urban-rural historical cultural landscape of Chenzhou based on three methods listed below, including the protection of a unique natural substrate, the shaping of three cultural landscapes, and the construction of four urban cultural landscape areas.

4.1 *The protection of a unique natural substrate*

The word "Chen" is used exclusively in the name of Chenzhou, it originally appeared in Qin dynasty, which means "the city in the forest". So Chenzhou is also called Linyi (which means a lot of trees), from its name we can say, there is no Chenzhou without "forest", the forest landscape possess not only a general ecological significance, but also an expression of the origin of Chenzhou. The nature landscape "city in the forest" is considered as an important support for the construction of historical culture landscape zone.

Man cannot talk about history without living space; the first step of construction urban-rural historical culture landscape area is to protect the living space of human beings-the land that support the landscape. Chinese people tend to think a place possessing water and mountains is a good place to live, so just like the water, mountain forest can not only satisfy people's living needs, but also comfort their spiritual needs. Chenzhou is located in a transitional zone to Hengshao hills, high in a mountain in the main complex, southeast and lower in the northwest. The landform is complex as well as diverse; its characteristic is the massif occupies most places. There are 16 mountains over 1500 meters, and 1135 mountains over 1000 meters. There are 71 mountain ridges over 1000 meters; its total length reaches 1234 kilometers. Mountains over 1500 meters are largely scattered in the south. Bamian Mountain, the peak elevation of 2042 meters, is one of the highest mountains in Hunan province. The relative height difference of the highest and lowest place is 1990.24 meters in this area; the topography grade is 30 per thousand. The various landform and beautiful natural defense is the typical advantage of building urban-rural special landscape.

4.2 The shaping of three culture landscape chains

Chenzhou is a small city with a long history. The carved bone awl state that, back to 10,000 years ago, there are already primitive man living on this land, the word "Chen" can be found in history record, Records of History written by Si Maqian, which said (Xiang Yu) "so that Xiangyu force messenger send the emperor Yi to Chen county, Changsha". Since then, the word "Chen" firstly appears on the paper and become known to the public. Since Qin dynasty, Chenzhou starts to Split the five ridges, build a network of roads, set up counties, with written records dating back over 2,000 years, and it has rich historical and cultural landscape resources.

According to the internal relations of historical and cultural landscape of Chenzhou and its historical origin, three obvious culture landscape chins can be found. The first one is the famous culture landscape chin of eight culture landscapes: Suxian Yun-song, the water and moon of North Lake, the sunset of South Tower, the sight of East Mountain, the flying snow of Yujiang, the waterfall of Xiang Mountain, Yuan spring and its tea, the smog of Longquan. The second is its culture landscape of gods and Fairy Mountains: there are many fairy mountains in Chenzhou, Many beautiful legends is still spreading now, with a certain social effect, which also conform to the worshiping nature philosophy of Chinese people since ancient times. The third one is the culture landscape of ancient dwellings or villages: Chenzhou is located in the south of Hunan province, there are plenty of ancient dwellings or villages with typical regional characteristics, such as Banliang village, Yangshan village, sanxing village, xiawan village, baijue village and so on, most of them are preserved in Ming and Qing dynasty, they all have a profound cultural accumulation.

Table 1. The table of historical culture landscape in Fu culture landscape.

Name	Location	Introduction	Content
White deer cave	Western foothills of Suxianling	Birthplace of Sudan	Cave and inscriptions
Orange well	Right side of No.1 middle school	A legend of orange well	Old well, inscriptions, the of relic of Guanshe
Suxian yun-song	Suxianling	One of the ancient eight sights of beautiful posture old pine trees	
Kuahetu	Feisheng pavilion in north of Suxianguan	Painted by wangzhen and caved on bluestone in 1995	Inscriptions
Chessboard stone	North of Suxianguan	The place where Sudan becoming a god riding a cran	Large stone inscriptions natural landscape
Eight words inscript	Wall of Suxianling	Inscript of Tang dynasty	
Taohuaju	Enter of Suxianling scenic spots	Qing Dynasty garden architecture dwellings	Garden
Suxian bridge	East side of Renmin east road	Built in 12th year of ZhengDe, only the stone pier left	Stone piers

The shaping of culture landscape chin is a way to line up the tangible landscape with a culture belongs to a uniform system; this method can avoid possible rejection because of its internal cultural attribute, and improving the internal culture radiation effect that similar to culture landscape, in the meanwhile, it can highlight the culture connotation of itself in a better way.

4.3 The construction of four urban cultural landscape areas

Urban culture area is the center of the whole urban-rural area, the improving of center landscape area will drive and direct the construction of the whole area. Thus, firstly we collect the major culture that can represent Chenzhou, and divide the city into four culture landscape areas: Fu culture area (the center is Suxianling) seen sheet 1; Liuxianling culture landscape; Yanquan landscape; Yidiling culture landscape. The four culture landscapes mainly covers most places of Chenzhou, which take the present historical culture landscape as the center, accomplishing the construction the culture landscape with a joint development mode of hard and soft culture information.

5 CONCLUSION

The integration of urban and rural historical and cultural landscape is not a specific mode of the landscape construction in some cities, it is a necessary social activity for every city, a lasting continuous improvement social activity, and a more of social activity that render our home everlasting glory and charm.

ACKNOWLEDGEMENT

2011 Hunan Soft Science Research Planning Project "Research on the Development Modes of Southern Hunan Rural Cultural Landscape (2011ZK3172)".

REFERENCES

Jin Jun. *Ideal landscape*: the system construction and integration design of urban landscape space [M]. Nangjing: Southeast University press, 2003.

Shi Zhouren. *Cultural gene pool*: A reflection on the function of humantics subject [A]. seen cross-cultural dialogue [C]. Shanghai: Shanghai Culture Press,1999.75–89.

The compiling committee of Chen county, *Chen county history* [M]. Beijing: China Social Sciences Press, 1995.12.

The compiling committee of Chenzhou. *Chenzhou history* [M]. Hefei: Huangshan books publishing house, 1994.6.

Yu Yiping. *The characteristics of urban landscape and its shaping* [J]. Chinese Garden, 2000, (4):53–55.

Hydraulic Engineering III – Xie (Ed)

Low beach guardian project of Lalinzhou

Xiao-Xin Fei, Zhang-Ying Chen & Xing-Nong Zhang
Nanjing Hydraulic Research Institute, Nanjing, China

ABSTRACT: Taipingkou waterway of middle Yangtze River is the first sand reach below the Three Gorges Dam, the influence of water releasing is distinct. Lalinzhou low beach, the right bank of Taipingkou waterway, is receding because of water scouring in recent years, which greatly threatens the channel condition's stability of Sanba Beach. Therefore, carrying on low beach guardian project of lalinzhou is the key measure on channel improvement. This article is based on generalized model experiment of low beach guardian project of Lalinzhou in Taipingkou waterway, aiming at the comparison of several plane layout and space structure projects. Experimental results show, after carrying out low beach guardian project of Lalinzhou, the effect of flow control is obviously improved, and Lalinzhou low beach can be restored in a certain extent, and flow connection "from south to north" can be intensified, so that the channel condition of Taipingkou waterway can be improved a lot.

1 INTRODUCTION

Taipingkou channel which located in upside of jingjiang reach in middle Yangtze River is the first sand reach below the Three Gorges Dam. It is well known for its hindering navigation. Although both sides of the coastline are stable, frequently changes of scouring and sedimentation on bed make the channel unstable. Lalinzhou low beach, the right bank of Taipingkou waterway, is receding because of water scouring in recent years, which greatly threatens the channel condition's stability of Sanba beach. Therefore, low beach guardian project of Lalinzhou is an important part of taipingkou waterway regulation (Zhang, 2013).

At present, research works related to waterway regulation of taipngkou has been carried out to establish the administering thinking of Lalinzhou, guard the low beach and maintain and strengthen the flow connection "from south to north". This project will make Lalinzhou beach to restore in a certain extent and guide the converging flow to scour main channel. The guard engineering not only protects Lalinzhou beach but also guides flow because of the head of protecting belt. This article based on generalized model experiment comprised several plan layout and space structure and then deeply analyzed impingement and diversion effect of protecting belt so that it can provide technical support for the preliminary design of waterway regulation of Taipingkou.

2 RIVER SITUATION AND RECENT EVOLUTION

Taipingkou waterway is a minor bend branching channel which controlled by artificial revetment. It locates from Chenjiawan to Yuheping in Jingzhou city, Hubei province, the total length is about 22 km. The waterway is divided into upper section and lower section by Yanglinji, the upper section is divided into north passage and south passage, there are Juzhang river feeds into this section in left bank and Taipingkou distributary in the right bank. The lower section is divided into north branch and south branch by Sanbba beach. From Chenguwan to Tangguqiao (Shashi reach) presents width in the middle and narrow in both sides of the river which have Taipingkou beach, Lalinzhou beach, Jingzhou Yangtze River highway bridge and Sanba beach in the width reach, shown in Figure 1.

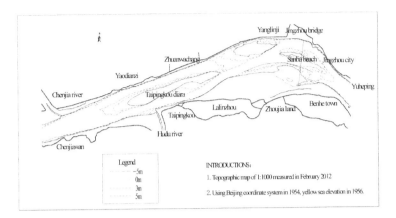

Figure 1. River regime of Taipingkou.

Taipingkou waterway has the significant characteristics of alluvial rivers. Although both sides of the river restrict by artificial revetment engineering and the outline is basically stable, the river bed evolution is often violent, with mainstream swing, growth and decline of the beach, alternating rise and fall as the main channel, etc. These three beaches associate with each other under the effect of different water and sediment conditions. Before 20th century 80 generation, the total area of Lalinzhou and Sanba beaches was large, so the mainstream was in left branch. After 80s, Taipingkou beach was deposited and Lalinzhou and Sanba beaches were scoured back by flow, especially in the flood year 1998 and 1999. Thus, the left branch silted seriously, the mainstream vacillate a little right. In 2000, Sanba beach broke up and formed a new Sanba beach. In 2001, the mainstream was in the central. In 2003, after the impoundment of the Three Gorges Hub, Lalinzhou and Sanba beaches were receding because of clear water scouring. In recent years, with the implementation of first-stage project of Shashi river waterway regulation and Lalinzhou high beach guard project, changing pattern of the river regime was under control and formed a relatively advantageous pattern of beach and channel, that is, water flowed from south passage to north branch. But unprotected beaches were still difficult to maintain its stability, such as Lalinzhou washed back, Sanba beach scoured seriously and so on, was still the major hidden danger of channel condition variation.

3 MODEL OVERVIEW

3.1 Model design

Considering protecting belt structure and deformation of riverbed sediment and three dimensional current around protecting belt, this model takes undistorted model to study the effect of erosion control and diversion of the beach protection belt and the model is made by river bed topographic map of 1:1000 measured in February 2012 (Huang, 2011). Model takes place on a flume with 60 m long and 14~18 m wide which is shown in Figure 2. Testing region included beach guardian Project of Lalinzhou and 3 ~ 6 km upstream and downstream of this waterways. The plane scale is 1:100 and vertical scale is also 1:100. According to the similar starting condition, this test selects plastic sand as model sand, its $\gamma_s = 1.15$ t/m^3 and $d_{50} = 0.24$ mm (Zhu, 2011). This sand meet the standards of 《Flow and sediment simulation regulations of inland waterway and port》 JTJ232-98 (TIWTE). Other related scales are shown in Table 1 (Li, 1981).

3.2 Simulation design of beach protecting structure

In this experiment, simulation design of low beach guardian project in Taipingkou waterway, including tetrahedron penetrating flame ballast and D-shape flexible mattress with concrete

Figure 2.   Plane layout of flume.

Table 1.   Main model scales.

Project	Name	Symbol	Scale
Fixed bed model	Velocity scale	λ_v	10
	Flow scale	λ_Q	100000
	Roughness scale	λ_n	2.15
Sediment model	Water movement time scale	λ_{t1}	10.0
	Velocity scale	λ_v	10.0
	Bed load size	λ_{d1}	0.953
	Bed material size scale	λ_{d2}	0.953
	Sediment initial velocity scale	λ_{v0}	10.0
	Transport scale	λ_{s*}	172
	River sediment time scale	λ_{t2}	189

Figure 3.   Simulation design of beach protecting structure.

blocks tied ballast and stone ballast, is carried out. Flexible mattress used to protect the bottom in river is D-shape flexible mattress with concrete blocks tied which consists of flexible mattress and concrete block ballast. Plan view scale of concrete block is $40 \times 26 \times 10$ cm (length × width × height). This test uses aluminum plate instead of concrete block. According to conversion scales, the scale of aluminum plate is $0.4 \times 0.26 \times 0.1$ cm. But it is so small that it is too difficult to produce. So, four concrete blocks are considered as a whole, then, the scale of aluminum plate becomes $0.8 \times 0.52 \times 0.1$ cm. Flexible mattress are made of aluminum plates paste on the cotton cloth to simulate the software in river. Tetrahedron penetrating flame in river uses 60 cm long stabs welded together. In the test, by filling the plastic processing of lead powder into all around tetrahedron penetrating flame is shown in Figure 3. Similarly, according to conversion scale, the length of tetrahedron penetrating flame uses in the experiment is 1.2 cm. The last ballast, stone ballast, its grain size in river is about 40 cm, test uses gravel, grain size is about 0.4 cm. These all guarantee geometric and gravimetric similarity of the model.

4 GUARDIAN PROJECT TEST

4.1 Project layout

At present, research works related to waterway regulation of Taipngkou has been carried out to establish the administering thinking of Lalinzhou, guarding the low beach and maintaining and strengthening the flow connection "from south to north". Based on the above management ideas, the measure of low beach guardian project of Lalinzhou is to arrange three bend-ended beach protecting belt to make Lalinzhou beach restore in a certain extent and guide the converging flow scour main channel. Test takes seven engineering proposal to determine weather the layout and façade structure of beach protecting belt is reasonable or not by comprehensive analyzing effect of diversion and protection against erosion, then, the recommended scheme which is shown in Figure 4 is put forward. The width of beach protecting belt is 180 m, belt structure uses stone ballast of 1.0 m height which fully paved on D-shape flexible mattress with concrete blocks tied. In every belt, there is 30 m wider strengthening ballast in the middle. The height of the strengthening ballast in 1# belt is 1.5 m and the others are controlled by ladder type, shown in Figure 5.

4.2 Analysis of diversion effect

Test uses different flow conditions to compare diversion effect before and after beach protecting belt project. Experimental results shows, after project, flow condition around beach protecting belt is smooth, deep groove velocity increases obviously on the left side of the dam in transition

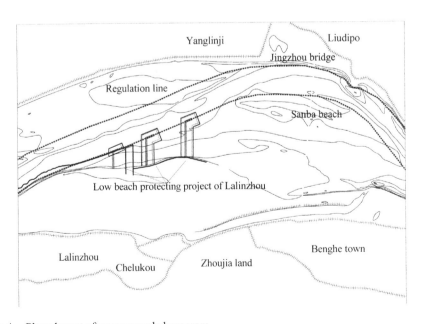

Figure 4. Plane layout of recommended program.

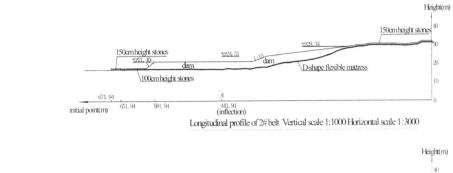

Longitudinal profile of 2# belt Vertical scale 1:1000 Horizontal scale 1:3000

Longitudinal profile of 3# belt Vertical scale 1:1000 Horizontal scale 1:3000

Figure 5.   The middle strengthening belt profile of recommended program.

Figure 6.   Local scour of edge (Q = 45000 m³/s).

section and flow direction turns left, velocity between belts is reduced. Relatively, flow field of low water period changes slightly larger than middle water period. In low water period (Q = 5500 m³/s), the biggest increase of velocity is 0.28 m/s and mainstream dynamic axis in transition section shift to the left of 100 ~ 150 m; In middle water period (Q = 16350 m³/s), the biggest increase of velocity is 0.16 m/s and mainstream dynamic axis in transition section shift to the left of 50 ~ 100 m; all results show that the recommended program has the function of protecting Lalinzhou low beach and enhancing diversion effect of transition section.

4.3   Analysis of scour prevention effect

After project, there is a certain intensity of local scour on beach edge, the scour range and intensity of erosion varies from the engineering plan layout and different hydrological conditions. Under the small and medium-sized flood, main scour occurs on the left side of belt head and the scouring depth is not large. But under the condition of floods especially flooding, the scour range and scouring depth are all increase around belt. Considering erosion intensity near belts, they are different from each other, scouring around 1 # and 2 #

Figure 7. Stability of the belt (Q = 45000 m³/s).

beach protecting belts are relatively strong and 3 # is smaller. Figure 6 shows local scour around the edge under the condition of flooding (Q = 45000 m³/s).

Although, there is certain intensity local scour around the beach protecting belt head, root segment erosion is not serious, phenomenon of collapse, bulge and concrete suspension are not occur. Even, a certain amount of silt covers upside the belt (Fig. 7). These all illustrate that, the beach protecting belt is stable, it plays an important role of protecting the beach, Lalinzhou beach can be guarded well.

5   CONCLUSIONS

1. Taipingkou waterway of middle Yangtze River is the first sand reach below the Three Gorges Dam, the influence of water releasing is distinct. Lalinzhou low beach, the right bank of Taipingkou waterway, is receding because of water scouring in recent years, which greatly threatens the channel condition's stability of Sanba Beach. Therefore, carrying on low beach guardian project of Lalinzhou is the key measure on channel improvement.
2. After project, deep groove velocity increases obviously on the left side of the dam in transition section, the flow direction turns left and velocity between belts is reduced. All results show that the recommended program has the function of protecting Lalinzhou low beach and enhancing diversion effect of transition section.
3. Considering stability and the effect of scour prevention of beach protecting belt, after project, flow condition around beach protecting belt is smooth. Although, there is certain intensity local scour around the beach protecting belt head, root segment erosion is not serious, phenomenon of collapse, bulge and concrete suspension are not occur. Even, a certain amount of silt covers upside the belt. These illustrate that, the beach protecting belt is stable and Lalinzhou beach can be guarded well.

REFERENCES

Cheng-Tao Huang, etc. Physical model research on Taipingkou waterway. Wuhan: Changjiang waterway planning design and research institute, 2011.
Chang-Hua Li, De-Chun Jin, Regulation for river model test, Beijing: China Communications Press, 1981.
Flow and sediment simulation regulations of inland waterway and port JTJ232-98, Beijing: China Communications Press.
Ling-Ling Zhu, etc. Experimental study on two-dimensional water and sediment mathematical model in Taipingkou waterway of middle Yangtze River. Wuhan: Changjiang waterway planning design and research institute, 2011.
Xing-Nong Zhang, etc. Scour prevention and diversion effect of beach protecting belt in Taipingkou waterway. Nanjing Hydraulic Research Institute, 2013.

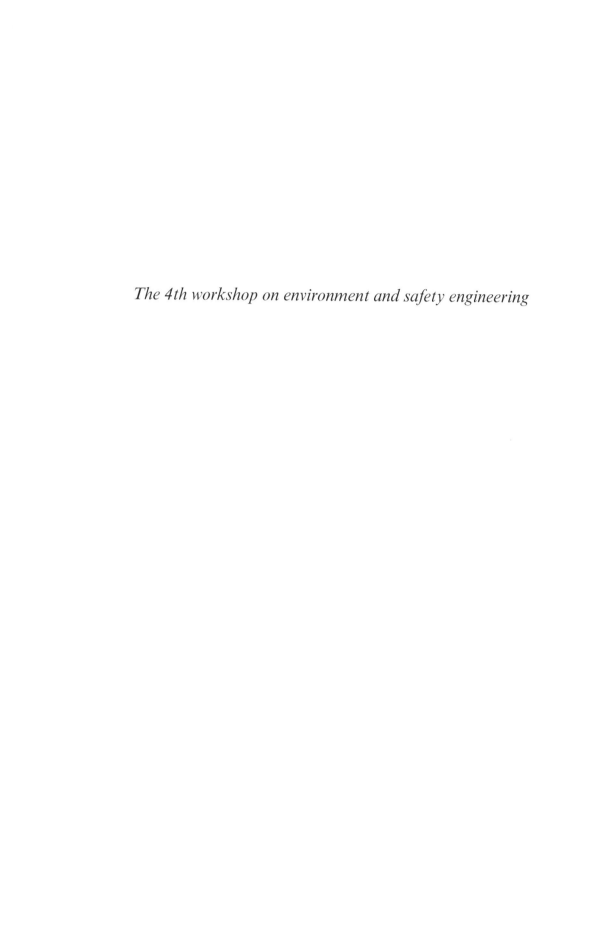

The 4th workshop on environment and safety engineering

Hydraulic Engineering III – Xie (Ed)
© 2015 Taylor & Francis Group, London, ISBN 978-1-138-02743-5

Study on the effect of Sanosil on removing algae

Ping Xia, Dong Zhang & Zheng Wang
Shanghai National Engineering Research Center of Urban Water Resources Co. Ltd., Shanghai, China

Jin-Rong Lu
Shanghai Waterworks, Shibei Co. Ltd., Shanghai, China

ABSTRACT: The removal of algae and chlorophyll a by Sanosil was studied at laboratory in different dosage and different reaction time. The comparison of NaClO pre oxidation and Sanosil pre oxidation on removal of algae and chlorophyll a was conducted in this paper. The results showed that effective components of Sanosil maintain stable concentration in the water, with persistent oxidation ability and Sanosil on algae and chlorophyll a removal effect was better than that of NaClO.

1   INTRODUCTION

Sanosil is a kind of disinfectant, launched by the German general Consulting Co. Ltd., which was first used in 1982 in Germany, Switzerland. Now Sanosil has been common used in the food industry, agriculture and animal husbandry, hospitals, public facilities, and etc in some European countries (Xiao Li, 1994). It is made by hydrogen peroxide and silver, which abandoned the disadvantages of hydrogen peroxide and silver when used alone, integrated their strong effect of sterilization, fast reaction, long-term effectiveness, less sensitive to light and heat (Rodriguez E, 2007).

In this paper, Sanosil010 is a type of disinfectant of drinking water with 5% hydrogen peroxide and 0.01% silver. Considering sanosil010 cannot produce residual chlorine in the outlet according to the National Drinking Water Standard (GB5749-2006). And no domestic application example in municipal water supply has been reported, so in our test, we use this disinfectant as oxidant to do some pre oxidation tests to remove algae in raw water (Von Gunten, 2003; Yue S-Y, 2006).

2   MATERIALS AND METHODS

2.1   *Materials*

Sanosil010 is composed of 5% hydrogen peroxide and 0.01% silver, purchased from a chemical factory in Shanghai. Water quality parameters measured for the raw water are summarized in Table 1. The water quality changed slightly over time.

Table 1.   Main water quality parameters of test raw water.

Water quality parameter	
Turbidity, NTU	5~10
COD, mg/L	2.0~3
DOC, mg/L	2.0~2.5
Fe, mg/L	0.05~0.15
Mn, mg/L	0.01~0.03

2.2 Methods

Stock Sanosil010 solution was made to obtain concentrations from 2 to 25 mg/L. Sanosil010 jar tests were conducted at room temperature. Experimental samples were collected at different time intervals.

Chlorophyll was measured by acetone extraction method and total algae density was measured by new method of detecting and counting—algae floating microscopic method.

3 RESULTS AND DISCUSSION

3.1 The active ingredient of Sanosil in water with time changing

The experimental, the effective components of Sanosil010 are composed of 5% hydrogen peroxide and 0.01% silver, when Sanosil010 dose was 2 mg/L and 25 mg/L, corresponding content of silver was 4 ug/L and 50 ug/L. Figures 1 and 2 show that stable performance of Sanosil010 product, whether it was low dose of 2 mg/L or high dose of 25 mg/L, after 24 h of reaction, the effective components of hydrogen peroxide and silver are not consumed, which concentration of reaction of after 24 h was close to the initial concentration.

3.2 Pre oxidation effect research of different concentrations of Sanosil

3.2.1 Effect of different dose of Sanosil on removing algae

As shown in Figure 3, removal efficiency of Sanosil010 on algae increased with the increasing of dose. When the Sanosil010 dose was 2 mg/L, 5 mg/L and 10 mg/L respectively, algae removal

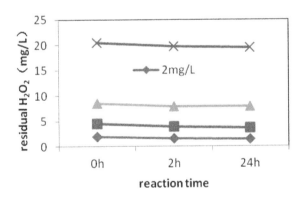

Figure 1.   The reduction of effective hydrogen peroxide in water with the time changing curve.

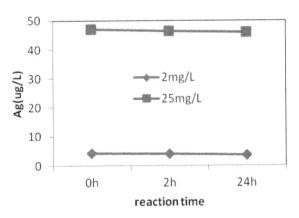

Figure 2.   The reduction of effective silver in water with the time changing curve.

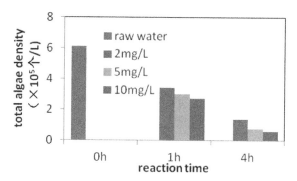

Figure 3. The removal of algae in different concentrations of Sanosil.

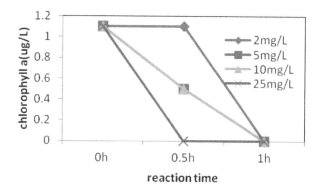

Figure 4. The removal of chlorophyll a in different concentrations of Sanosil.

rate was ranging from 43.9% to 55.3% after reaction of 1 h. And after reaction of 4 h, algae removal rate was in 77.3%~90.4%. Therefore, Sanosil010 had a good effect on removing algae.

3.2.2 Effect of different dose of Sanosil on removing chlorophyll a

Raw water chlorophyll a concentration was 1.1 μg/L as shown in Figure 4. Experimental results showed that, Sanosil on chlorophyll a had better removal effect. When the dose increased from 2 mg/L to 10 mg/L, with the reaction time prolonged, the chlorophyll a removed significantly. After the reaction of 1 h, chlorophyll a concentrations were below 0.1 μg/L.

3.3 Comparison of Sanosil and NaClO on pre oxidation effect

3.3.1 Removing effect of algae

As can been seen in Figure 5, when the Sanosil and NaClO in the same effective ingredient dose of 2.0 mg/L, we compared their pre oxidation effect of removing algae. The original algae density is 6.11×10^5 /L. As shown in Figure 6, Sanosil and NaClO of pre oxidation removal effect in the initial 1 h was similar, after reaction of 1 h, the total density of algae from the initial of 6.11×10^5/L respectively to 3.43×10^5/L and 2.89×10^5/L. Along with the extension of time, NaClO had obvious advantages of algae removing, after reaction of 4 h, algae removal rate of NaClO was 93.4% and that of Sanosil was 77.3%.

3.3.2 Removing effect of chlorophyll a

Chlorophyll a concentration in raw water was 1.1 μg/L. The experimental results in Figure 6 showed that, after reaction of 1 h, Sanosil on chlorophyll a removal rate was 100% and NaClO on chlorophyll a removal rate was 54%, so Sanosil on chlorophyll a removal effect was better than that of NaClO.

153

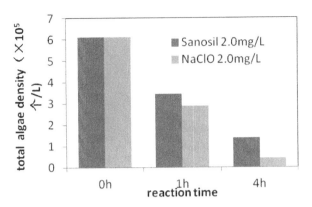

Figure 5.   The comparision of Sanosil and NaClO of removing total algae density.

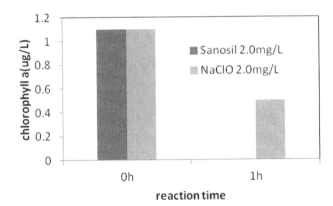

Figure 6.   The comparision of Sanosil and NaClO of removing chlorophyll a.

4   CONCLUSIONS

1. Effective components of Sanosil decay slow in the water, which can maintain stable concentration in the water, with persistent oxidation ability.
2. Sanosil on algae and chlorophyll a removal effect is better than that of NaClO.

ACKNOWLEDGEMENTS

The project was financially supported by Shanghai Natural Science Foundation (12ZR1410000), National Science and Technology Major Project of water pollution control and treatment (2012ZX07403-002), Shanghai Rising-Star Program (14QB1400800) and Shanghai Municipal Science and Technology Commission Project (11231200300).

REFERENCES

Rodriguez E, Onstad G, Kull T, et al. Oxidative elimination of cyanotoxins: Comparison of ozone, chlorine, chlorine dioxide and permanganate [J]. Water Res, 2007, 41(15): 3381–3393.

Von Gunten, U. (2003), Ozonation of drinking water: Part I. Oxidation kinetics and product formation, Wat. Res., 37, pp. 1443–1467.

Xiao Li. Ideal disinfectant: Sanosil [J]. JIANKANG, 1994, 12(15): 44.

Yue S-Y. Exploration on the algae removal effect with composite water treatment process, Journal of Water Purificaition Technology, Vol. 25 (2006), 1–5.

Hydraulic Engineering III – Xie (Ed)
© 2015 Taylor & Francis Group, London, ISBN 978-1-138-02743-5

Indoor air quality of the storage of semi-active records in Oman

N. Salem
Department of Library and Information, Sultan Qaboos University (SQU), Sultanate of Oman

S.A. Abdul-Wahab
Department of Mechanical and Industrial Engineering, Sultan Qaboos University, Sultanate of Oman

ABSTRACT: This paper addressed the impacts of environmental factors on the storage of semi-active records in the Sultanate of Oman. To this effect, the measurements of indoor environmental parameters were conducted in documents storage buildings of the Ministry of Manpower, and the Ministry of Municipalities and Water Resources. The recorded measurements were then compared with the available international standards/guidelines. The results indicated that most of the environmental factors inside these two buildings were found not within the stipulated standards.

1 INTRODUCTION

The storage media can be seriously declined by range of environmental parameters such as temperature, humidity, particulates, and gaseous pollutants (Karbowska-Berent et al. 2011). In order to extend the lives of these storage media, it is important to monitor regularly the indoor air quality parameters in them. Forde (2002) and Peters (1996) stated that environmental control and monitoring is one of the means to sustainable collection management in libraries and archives. In the present study, two important archives and record storage locations were selected for the assessment of indoor environmental conditions. The first location was in the Ministry of Manpower whereas the second location was in the Ministry of Municipalities and Water Resources. The aim of the study was to assess the indoor environmental factors in the storages of semi-active records of these two stated Ministries. In order to achieve this, the values of environmental factors inside the storages of semi-active record were logged for a period of at least 24 hours. The assessment of indoor air quality was carried out by comparing the measured levels of environmental parameters with that of international standards or guidelines for averaging periods of 1-hour, 8-hours and 24-hours. The basis of the standards/guidelines values was on the effects on archival and record collections in storage building.

2 MATERIALS AND METHODS

2.1 Description of sampling sites

The Ministry of Manpower covers industrial and technical education and vocational training sectors. The Department of the documents of the Ministry was established to protect and preserve the educational, industrial and manpower record of the ministry for a period of at least fifty years through digitization. The record department building of the Ministry of Manpower contains huge collections of documents of very important nature. The area of the record storage building is about 312 m² with a height of 4.5 m. On the other hand, the record storage building of the Ministry of Municipalities and Water Resources is dedicated to store the document produced by the ministry since 1990. It has two large rooms for record storage along with the offices of administration. The area of first hall is 168 m² while that of second hall is 282 m² with ceiling heights of 10 m. The building housed 94 racks to place the record documents.

2.2 Monitoring

The levels of indoor environmental factors in storages of semi-active record of the two Ministries were carried out by using GrayWolf Sensing Solutions monitoring equipment (Pegas et al. 2010; Muhamad-Darusa et al. 2011; Hariri et al. 2012). GrayWolf is a fully integrated system for simultaneous measurements of IAQ parameters, toxic gases and air speed. In the present study, the WolfPack-Modular Area Monitorwere used in present study to measure carbon dioxide (CO_2), carbon monoxide (CO), Relative Humidity (%RH), temperature, ammonia (NH_3), nitric oxide (NO), nitrogen dioxide (NO_2), sulfur dioxide (SO_2), hydrogen sulfide (H_2S), ozone (O_3), Total Volatile Organic Compounds (TVOCs). The indoor air speed was measured with the help of AS-202 A hot anemometer probe. The size distribution and number concentration of airborne particles were recorded using a 6 channel GW-3016 particle counter.

3 RESULTS AND DISCUSSION

3.1 Ministry of Manpower

Table 1 shows the levels of environmental parameters in the records storage building of the Ministry of Manpower for averaging periods of 1 hour, 8 hours and 24 hours, along with the relevant standards/guidelines.

Table 1. Levels of environmental parameters in the Ministry of Manpower's records department.

Parameter		1-hour parameter averaging	8-hours parameter averaging	24-hours parameter averaging	Standard/ guideline
Sulfur	Max	0.0	0.0	0.0	0.0004–0.002[a]
dioxide	Min	0.0	0.0		(ASHRAE 2011)
(ppm)	Ave	0.0	0.0		
Ozone	Max	0.026	0.021	0.020	0.0005–0.005[a]
(ppm)	Min	0.02	0.020		(ASHRAE 2011)
	Ave	0.020	0.020		
Nitrogen	Max	0.0	0.0	0.0	0.002–0.01[a]
dioxide	Min	0.0	0.0		(ASHRAE 2011)
(ppm)	Ave	0.0	0.0		
Hydrogen	Max	0.03	0.0180	0.008	<0.0001[a]
sulfide	Min	0.0	0.0016		(ASHRAE 2011)
(ppm)	Ave	0.008	0.008		
TVOC	Max	570.40	531.33	82.70	<100[a]
(µg/m³)	Min	409.60	428.50		(ASHRAE 2011)
	Ave	482.27	481.59		
Temperature	Max	18.46	17.11	16.88	15–25[a]
(°C)	Min	16.10	16.25		(ASHRAE 2011)
	Ave	16.86	16.74		
Relative	Max	69.64	66.17		30–50[a]
humidity	Min	51.32	53.88		(ASHRAE 2011)
(%)	Ave	60.17	60.36		
PM₂.₅	Max	15.44	–	–	1–10[a]
(µg/m³)	Min	13.53	–		(ASHRAE 2011)
	Ave	14.80	–		
TPM	Max	108.77	–	–	≤75[a]
(µg/m³)	Min	57.83	–		(PURAFIL 2004)
	Ave	85.20	–		

[a]Based on protection and preservation of archives.

It is clear from Table 1 that the oxides of sulfur and nitrogen were not detected. The temperature in the storage building was found within the allowable range. The average concentration of O_3 was 0.02 ppm and it was indicated for all averaging periods. The measured O_3 was higher than the specified range of 0.5 to 5 ppb. The average concentration of H_2S was 0.008 ppm for all selected intervals of time. This value of H_2S was higher than that of recommended value (<0.1 ppb) suggested by ASHRAE (2011). Also, the concentrations of TVOCs were found exceeding the permitted value (<100 µg/m³) with magnitudes of 482.27 µg/m³, 481.59 µg/m³ and 482.70 µg/m³ for all periods respectively. The observed relative humidity was 60.17%, 60.36% and 60.19% as averaged over 1-hour, 8-hours and 24-hours respectively. These levels of relative humidity were surpassing the ASHRAE recommendation of 15% to 25%. This was in line with the results of Chang and Falk (2009) who also reported in their study higher relative humidity on rare books archives. High relative humidity levels can result in the growth of mould, increased chemical deterioration, cockling of paper and parchment, warping of books and increase the likelihood of pest infestations (CCA 2003). The particulate matter along with raised concentrations of other environmental parameters, as mentioned above, might be damaging factors for the documents storage of at least 50 years, which is the ultimate preservation goal of the Ministry of Manpower.

3.2 *Ministry of municipalities and water resources*

Table 2 shows the measured levels of indoor air quality parameters at the record storage building of the Ministry of Municipalities and Water Resources.

Table 2. Levels of environmental parameters in the Ministry of Regional Municipalities and water resources' records department.

Parameter		1-hour parameter averaging	8-hours parameter averaging	24-hours parameter averaging	Standard/ guideline
Sulfur	Max	0.12	0.019	0.01	0.0004–0.002[a]
dioxide	Min	0.08	0.01		(ASHRAE 2011)
(ppm)	Ave	0.01	0.01		
Ozone	Max	0.070	0.67	0.065	0.0005–0.005[a]
(ppm)	Min	0.040	0.063		(ASHRAE 2011)
	Ave	0.065	0.065		
Nitrogen	Max	0.038	0.023	0.024	0.002–0.01[a]
dioxide	Min	0.038	0.020		(ASHRAE 2011)
(ppm)	Ave	0.022	0.022		
Hydrogen	Max	0.026	0.020	0.008	<0.0001[a]
sulfide	Min	0.026	0.002		(ASHRAE 2011)
(ppm)	Ave	0.008	0.009		
TVOC	Max	570.40	472.73	323.12	<100[a]
(µg/m³)	Min	95.20	238.14		(ASHRAE 2011)
	Ave	323.04	334.68		
Temperature	Max	31.38	30.51	30.41	15–25[a]
(°C)	Min	29.32	30.29		(ASHRAE 2011)
	Ave	30.41	30.39		
Relative	Max	69.64	65.71	63.42	30–50[a]
humidity	Min	52.04	61.43		(ASHRAE 2011)
(%)	Ave	63.32	63.24		
PM$_{2.5}$	Max	48.87	–	–	1–10[a]
(µg/m³)	Min	41.39	–		(ASHRAE 2011)
	Ave	50.12	–		
TPM	Max	465.31	–	–	≤75[a]
(µg/m³)	Min	126.87	–		(PURAFIL 2011)
	Ave	302.10	–		

[a]Based on protection and preservation of archives.

157

It is obvious from Table 2 that the levels of all the environmental parameters are exceeding their acceptable limits. Sulfur dioxide and ozone with average concentrations of 10 ppb (Allowable range: 0.4–2 ppb) and 65 ppb (Allowable range: 0.5–5 ppb) were observed for all periods. The noted average concentrations of nitrogen dioxide were 22 ppb for 1-hour and 8-hours while it was 24 ppb (Allowable range: 2–10 ppb) for 24-hours period. Hydrogen sulfide was observed as 8 ppb for averaging periods of 1-hour and 24-hours while its concentration was 9 ppb (Allowable range: < 0.10 ppb) for 8-hours period. The high concentrations of gaseous pollutants such as SO_2, O_3 and NO_2 can seriously endanger the preservation of record collections. The total VOCs can also have deteriorating effect on paper (Menart et al. 2011). The most critical factors for the storage of paper record are temperature and humidity. The measured values of temperature and relative were also greater than the limiting values (ASHRAE 2011). High humidity levels can cause the swelling of paper and parchment materials resulting in planar distortions, coated papers to stick together, the transfer of inks from one surface to another, and mold growth in levels above 60%. The higher levels of temperature in the building under study were may be due to non-availability of air-conditioning system in the store, which was seen during sampling campaign.

4 CONCLUSIONS

The results of the study showed that most of the environmental factors measured at the storages of semi-active record of the Ministry of Manpower and the Ministry of Municipalities and Water Resources were not within the stipulated values defined in international standards. The indoor air quality of the two buildings was not suitable for protection and preservation of document collections. Therefore, some preventive measures must be taken to safeguard the documentary assets. The document collection of the two Ministries will be damaged if remedial measures were not taken timely.

REFERENCES

ASHRAE. 2011. ASHRAE Handbook HVAC Applications. Museum, galleries, archives and libraries. Atlanta, GA: 23.21–23.22.

CCA. 2003. Basic Conservation of Archival Materials: Chapter 3: Environment, Canadian Council of Archives(CCA), Ottawa, Canada.

Chang, J.D. & Falk, B. 2009. Experimental and computer simulation approach to investigating the environmental performance of a rare books archive. *E-Preservation Science* 6: 180–185.

Forde, H. 2002. Overview of Collections Information and Advice in the Archives Domain. Council on Library and Information Resources. URL: http://www.clir.org/pubs/reports/pub89/contents.html.

Hariri, A., Leman, A.M., Yusof, M.Z.M., Paiman, N.A. & Noor, N.M. 2012. Preliminary measurement of welding fumes in automotive plants. *International Journal of Environmental Science and Development* 3(22): 146–151.

Karbowska-Berent, J., Górny, R.L., Strzelczyk, A.B. & Wlazło, A. 2011. Airborne and dust borne microorganisms in selected Polish libraries and archives. *Building and Environment* 46(10): 1872–1879.

Menart, E., De Bruin, G. & Strlič, M. 2011. Dose-response functions for historic paper. *Polymer Degradation and Stability* 96(12): 2029–2039.

Muhamad-Darusa, F., Zain-Ahmeda, A. & Latifc, M.T. 2011. Preliminary assessment of indoor air quality in terrace houses. *Health and the Environment Journal* 2(2): 8–14.

Pegas, P.N., EvtyuginaI, M.G., AlvesI, C.A., NunesI, T., CerqueiraI, M., FranchiI, M., PioI, C., AlmeidaII, S.M. & FreitasII, M.D.C. 2010. Outdoor/indoor air quality in primary schools in Lisbon: A preliminary study. *Quim Nova* 33 (5): 1145–1149.

Peters, D. 1996. Our environment ruined? Environmental Control Reconsidered as a Strategy for Conservation. Journal of Conservation and Museum Studies No. 1. URL: http://www.ucl.ac.uk/archaeology/conservation/jcms/.

PURAFIL. 2004. Environmental control for museums, libraries and archival storage areas, Technical Bulletin-600 A. 2004. Purafil, Inc. 2654 Weaver Way Doraville, Georgia 30340.

Hydraulic Engineering III – Xie (Ed)
© *2015 Taylor & Francis Group, London, ISBN 978-1-138-02743-5*

Modeling of CO, NO$_x$ and CO$_2$ due to vehicle emissions from Sultan Qaboos University in Oman

S.A. Abdul-Wahab & S.O. Fadlallah
Department of Mechanical and Industrial Engineering, Sultan Qaboos University, Sultanate of Oman

ABSTRACT: The objective of this study was to use the CALPUFF modeling system to predict the concentrations of CO, NO$_x$ and CO$_2$ released, due to the traffic, from the Sultan Qaboos University (SQU) facility at its location in Muscat, Sultanate of Oman. These pollutants were selected as the indicator since they were the most dominant products of vehicle emissions. The study focused on investigating the dispersion of CO, NO$_x$ and CO$_2$ by considering the SQU as an area source. The results from the CALPUFF model clearly demonstrated that the levels of the three targeted pollutants, emitted from the vehicles, were found to significantly exceed the allowable limits defined by regulated standards.

1 INTRODUCTION

CALPUFF is as an effective and reliable atmospheric modeling tool. It is non-steady-state modelling software produced by the Atmospheric Studies Group and distributed by TRC. Due to its flexibility, CALPUFF have attracted several research studies. Over the past years, several research papers have been published with different case studies (Abdul-Wahab 2003; Abdul-Wahab et al. 2011; Abdul-Wahab et al. 2012; Abdul-Wahab et al. 2014; Ghannam & El-Fadel 2013; Tian et al. 2013; Yu & Stuart 2013). Although a significant amount of research papered related to CALPUFF modelling software were published, few studies considered the effect of vehicle emissions and investigated the dispersion of CO, NO$_x$ and CO$_2$ concentrations. It is noteworthy to emphasize that there have not yet been any comprehensive studies published which involves using CALPUFF to predict and model concentration dispersions of these pollutants from vehicles. Hence, this present study aims to contribute to this field by focusing on a computational-based CALPUFF investigation of the significant pollutants emitted from traffic in the SQU facility. The concentration and dispersion of CO, NO$_x$ and CO$_2$ in and around the SQU facility were modelled for the year 2014. The CALPUFF dispersion modeling software was used along with the anticipated emission data of the SQU facility to determine the general air quality impacts of CO, NO$_x$ and CO$_2$ during this year.

2 MATERIALS AND METHODS

2.1 *Area of study description*

Sultan Qaboos University (SQU) is considered as the first public university in the Sultanate of Oman. The facility of the SQU is located in Muscat, Oman. SQU was constructed in the year 1982 with a total land area of 6.8 km^2 and the first batch of students was enrolled in 1986. The SQU Hospital (area of 40,000 m^2) is also located within the SQU boundary. The hospital officially started in the year 1990 and it is considered as an educational as well as medical institution. Ever since its establishment, the enrollment of students has significantly increased within the years. SQU started with 557 students as first batch and currently it has around 15,000 students. Due to this rapid increase, it has been noticed that the number of vehicles within SQU has significantly increased which causes as a result a traffic jam. It is worth to mention that female students who are basically living away from Muscat are provided with a living inside the university campus. Not to mention as well the academic staff which are living inside the domain of study. On the other hand, male students are required to live outside the campus. As for the traffic jam, it has been observed that during morning period, between 7:30 and 8:30, SQU is suffering from a massive flow of traffic due to the number of cars entering the SQU from the main three university gates including staff and male students. The study will consider taking the SQU domain as an area source.

2.2 CALPUFF modelling system

CALPUFF is considered as a meteorological and air quality modelling software. The software can be used to investigate air pollution dispersions and to predict the concentration of air pollutants which are emitted from different types of sources. Additional details about the model can be found in Scire et al. (2000) and Abdul-Wahab et al. (2014). Prior to running the CALPUFF pre-processors, shared information on the meteorological grid was entered into a common file using the Identify Shared Information module. This information was shared amongst all CALPUFF processors (Table 1).

2.3 Surface and upper air meteorological data

The surface meteorological data used in this study was obtained from the Directorate General of Meteorology and Air Navigation (www.met.gov.om). The hourly meteorological data consists of the following information: temperature (°C), precipitation (mm), station pressure (mbar), relative humidity (%), wind direction (°), wind speed (m/s), cloud cover (tenths), and cloud height (ft). Data was prepared in a format that could be run using CALPro's SMERGE program to produce a SURF.DAT file suitable for input into CALMET. On the other hand, the upper air meteorological data information was retrieved from the Radiosonde Database website that is developed and run by the Earth System Research Laboratory at the National Oceanic and Atmospheric Administration (esrl.noaa.gov/raobs/). Twelvehour interval data was obtained from the upper air climate station for a full period similar to that of the surface data period. Hourly upper air meteorological data was prepared in a format that could be run using CALPro's READ62 program to produce UP.DAT file

Table 1. Model input information for the study area.

Parameter	Sultan Qaboos University
Projection type	LLC
LCC latitude of origin	23.590828 °N
LCC longitude of origin	58.163697 °E
Latitude 1	5 N
Latitude 2	45 N
False Easting	0
False Northing	0
Continent/ocean	Global
Geoid-Ellipsoid	WGS-84:WGS84
Region	WGS-84
DATUM Code	WGS-84
X (Easting)	−100 km
Y (Northing)	−100 km
Number of grid cells (X)	200
Number of grid cells (Y)	200
Grid spacing	1 km
Number of vertical layers	9
Cell face heights	0, 20, 50, 100, 150, 200, 300, 500, 1000, 2000 m
Base time zone	UTC + 04:00

suitable for input into CALMET. Information pertaining to the surface and upper stations is summarized in Table 2.

2.4 *Emission data*

Emission factors are required to calculate the emission rates for area source. Despite the fact that vehicles come in different types, starting with motorcycles and ending with heavy-duty trucks, emission rates differ in terms of vehicle specifications (i.e., weight, engine capacity, etc.), vehicle speed, and fuel type. Many research papers and environmental institutes focused their resources to calculate vehicle emission rates. In this current study, the emission factors were taken from Transport Research Laboratory (2009). Transport Research Laboratory (2009) has developed an excel sheet that categorizes vehicles based on weight, engine capacity, emission standard, and vehicle average speed for various pollutants (https://www.gov.uk/government/publications/road-vehicle-emission-factors-2009). Table 3 shows the values of the emission factors used in this study.

2.5 *Domain area*

The study focused on studying the effect of vehicle emissions by taking the SQU facility as an area source. The domain area was considered as a polygon with an area of 5.9 km². Table 4 represents statistical data collection for average number of vehicles entering the SQU

Table 2. Information about surface and radiosonde stations.

Parameter	Surface station (surface meteorological data)	Radiosonde station (upper air meteorological data)
Station name	SEEB INTL/MUSCAT 99 OM	ABU DHABI INTL 99 AE
INIT	OOMS	OMAA
UTM latitude	23.58 °N	24.43 °N
UTM longitude	58.28 °E	54.65 °E
Location X on grid	10 km	−400 km
Location Y on grid	12 km	60 km
Station elevation	17 m	27 m
WMO ID	41256	41217
WBAN	99999	99999

Table 3. Emission factors (Transport Research Laboratory 2009).

Pollutant	Emission factor	Additional specifications
CO (g/veh.km)	15.74	Vehicle speed at 34.5 km/hr,
NO_x (g/veh.km)	1.29	petrol, pre-euro standard,
CO_2 (g/veh.km)	182.3	1400 cc, 2.5t

Table 4. Statistical data collection for average number of cars entering the domain within selected timing.

	Timing 7:30 am–8:30 am				
	Entrance gate 1	Entrance gate 2	Entrance gate 3	Staff living in campus	Total
Day 1 (1/4/2014)	5423	5386	4732	799	16,340
Day 2 (2/4/2014)	5671	5495	5163	799	17,128
Day 3 (3/4/2014)	5529	5537	4914	799	16,776

Table 5. Area source emission rate calculation for the desired pollutants on April 2, 2014.

Pollutant	Emission factor (g/veh.km)	Average speed (km/h)	Number of vehicles for selected day (veh)	Case study total area (m²)	Emission rate (g/m²/s)
CO	15.74	34.5	17,128	5,972,922	157.19
NO$_x$	1.29	34.5	17,128	5,972,922	12.928
CO$_2$	182.35	34.5	17,128	5,972,922	1820.9

Table 6. CALPUFF area source input parameters.

Parameter	Sultan Qaboos University
Point 1 (upper left)	0.14408 km, 1.734 km
Point 2 (upper right)	1.9473 km, 0.10205 km
Point 3 (lower left)	-1.8995 km, -0.7709 km
Point 4 (lower right)	1.16538 km, -1.13964 km
Effective height	0.35 m
Base elevation	55.4736 m
Initial sigma z	0.2

facility from the three main gates and vehicles owned by SQU staffs living in the SQU campus. According to the SQU administration, the numbers of cars owned by the university staffs that are already living inside the campus are around 799 vehicles. This data collection was conducted for three continues days starting from April 1, 2014 to April 3, 2014 for the period between 7:30 am and 8:30 am. Based on the data, the maximum number of vehicles entering the domain of study was seen on April 2, 2014 with a total number of vehicles equal approximately to 17128. Based on the number of vehicles obtained from statistics and the total area calculated using Google Earth, Table 5 summarizes emission rate calculation for CO, NO$_x$ and CO$_2$ taking into account the assumed average vehicle speed, the domain total area, and the total number cars within the domain for the specified modelling day, which is April 2, 2014. Table 6 lists CALPUFF area source input parameters. The parameters include the polygon coordinates with respect to the specified origin, the effective height, and initial sigma (z) which is calculated by dividing the effective height by 2.15.

3 RESULTS AND DISCUSSION

Since Muscat is considered as a coastal city, the meteorological conditions of the land are strongly affected by the sea. Table 7 lists the top 5 highest 1-hour average concentrations of CO and NO$_x$ and the top 5 highest 0.5-hour average concentrations for CO$_2$ simulated for April 2, 2014 from 00h00 to 23h00. These top 5 highest average concentrations all occurred at a location 0.5 km east and 0.5 km south of the SQU. Thus, all of the top 5 highest CO, NO$_x$ and CO$_2$ concentrations occurred at the same locations nearby the SQU. They occurred near the beginning of the day between 03h00 to 07h00. The first highest concentration was occurring at 06h00, the second highest concentration occurring at 05h00, the third highest concentration occurring at 03h00, the fourth highest concentration occurring at 04h00 and the fifth highest concentration occurring at 07h00.

Table 7 also shows a comparison of the maximum highest average CO, NO$_x$ and CO$_2$ concentrations for the day examined in this study against their respective criterion limits. It can be seen that CO and NO$_x$ have 1-hour maximum concentration criterions of 40096.1 μg/m³ (35 ppm) and 188.2 μg/m³ (100 ppb) respectively (U.S. EPA 2012). On the other hand, CO$_2$ has 0.5-hour maximum concentration criterion of 63000 μg/m³ (Ontario Ministry of the Environment 2008). Further examination of Table 7 illustrates that all the simulated 1-hour average CO and NO$_x$ concentrations were found to significantly exceed their respective 1-hour

Table 7. List of the top 5 average CO, NO$_x$, and CO$_2$ concentrations simulated on April 2, 2014.

Coordinates (km)	Time (HH:MM)	Average peak values (µg/m³)		
		CO* 1-hour	NO$_x$** 1-hour	CO$_2$*** 0.5-hour
0.5, −0.5	06:00	91,283	7505.0	1,215,044.8
0.5, −0.5	05:00	86,460	7108.5	1,150,886.7
0.5, −0.5	03:00	84,027	6908.5	1,118,504.1
0.5, −0.5	04:00	77,331	6357.9	1,029,382.2
0.5, −0.5	07:00	73,926	6078.0	984,056.3

*allowable 1-hour average concentration is 40096.1 µg/m³ (35 ppm) according to U.S. EPA (2012).
**allowable 1-hour average concentration is 188.2 µg/m³ (100 ppb) according to U.S. EPA (2012).
***allowable 0.5-hour average concentration is 63000 µg/m³ (Ontario Ministry of the Environment 2008).

criterion limits. Similarly, the simulated 0.5-hour average CO$_2$ concentrations were found to be above its 0.5-hour limit. Moreover, given that SQU emission levels will vary on a day-to-day basis, changes in the seasonal weather may cause CO, NO$_x$ and CO$_2$ concentration levels to even exceed the highest concentration levels determined in this study.

4 CONCLUSION

The CALPUFF dispersion modelling system was utilized to assess the maximum ground level concentrations of CO, NO$_x$ and CO$_2$ emitted from traffic emissions in the SQU. The results of the dispersion models were evaluated against the CO, NO$_x$ and CO$_2$ concentration limits. All simulated 1-hour average CO and NO$_x$ concentrations and 0.5-hour average CO$_2$ concentrations were found to significantly exceed their respective criterion limits. The maximum concentrations were found to be occurred at the same locations nearby the SQU. This can be related to the wind blows that come from the sea toward the land which directly influenced the dispersion of the three pollutants. As a result, the pollutants accumulated and gathered nearby. It is recommended that external parking slots away from the SQU campus must be provided for student and from there, students can take a bus to their classes and lecture theaters. These buses need to be in a good condition with low pollutant emissions.

REFERENCES

Abdul-Wahab, S.A. 2003. SO$_2$ dispersion and monthly evaluation of the Industrial Source Complex Short-Term (ISCST32) model at Mina Al-Fahal refinery, Sultanate of Oman. *Environmental Management* 31(2): 276–291.
Abdul-Wahab, S.A., Ali, S., Sardar, S., Irfan, N. & Al-Damkhi, A. 2011. Evaluating the performance of an integrated CALPUFF-MM5 modeling system for predicting SO$_2$ emission from a refinery. *Clean Technologies and Environmental Policy* 13(6): 841–854.
Abdul-Wahab, S., Ali, S., Sardar, S. & Irfan, N. 2012. Impacts on ambient air quality due to flaring activities in one of Oman's oilfields. *Archives of Environmental and Occupational Health* 67(1): 3–14.
Abdul-Wahab, S.A., Chan, K., Elkamel, A. & Ahmadi, L. 2014. Effects of meteorological conditions on the concentration and dispersion of an accidental release of H$_2$S in Canada. *Atmospheric Environment* 82: 316–326.
Ghannam, K. & El-Fadel, M. 2013. A framework for emissions source apportionment in industrial areas: MM5/CALPUFF in a near-field application. *Journal of the Air and Waste Management Association* 63(2): 190–204.

Ontario Ministry of the Environment. 2008. Jurisdictional Screening Level (JSL) List-A Screening Tool for Ontario Regulation 419: Air Pollution—Local Air Quality. Ontario, Canada.

Scire, J.S., Strimaitis, D.G. & Yamartino, R.J. 2000. A User's Guide for the CALPUFF Model (Version 5.0). Concord, MA: Earth Technologies Inc.

Tian, H., Qiu, P., Cheng, K., Gao, J., Lu, L., Liu, K. & Liu, X. 2013. Current status and future trends of SO_2 and NO_x pollution during the 12th FYP period in Guiyang city of China. *Atmospheric Environment* 69: 273–280.

Transport Research Laboratory. 2009. A review of Available Road Traffic Emission Models. UK. Prepared for Charging and Local Transport Division, Department of the Environment, Transport and the Regions.

U.S. EPA. 2012. Air and Radiation: National Ambient Air Quality Standards (NAAQS).

Yu, H. & Stuart, A.L. 2013. Spatiotemporal distributions of ambient oxides of nitrogen, with implications for exposure inequality and urban design. *Journal of the Air and Waste Management Association* 63(8): 943–955.

Hydraulic Engineering III – Xie (Ed)
© 2015 Taylor & Francis Group, London, ISBN 978-1-138-02743-5

Application of countermeasure research on rain adaptative landscape in Fujian coastal cities

Yao Wang
Fine Arts Design and Clothing Engineering College, Xiamen University of Technology, Xiamen, Fujian, China

Jiping Wang
College of Environmental Science and Engineering, Xiamen University of Technology, Xiamen, Fujian, China

ABSTRACT: Coastal cities of Fujian locating in southeast coast of China have abundant rainfall annually. Latosolic red soil is zonal soil. With urbanization development, largely using of waterproof paving in cities enlarge the storm runoff, which result in urban inland inundation, water pollution, fresh water resource deficiency and other problems. The urgent issues are that urban subsurface drainage infrastructure should be reconstructed and extended in large scale. This paper discusses the way of solving urban rainwater problem by creating rain adaptative landscape. Linear retention and dredge landscape skeleton is established combining with dotted water landscape and surface permeable elements for reference foreign rainwater management experiences. Rainwater adaptative landscape network system is composed of diminishing hierarchies to achieve ecological goals of controlling and utilizing urban rainwater effectively, economic goals of reducing infrastructure construction and maintenance cost, and entertainment goals of meeting citizen's recreation requirements.

1 INTRODUCTION

Some Chinese cities suffer from serious water scarcity, water pollution, flood and other environmental problems related to water, so it's an important development strategy for China to increase water sources, save water and energy, reduce emission, protect and restore environment. Coastal cities of Fujian Province are subject to subtropical oceanic monsoons, so rain mainly comes in spring and summer, especially in midsummer and early autumn which are often hit by thundershower, typhoon rain and long and powerful rainstorm. Take 5·16 rainstorm of Xiamen in 2013 for example, maximum precipitation contributed by 6 hours reached 185.5 mm (Chen, 2013) and caused extreme causalities and economic losses. The region is covered by latosol, so total reservoir capacity, capacity of single reservoir and effective reservoir capacity are all small. Scoured by rainstorm, the local infiltration rate and water retention rate are decreased, subsequently surface runoff is increased (Quan, 2004). In addition, as an urgent task for coastal cities, it's also a good opportunity to solve the rainwater problem when speeding urbanization requires more investment for infrastructure and water conservancy project which usually are faced with difficult maintenance.

1.1 *The current predicament*

Present water problems suffering coastal cities in Fujian include: (1) "Excessive water". Serious city waterlogging often emerges. Because of speeding urbanization, water-proof paving is freqently applied for efficiency and it results in higher surface runoff coefficient and shorter concentration time, so during heavy rain period rainstorm runoff, peak discharge and frequency are increased, and city drainage capacity becomes insufficient against seri-

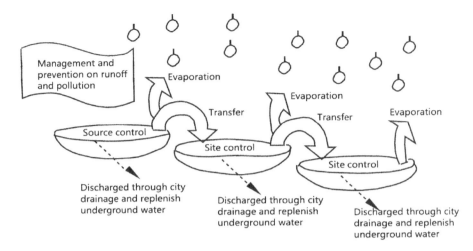

Figure 1.   Stormwater runoff management chain of SUDS.

ous waterlogging. For example, the typhoon Longwang of 2005 brought heavy rainfall to Fujian coast. Fuzhou City received a maximum precipitation of 118 mm in one hour. For then was time for flood tide, rainwater in the city failed to be discharged, resulting in an 8-square-kilometer flooded area with maximum depth of 1.9 m (Nan, 2005); (2)"Insufficient water". Though located in southeast coast and granted with rich rainfall, the region has greater evaporation, so seasonal drought is seen often; lower infiltration of latosol makes rainfall difficult to be preserved, so the region is short on fresh water. Xiamen is one of the typical coastal cities without sufficient such resources (Lin, 2009); most of coastal cities of the province still discharge rain as waste water without reasonable retention for use, only focusing on flood prevention and drainage. Such treatment leads to lower level of underground water. These problems together with shapely increased extremely water-consuming enterprises intensify city's water crisis and environmental problems. (3) "Dirty water". There is few concern raised for quality, amount and pollution of rainwater by citizens. However, pollutions of surface runoff like nitrogen, phosphorus, and heavy metals etc. remain in city, especially in industrial and commercial areas and streets due to they are frequently visited by vehicles and passengers. Pollutions in surface runoff are discharged directly or through drainage into nearby river or sea, severely polluting city water system and ocean.

As a key element in natural circulation, rainwater plays an important role in sustainable water circulation. For above mentioned problems of stormwater suffer these cities, rain adaptative landscape is a good solution for such excessive water, insufficient water and dirty water. Rain adaptative landscape is a way of control and utilization including stormwater's infiltration, regulation and storage, purification, use and discharge. Natural water circulation restoration, pollution elimination, surface runoff period delay, peak discharge reduction, waterlogging mitigation, underground water replenishment, land subsidence alleviation, oceanic environmental protection, city landscape enrichment, biodiversity preservation, public environmental awareness improvement and pressure relief for city underground pipelines feature the solution which aims to decrease stormwater's pollution and damage in city, realize management on stormwater, create open space in city and lower investment on underground infrastructure etc. And these jobs will produce synergy.

1.2   Development of domestic and foreign

Some developed countries used to be faced with the same stormwater problems as China do today. They have developed approved theories and technologies and put them into wide use after decades of study. American BMPs (Best Management Practices), a typical system of 1970s, solves problems related to water quality, water amount and environmental protection by focusing on end control of stormwater. LID (Low Impact Development) which has evolved since 1990s controls stormwater source based on small-scale and dotted landscape. English SUDS (Sustainable Urban Drainage Systems) has a comprehensive consideration on water quality, amount, possibility of recreating landscape, ecological value of surface runoff and performs a grading reduction and control throughout since runoff inception (Illustration 1). WSUD (Water Sensitive Urban Design) of Australia, giving consideration to landscape and environment, focuses on water circulation to protect water ecosystem for urban plan and design, so stormwater control is only a part of it. LIUDD (Low Impact Urban Design and Development) applied by New Zealand integrates low impact development, small area protection, comprehensive watershed management, green architecture etc. and is applicable to both urban and rural area (Che, 2005).

2 THE APPLICATION COUNTERMEASURES

The solution proposed in the paper uses experience of approved rainwater and flood management of developed countries, mainly from stormwater management of SUDS and rest from LID and BMPs, with consideration of Fujian's actual conditions, to design rain adaptative landscape system for city. Based on linear landscape of retention and drainage which leads water into dotted water landscape of reservoirs, combined permeable paving, it forms primary, intermediate and advanced stormwater reducer made by rain adaptative landscape to perform source, midway and end controls.

2.1 Advanced rain adaptative landscape

1. Primary linear element: ditch of green plant. Road landscape is created considering conditions of low-grade road and basin as well as road in residence area. Sunken ditch under landscape plant can be done for new district of city, and overflow outlets are added for raised roadside greenbelt in old city district to make the greenbelt underground, so it can gather rainwater runoff to greenbelts and replenish underground water.
2. Primary dotted landscape: It mainly consists of bioretention including small rainwater park, roof park, rainwater pond etc. dotted in roadside greenbelts, and affiliated green space of residence areas, commercial areas and educational areas to gather rainwater from nearby hard paving or roadside green ditch.
3. Primary patch element: Permeable paving is made on sidewalk of residence area and small court to facilitate nature infiltration of rainwater.

Major purpose of primary rain adaptative landscape is to solve problems arising from light and moderate rain by focusing on rain source control through dotted small scale landscape and assistance of LID theory. The region lightly suffers from non-point source pollution so rainwater interception and infiltration are major issues. Therefore, it requires small investment and low cost as well as enjoys easy execution and good visual effect. In case rainfall exceeds its designed capacity, stormwater can be discharged though rainwater pipeline.

I. Intermediate rain adaptative landscape:
1. Intermediate linear elements: Retention landscape. Road greenbelt and green space in city may be used to form a middle-size city roadside green landscape. Roadside landscape also applies sunken model with area decided according to rainfall and served area.
2. Intermediate dotted landscape: bioretention landscape. Combined with public green space and park etc., it forms a park landscape system incorporating functions of rainwater interception, retention, collection and usage. Small and middle-size ground rainwater purification area, sunken green land or square and underground rainwater sedimentation basin may be made to recycle rainwater assisted by infiltration measurement and water quality assurance.
3. Intermediate patch element: Permeable paving used for scenic area, small square, small outdoor parking lots, sidewalk and light vehicle lane.

Intermediate rain adaptative landscape is used to solve problems arising from large amount of rainwater and those can't be solved by primary landscape. This part has more serious problems due to non-point source pollution, so it shall perform rainwater interception—infiltration—purification—reuse by integration of source control and midway control. Collected rainwater can be used for irrigation, hard ground cleaning, and replenishment for water consumption in greenbelt and water landscape. Intermediate rain adaptative landscape is also linked with city underground pipelines to treat overflow stormwater.

II. Advanced rain adaptative landscape
    1. Advanced liner element: large rain retention landscape. It can be located near natural water system, or create landscape along bank taking use of existing river or lager channel to conduct purification and water diversion.
    2. Advanced dotted landscape: large bioretention landscape. It combines large city public green land and ecological wetland park. Existing lake, wetland, artificial lake or sunken square can be used for water regulation and storage to cut down runoff discharge amount and reduce peak discharge amount.
    3. Advanced patch element: large square, garden path, large ground parking lot, and even city first-grade road can apply permeable concrete and permeable pitch for paving.

Advanced rain adaptative landscape is made to solve problems resulted from heavy or extremely heavy rain, typhoon rain, persistent heavy rain and rainwater can't be treated by intermediate landscape. It mainly draws on experience of BMPs to control midway and end rainwater. Regulation and storage which is served as major role, together with purification and reuse are performed for city control on drainage, erosion and waterlogging. It has similar frame with intermediate counterpart. According to designed standard for rainwater treatment, area and scale may be larger. Advanced rain adaptative landscape retains great amount of rainwater and produces large ecological city landscape after overflow rainwater is purified and discharged through city rainwater pipeline with plant purification, square ditch containing sand, soil for infiltration which assist underground water replenishment, gradually evaporation. Purified rainwater is reusable.

These three levels link each other and each works according to its level to shape a city rain adaptative landscape system. Such system, working with city rainwater pipelines, conducts treatment on overflow rainwater as well as relieves pressure of city pipelines, improves its drainage standard to reduce investment amount and maintenance cost for underground drainage.

2.2 *Macro and micro level*

During the operation of rain adaptative landscape, in terms of macro jobs, first people shall alter traditional idea and discharge of city rainwater. So a comprehensive model for rainwater control and usage needs to be established by taking water quality, water amount, natural water system and city landscape into consideration. Second, intensively cooperative work involving multiple disciplines shall be conducted to build city rain adaptative landscape. For instance, talents of architecture, landscape, city planning, water conservancy, environment project, municipal engineering, and city management shall be involved for it. Third, relevant authorities shall prepare or revise laws, policies, management mechanism in time according to Fujian's local conditions, such as local standard for environment-friendly building and rainwater usage, completed frame of rainwater management and appraisal, and control and assessment for runoff. The last one, related education and promotion for the public are needed.

In terms of detailed jobs, first, attention shall be paid to soil function as it also regulates water amount and quality in the field of hydrology. It can reduce runoff amount, replenish underground water, increase river's base flow and deal with runoff pollution by filtration, adsorption and biological effect etc. According to Fujian's actual conditions, soil improvement for landscape shall be done at very beginning. Latosol with good adsorption for phosphorus is suitable for control of rainwater's non-point source pollution, but it has bad permeability, so not good for rainwater infiltration (Zhu, 2013). Therefore, soil improvement requires right amount of sand and humus to facilitate infiltration, in addition, alimentation for plant. Second, excessively tamped soil shall be prevented as it produces negative effect on infiltration and increases surface runoff; original organics shall be preserved to maintain its function of water quality improvement (Zhang, 2013). Second, select plant. For rainwater retention and infiltration in green land, amphibian plants adaptable to Fujian's climate shall be selected for it withstands short-term water immersion and grows without water; besides, these selected plants shall be of functions of pollution adsorption and water purification. Third, when on site investigation is performed, pay attention to underground water's level to prevent construction executed above high level underground water, which would pollute such underground water.

3 CONCLUSION

The paper proposes rain adaptative landscape as a solution to city rainwater. As for problems of "excessive water", such landscape is used for interception, regulation, storage and infiltration to deal with heavy rain and waterlogging; to alleviate problems of "insufficient water", retained rainwater is purified for reuse; to solve problems of "dirty water", pollution of rainwater runoff is controlled by filtration offered by plant, soil or by other deposit. These ways are integrated in each level of landscape to solve problems resulted from frequent light and moderate rain as well as waterlogging in city made by typhoon rain and heavy rain as an environment-friendly, energy-saving and highly acceptable system to realize control and use of rainwater. The study is made according to Fujian's local conditions. Hopefully it offers a suitable proposal for Fujian's ecological civilization development, and provides useful information for rainwater management of other areas in China.

ACKNOWLEDGEMENT

This work are funded by the Natural Science Foundation of Fujian Province (Grant No. 201301211) and the Social Science Foundation of Fujian Provincial Education Department (Grant No. JA1228S).

REFERENCES

Che Wu, Lu Fangfang, Li Junqi, Li Haiyan, Wang Jianlong, 2009, Typical Rainwater and Flood Control System of Developed Countries and Inspiration, *China Water Supply and Drainage*, (10)1–6.
Chen Ying, 2013, Water Rises at Night. Natural Disaster Tests Xiamen's Wisdom, *Southeast Morning Post*, (20/05/2013): Monday version for Xiamen, Zhangzhou and Jinmen.
Lin Weihong, 2009, Xiamen Research and Development of Rainwater Usage, *Water Supply, Drainage and Building Electricity*, (1)66–82.
Nan Guang, 2005, Typhoon Longwang Tests City Function Fuzhou, *South Reviews*, (20)14–15.
Quan Bin, Chen Jianfei, Zhu Hejian, Guo Chengda, 2004, Water Problems and Management of Latosol—Take Zhangpu County in South Fujian as Example, *Soil*, (5)532–537.
Zhang Zhiwei, 2013, Soil: A Neglected Factor in Water System, *Landscape Design*, (4)70–77.
Zhu Mulan, Liao Jie, Chen Guoyuan, Wang Jiping, She Nian, 2013, Soil Infiltration Improvement for LID Roadside Greenbelt, *Water Resources Preservation*, (3)25–28.

Hydraulic Engineering III – Xie (Ed)
© *2015 Taylor & Francis Group, London, ISBN 978-1-138-02743-5*

An evaporation duct prediction model based on the MM5

Lin Jiao
Department of Military Oceanography, Dalian Naval Academy, Dalian, China

Yong Gang Zhang
Research Centre of Military Oceanography, Dalian Naval Academy, Dalian, China

ABSTRACT: This paper introduces the flux profile relations of Zhang and Hu, namely the nonlinear factor α_v, and the gust wind item w_g into the Babin model, and thus extends the evaporation duct diagnosis model based on the Babin model to the offshore coast under extremely low wind speed. In addition, an evaporation duct prediction model is designed and coupled with the fifth generation Mesoscale Model (MM5). The tower observational data and radar data at the Ping tan island of Fujian Province on May 25–26, 2002 are used to validate the forecast results. The outputs of the prediction model agree with the observations from 0 to 48 hours. The error of the predicted evaporation duct height is 0.193 and the prediction results are in accordance with the radar detection.

1 INTRODUCTION

Evaporation duct is a kind of abnormal refractive structure in marine boundary layer; it affects the propagation of electromagnetic wave on the sea. The study shows that the propagation characteristics of electromagnetic wave will be changed and the propagation loss will be decreased and the propagation path will be bended in the evaporation duct, the part of electromagnetic wave will be trapped into the duct layer and be formed the over-the-horizon propagation. Because of the turbulent characteristics in marine boundary layer, the instantaneous value of meteorological element is difficult to measure. As a result, it is not recommended to adopt the method of building the profile of the atmospheric refractive index to determine the evaporation duct height. Based on similarity theory in the surface layer, the macroscopic measurement method of the hydrological and meteorological elements is used to calculate the evaporation duct height. However, due to the lack of initial field data and vertical resolution in the past weather forecasting model, which output of the bottom level is higher than the surface layer, it is hard to find a method to solve the problem of the evaporation duct forecast. With the development of mesoscale meteorology, the mesoscale model has a higher vertical resolution, which can be used to forecast and simulate the mesoscale phenomena in the surface layer. Base on the fifth generation mesoscale model, an evaporation duct prediction model is designed and established in this paper.

2 FORMATION MECHANISM AND CHARACTERISTICS CALCULATION OF EVAPORATION DUCT

2.1 The formation mechanism of evaporation duct

Because of the different thermodynamic properties between sea and atmosphere, an imbalance thermal structure will be formed on the air-sea interface under the influence of solar radiation. Based on the structure, the atmosphere will be driven through the latent heat of

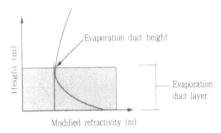

Figure 1. The refractive profile of an evaporation duct.

the sea evaporation. A large number of water vapor was attached in the sea surface layer due to the evaporation of the sea. With the help of vertical transport of the turbulent, the water vapor will be raised to a certain height and formed the interface. Above the interface the water vapor content is low and the water vapor content is higher under the interface, on the sea surface the water vapor content is even saturated. When the water vapor content decreased with increase in height from the sea surface to the interface, the normal distribution of the atmospheric temperature and humidity in the sea surface layer was changed and the atmospheric refraction structure was changed. When the water vapor content sharply decreased, the extreme super refraction appeared, it is evaporation duct. Its refraction structure is shown in Figure 1.

2.2 The calculation of evaporation duct height

The evaporation duct height is an important parameter of the evaporation duct, which describes the intensity of duct and determines the effect on the electronic devices. There are many numerical models for calculating the evaporation duct height at present, but these models are all based on the surface layer similarity theory, only the methods for calculating the surface layer flux and the characteristic scale are different. This paper uses the Babin model (1996) based on the flux-algorithm as the method of calculating the evaporation duct height, in addition, the flux profile equation of Zhangqiang (1995) is added the model. As a result, the theory of Monin-Obukhov is extended to very low wind speed and the coastal areas.

For microwave band, troposphere atmospheric refractive index N is:

$$N = 77.6\frac{P}{T} - 5.6\frac{e}{T} + 3.73 \times 10^5 \frac{e}{T^2} \tag{1}$$

In the formula, P is atmospheric pressure, e is vapor pressure, and T is atmospheric temperature.

Take the height derivative of both sides in the last-written, replace the atmospheric temperature T with potential temperature θ and replace the vapor pressure e with specific humidity q, then

$$\frac{dN}{dz} = C_1 + C_2\frac{d\theta}{dz} + C_3\frac{dq}{dz} \tag{2}$$

In the formula, C_1, C_2, C_3 are related parameters, the expressions are as follows:

$$C_1 = (-\rho g \times 0.01) \cdot \left(\frac{A}{T} + \frac{ABq}{T^2(\varepsilon + (1-\varepsilon)q)}\right)$$
$$- \frac{g(p - (1-\varepsilon)e)}{C_{pa}}\left(-\frac{A}{T^2} - \frac{2ABq}{T^3(\varepsilon + (1-\varepsilon)q)}\right)$$

172

$$C_2 = \left(\frac{p}{p_0}\right)^{\frac{R_d}{C_{pa}}}\left(-\frac{Ap}{T^2} - \frac{2ABpq}{T^3(\varepsilon+(1-\varepsilon)q)}\right)$$

$$C_3 = \left(\frac{ABp\varepsilon}{T^3(\varepsilon+(1-\varepsilon)q)^2}\right)$$

$A = 77.6$ K/hPa, $B = 4810$ K^2/hPa; P is atmospheric pressure, e is vapor pressure, the unit is hPa; T is atmospheric temperature, the unit is K; P_0 is reference pressure, its value is 1000 hPa; R_a is the specific gas constant of dry air, its value is 287.05 J/(kg·K); C_{pa} is the specific volume of dry air in constant pressure, its value is 1004 J/(kg·K); $\varepsilon = 0.62197$; ρ is air density, its unit is kg/m^3; g is acceleration of gravity, its value is 9.8 m/s^2. C_1, C_2, C_3 can be calculated by atmospheric temperature, atmospheric humidity, atmospheric pressure and sea surface temperature in the reference height.

$d\theta/dz$ and dq/dz can be determined by the similarity theory of Monin-Obukhov.

$$\frac{d\theta}{dz} = \frac{\theta_*}{kz}\varphi_H\left(\frac{z}{L}\right) \tag{3}$$

$$\frac{dq}{dz} = \frac{q_*}{kz}\varphi_V\left(\frac{z}{L}\right) \tag{4}$$

θ_* is the Monin-Obukhov characteristic scale parameter of potential temperature, q_* is the Monin-Obukhov characteristic scale parameter of specific humidity; φ_H is the Monin-Obukhov dimensionless profile function of the atmospheric temperature, φ_V is the Monin-Obukhov dimensionless profile function of the atmospheric humidity; L is the Monin-Obukhov Length; k is Karman constant, its value is 0.4.

Let's plug equation (3), (4) into equation (2), then

$$\frac{dN}{dz} = C_1 + C_2\frac{\theta_*}{kz}\varphi_H\left(\frac{z}{L}\right) + C_3\frac{q_*}{kz}\varphi_V\left(\frac{z}{L}\right) \tag{5}$$

Here a nonlinear correction factor α_V, which proposed by Zhangqiang (1995), is introduced in this paper.

$$\alpha_V = \frac{\varphi_V}{\varphi_H} \approx \frac{1+2z/L}{1+5.4z/L} \tag{6}$$

By the definition of atmospheric duct, when the vertical gradient of atmospheric refraction index is equal to -0.157 m^{-1}, the atmospheric duct is formed, at the same time; corresponding height z is duct height z_d, so

$$z_d = \frac{-(C_2\theta_* + \alpha_V C_3 q_*)}{k(C_1+0.157)}\varphi_H\left(\frac{z_d}{L}\right) \tag{7}$$

In order to calculate the evaporation duct height z_d, L, θ_*, q_*, φ_H must be determined. According to the relations which was deduced by Liu (1979), L, θ_*, q_*, φ_H can be calculated.

$$\theta_* = k(\theta_z - \theta_{z0})[\ln(z/z_0) - \psi_H(z/L)]/0.74 \tag{8}$$

$$q_* = k(q_z - q_{z0})[\ln(z/z_0) - \psi_V(z/L)]/0.74 \tag{9}$$

Here,

$$\begin{cases} \psi_H(z/L) = 2\ln[(1+0.74\varphi_H^{-1})/2] & \text{当} z/L \leq 0 \\ \psi_H(z/L) = -6.35z/L & \text{当} z/L > 0 \end{cases}$$

173

$$\begin{cases} \psi_V(z/L) = 2\ln[(1+0.74\alpha_V\varphi_H^{-1})/2] & \text{when } z/L \le 0 \\ \psi_V(z/L) = -6.35z/L & \text{when } z/L > 0 \end{cases}$$

$$\varphi_H \approx \begin{cases} 0.74(1-9z/L)^{-0.5} & \text{when } z/L \le 0 \\ 0.74+4.7z/L & \text{when } z/L > 0 \end{cases}$$

The Monin-Obukhov Length L is expressed as follows:

$$L = \frac{U_*^3}{kg\left(\dfrac{\theta_*}{T_0}+0.608q_*\right)} \tag{10}$$

In the formula,

$$U_* = kV_z[\ln(z/z_0) - \psi_M(z/L)]/0.74$$

$$\begin{cases} \psi_M(z/L) = 2\ln[1+\varphi_M^{-1})/2] + \ln[(1+\varphi_M^2)/2] - 2\arctan\varphi_M^{-1} + \pi/2 & \text{when } z/L \le 0 \\ \psi_M(z/L) = -4.7z/L & \text{when } z/L > 0 \end{cases}$$

$$\varphi_M \approx \begin{cases} (1-15z/L)^{-0.25} & \text{when } z/L \le 0 \\ 1+4.7z/L & \text{when } z/L > 0 \end{cases}$$

Here, V_z is the full wind speed at the height of z, T_0 is the average temperature in the surface layer.

In order to extend the similar theory to the low wind speed condition, the gust w_g was introduced; we replaced the wind speed value with the square root of the new wind speed S, here $S = u_x^2 + u_y^2 + u_z^2 + w_g^2 = u^2 + w_g^2$

$$w_g = \beta w_* = \beta(F_b z_i)^{1/3} \tag{11}$$

$$F_b = -\left(\frac{g}{T}\right)\overline{w'T'}_v \tag{12}$$

In the formula, w_g is the gustiness wind speed, w_* is the characteristic scale of the free convection velocity, β is the empirical constant, its order is 1.0, usually takes 1.25 (Fairall, 1996), F_b is the buoyancy flux terms, which reflects the turbulent fluctuations, $\overline{w'T'}_v$ is the vertical flux of the virtual temperature, z_i is the mixing height in the troposphere.

3 CONSTRUCTION OF THE EVAPORATION DUCT PREDICTION MODEL

Based on the formation mechanism of evaporation duct and the calculation of evaporation duct height, this paper proposes a new model which uses the mesoscale model MM5 coupling the improved Babin model for the prediction of the evaporation duct. The process is illustrated in Figure 2.

In the prediction model, the TERRAIN module horizontally interpolates the regular latitude-longitude terrain elevation, and vegetation onto the chosen mesoscale domains. The purpose of REGRID module is to read archived gridded meteorological analyses and forecasts on pressure levels and interpolate those analyses from some native grid and map projection to the horizontal grid and map projection as defined by the TERRAIN module. The RAWINS module is to improve meteorological analyses on the mesoscale grid by incorporating information from observations, the analyses input to RAWINS as the first guess are usually fairly low-resolution analyses output from REGRID module. The INTERPF module handles the data transformation required to go from the analysis programs to the mesoscale model. This entails vertical interpolation, diagnostic computation, and data reformatting.

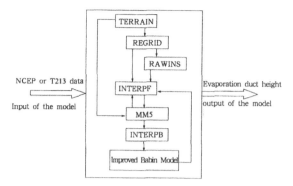

Figure 2. The framework of the evaporation duct prediction model.

INTERPF module takes REGRID, RAWINS, INTERPB output data as input to generate a model initial, lateral boundary condition and a lower boundary condition. The MM5 module is the core of the prediction model; it is the numerical weather prediction part of the modeling system. The INTERPB module handles the data transformation required to go from the mesoscale model on σ coordinates back to pressure levels. The improved Babin model is used to calculate the evaporation duct height. The fifth generation mesoscale model is coupled with the improved Babin model through related parameters. These parameters include the air temperature, the air humidity, the sea surface temperature, the sea surface wind, the sea level pressure, the roughness, the mixing height and Monin-Obukhov length. The MRF parameterization scheme of MM5 is adopt as Planet Boundary Layer scheme, which includes the boundary layer physical process and the free atmosphere physical process. Its resolution is higher.

4   THE NUMERICAL SIMULATION AND THE VERIFICATION OF THE PREDICTION MODEL

In order to testify the accuracy of the evaporation duct prediction model, this paper utilized NCEP data of May 25 in 2002 on the numerical simulation. Moreover, the simulation results were verified with the measured data of Ping Tan Island on May 25, 2002. The observation site of this experiment was in Ping Tan Island Fujian province. The main content of this experiment was to calculate the evaporation duct height by the measured data of the iron tower; these data include the air temperature, the air humidity, the sea level pressure, the wind speed, the wind direction and the sea surface temperature. Then the evaporation duct diagnostic accuracy was verified by comparing the measured radar data.

4.1   The design scheme of the simulation experiment

The initial field data of the prediction model is the $1^\circ*1^\circ$ NCEP data. Two layer nested area is chosen as the forecast area. The horizontal network points of the first area is 75×75, the grid point distance is 81 kilometers; the horizontal network points of the second area is $31*31$, the grid point distance is 27 kilometers. The center point of the forecast area is located in 32°N and 118°E. The model vertical resolution selection is the inhomogeneous 31 layers (two layers in the surface layer), the iron tower and the radar observation point are both located in the second forecast area, its horizontal coordinate point is 22; its vertical coordinate point is 16. The initial time is at 00:00 GMT, 25 May 2002, the forecast time is 48 hours. For the stabilization of the calculation, the split-time-step method is adopt to cope with the acoustic item (u, v, w) in the forecast equation, which uses several short time step to replace the long time step for predicting the velocity field and the pressure field. The forecast areas are illustrated in Figure 3.

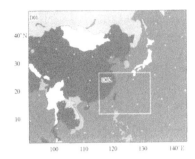

Figure 3.　The model domain.

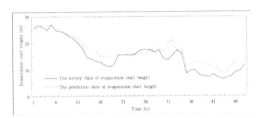

Figure. 4.　The comparison between the predicted and the observed evaporation duct height.

Table 1.　A contrast between the model prediction and the radar detection.

Time (UTC)	The evaporation duct prediction height (m)	Radar detection distance (n mile)	Detection results	Comparison between the duct heights and the radar observation
2002/05/25/01	24.048	37.4	Over-the-horizon	Consistent
2002/05/25/04	26.640	45.5	Over-the-horizon	Consistent
2002/05/25/07	25.070	65.3	Over-the-horizon	Consistent
2002/05/25/12	20.177	38.2	Over-the-horizon	Consistent
2002/05/25/15	16.705	62.4	Over-the-horizon	Consistent
2002/05/26/00	15.720	61.7	Over-the-horizon	Consistent
2002/05/26/03	17.280	54	Over-the-horizon	Consistent
2002/05/26/09	22.163	88.8	Over-the-horizon	Consistent
2002/05/26/12	25.197	79.2	Over-the-horizon	Consistent
2002/05/26/15	25.800	78.5	Over-the-horizon	Consistent

4.2　*Comparison between the prediction results and the measured results*

Based on the outputs of the evaporation duct prediction model, we can get the evaporation duct heights from May 25 to 27, 2002. The measured data from iron tower was used to validate the prediction model. Compared with the prediction outputs and the survey data, the result indicates that the predicted results are good agreement with the measured values within 48 hours and the changing rules are basically identical. As is shown in Figure 4. The average error calculation formula is as follows:

$$\sigma = \frac{\sum\limits_{n=1}^{48} abs(P-O)/O}{48}$$

(13)

Here σ is the average error; P is the predicted results; O is the observed data.

176

Using the above formula, we get the average relative error of the evaporation duct heights prediction is about 19.3%.

4.3 *Comparison between the prediction results and the radar observations*

The radar installation height is 14.8 meters; its scope of observation is from 79° to 232°. Thus as long as the evaporation duct height is higher than 14.8 meters, the radar may find the over-the—horizon targets. Based on the comparison between the evaporation duct heights and the radar observations, we further check up the accuracy of the evaporation duct prediction model, as shown in Table 1.

It should be noted that the last two sets of results in the table are caused by the surface duct, because the evaporation duct height is less than 14.8 meters at that time.

5 CONCLUSION AND DISCUSSION

Based on the numerical simulation and the prediction results analysis, we can come to the conclusion:

1. The Babin model was improved with the flux profile relations (Zhangqiang and Huyinqiao, 1995) and the gust wind item, thus the evaporation duct diagnosis model will be extended to the offshore coast under extremely low wind speed.
2. Base on the fifth generation Mesoscale Model (MM5), coupling with the improved Babin model, an evaporation duct prediction model is designed and established in this paper.
3. The outputs of the prediction model agree with the observations from 0 to 48 hours. The error of the predicted evaporation duct height is 19.3%.
4. If more real-time data can be assimilated in the evaporation duct prediction model, the forecast accuracy may be improved. Furthermore we can use the inversion method of satellite remote sensing to study the evaporation duct prediction.

REFERENCES

Babin S M. 1996. A new model of the oceanic evaporation duct and its comparison with current models [D]. University of Maryland at College Park, 189–190.

Brutsaert W. 1991. Evaporation into the Atmosphere Theory, History, and Applications. Kluwer Academic Publishers, 298–299.

Burk S D, Thompson W T. 1997. Mesoscale modeling of summertime refractive conditions in the southern California Bight. J Appl Meteor, 36: 22–31.

Dai Fushan, Li Qun. 2002. The Atmospheric Duct and Its Military Application (in Chinese). Beijing: Peoples Liberation Army Press, 212–220.

Edson J B, Fairall C W. 1998. Similarity relationships in the marine atmospheric surface layer for terms in the TKE and scalar variance budgets. J Atmos Sci, 55: 2311–2328.

Hu Yingqiao, Ji Yuejin. 1993. The combinatory method for determination of the turbulent fluxes and universal functions in the surface layer. Acta Meteor Sinica, 7(1): 101–109.

Liu W T, Katsaros K B, Businger J A. 1979. Bulk Parameterization of air-sea exchanges of heat and water vapor including the molecular constraints at the interface. J Atmos Sci, 36: 1722–1735.

Li Xiaoli, He Jinhai, et al. 2003. The design of urban canopy parameterization of MM5 and its numerical simulations. Acta Meteor Sinica (in Chinese), 61(5): 526–538.

Lv Meizhong, Peng Yongqing. 1990. Tutorial of the Dynamical Meteorology (in Chinese). Beijing: China Meteorology Press, 120–128.

Hydraulic Engineering III – Xie (Ed)
© 2015 Taylor & Francis Group, London, ISBN 978-1-138-02743-5

Adsorption of Cu²⁺ from aqueous solution by aminated ephedra waste

N. Feng, Y. Shi & J. Cao
School of Pharmacy, Ningxia Medical University, Yinchuan, Ningxia, China

ABSTRACT: Ephedra waste was modified by epichlorohydrin and diethylenetriamine to obtain aminated ephedra waste biosorbent. The factors affecting the adsorption efficiency, such as pH and contact time were investigated. The results showed that the optimum absorption conditions of aminated ephedra waste: pH was 4.7; contact time was 3 h; equilibrium was well described by Langmuir isotherms and kinetics was found to fit pseudo-second order type. According to the Langmuir equation, the maximum adsorption capacities of modified adsorbent for Cu²⁺ are 93.11 mg/g, which are higher than untreated adsorbent (17.61 mg/g). The aminated ephedra waste biosorbent had excellent absorbability toward heavy metal ions Cu²⁺.

1 INTRODUCTION

As a result of rapid industrialization, the disposal of heavy metals into the environment has increased. Many industries discharge aqueous effluents that contain heavy metals, such as copper, cadmium, zinc and mercury (Saygideger et al. 2005). These metal ions can be harmful to aquatic life, and water contaminated by toxic metal ions remains a serious public health problem for human health (Pehlivan et al. 2008). Therefore, it is of great practical interest to explore ways to effectively remove these heavy metal ions from the wastewaters before their discharge, and to possibly separate them for recovery and re-use.

The main techniques that have been used to reduce the heavy metal content of effluents include chemical precipitation, liquid–liquid extraction and resins, membrane processes, adsorption onto activated carbon, ion exchange, and electrolytic methods (Seco et al. 1999). These methods have been found to be limited, because they often involve high operational costs or may also be insufficient to meet strict regulatory requirements as for chemical precipitation. The biosorption which uses cheap adsorption materials may be an alternative wastewater technology, and this method can avoid the generation of secondary waste and used easily on an industrial scale (Feng & Aldrich 2004).

Ephedra sinica Stapf is a traditional Chinese medicine; its main active constituents are alkaloids, primarily ephedrine and pseudoephedrine which has the functions of sweating, diuresis and detumescence and so on. Beside the production of ephedrine and pseudoephedrine, a large amount of solid waste–ephedra waste is also an output from Chinese traditional medicine industry. Currently, they are disposed through combustion or land filling, which produces secondary pollution easily. The main components of the ephedra waste are cellulose, hemicellulose and lignin (Feng & Zhang 2013). We can have a chemical modification of ephedra waste, then use it as biosorbent of Cu²⁺.

In this study, ephedra waste was modified by epichlorohydrin and diethylenetriamine to obtain aminated ephedra waste biosorbent and absorbed for Cu²⁺. By dint of static absorption test, the effects of pH and contact time on absorption were investigated. Biosorbent characterization was determined with FTIR analysis. Langmuir adsorption isotherm was applied to the experimental data. The pseudo second-order model was used for determining of the adsorption kinetics.

2 MATERIALS AND METHODS

2.1 *Materials and instruments*

The ephedra waste was obtained from a traditional Chinese medicine factory. $CuSO_4 \cdot 5H_2O$ were used as copper source. Epichlorohydrin, diethylenetriamine and sodium carbonate were the analytic grade reagents commercially available. A pH meter (PHS-3C) was used to measure pH of the suspensions. FTIR spectroscopy (Nicolet-6700) was used to identify the chemical groups in the biosorbent. Atomic absorption spectrometer (Persee A3 series) was used to analyze the concentration of residual copper ions. The double-distilled water was used for the entire experiments.

2.2 *Methods*

2.2.1 *Modification of aminated ephedra waste adsorbent*

Raw material was washed over and over again with tap water, then twice with double-distilled water. After a drying at 60 °C, the ephedra waste were crushed, sieved, and under the 0.42 mm particle size fraction was kept for the experiments.

5 g of ephedra waste was added to 50 mL 5% of sodium hydroxide, the mixed solution was added to 15 mL of epichlorohydrin, followed by stirring at 30 °C for 1 h. The product was filtered through filter paper, the filtered precipitate was washed with double-distilled water until pH is equal to 7.0, and then was washed twice with acetone, and the product was drying at 70 °C for 8 h. By this method we can get the ephedra waste epoxy group cellulose ether.

2.5 g of ephedra waste epoxy group cellulose ether and 2.5 g of sodium carbonate were added to 225 mL double-distilled water, and then added 2.5 mL of diethylenetriamine to the mixed solution, followed by stirring at 100 °C for 2 h. The product was filtered through filter paper, the filtered precipitate was washed with double-distilled water until pH is equal to 7.0, and then was washed twice with ethanol, and the product was drying at 70 °C for 8 h. By this method we can get the aminated ephedra waste.

2.2.2 *Adsorption experiments and analytical method*

The adsorption experiments were conducted by batch process. Fixed amount of adsorbent was added in the stoppered conical flasks containing 25 mL of Cu^{2+} ion solution of desired concentration, the initial pH of the Cu^{2+} solution was changed by adding 0.1 mol/L HCl or 0.1 mol/L NaOH solutions as required. Then, the content was shaken for the desired contact time in a thermostated reciprocating shaker at the rate of 200 rpm at 25 °C. At the regular interval of time the conical flask from the shaker was withdrawn and then the adsorbent was separated from the Cu^{2+} ion solution by filter paper, and the filtrate was analyzed using AAS. The amount of Cu^{2+} (q_e, mg/g) retained in the adsorbent phase was estimated by the following equation:

$$q_e = \frac{(c_0 - c_e)V}{W}$$

(1)

where c_0 is the initial concentration of the solution in mg/L, c_e is the equilibrium concentration of the solution in mg/L, V is the volume of the solution in L and W is the mass of sorbent in g.

The absorption rate of Cu^{2+} ion in solution was calculated using the following equation:

$$\text{Absorption rate (\%)} = \frac{c_0 - c_e}{c_0} \times 100\%$$

(2)

3 RESULTS AND DISCUSSION

3.1 *FTIR analysis*

The FTIR spectra of unmodified (a) and modified (b) ephedra waste are shown in Figure 1. The broad and intense absorption peaks at 3417 cm⁻¹ correspond to O-H stretching vibrations

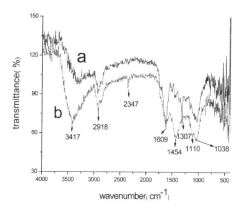

Figure 1.   FTIR spectra of unmodified (a) and modified (b) ephedra waste.

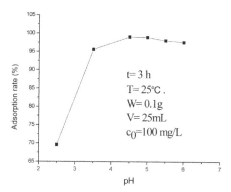

Figure 2.   Effect of pH on biosorption to Cu^{2+}.

of cellulose, hemicellulose and lignin. The peaks at 2918 cm^{-1} can be attributed to C-H stretching vibrations of methyl, methylene and methoxy groups. The absorption bands at 1609 cm^{-1} are characteristic of the elongation of the aromatic $-C=C-$ bonds (Gnanasambandam & Protor 2000). 1307 cm^{-1} is assigned to the bending vibration of the hydroxyl group (Pan et al. 2013). The absorption peaks at 1038 cm^{-1} can be assigned to C-O stretching vibrations of carboxylic acids and alcohols (Feng et al. 2009).

The IR spectra of modified (b) ephedra waste were compared with the IR spectrum of the unmodified ephedra waste. The absorption peaks at 1307 cm^{-1} was disappeared and three new absorption peaks at 2347 cm^{-1}, 1454 cm^{-1} and 1110 cm^{-1} were occurring. The absorption peaks at 1454 cm^{-1} correspond to N-H in-plane bending vibrations; the absorption peaks at 2347 cm^{-1} can be attributed to N-H stretching vibrations of amines (Argun & Dursun 2008). The absorption peaks at 1110 cm^{-1} correspond to C-O stretching vibrations of ethers and alcohols (Yu et al. 2010); the disappearing of absorption bands at 1307 cm^{-1} indicated that the hydroxyl in the raw material was removed by treating with epichlorohydrin. Comprehensive analysis shows that the epoxidation and amination reaction were happened and the groups were changed.

3.2   *Effect of pH on biosorption*

Solution pH is an important parameter for the adsorption experiments. The effect of pH on adsorption was given in Figure 2. At lower pH, the amount of biosorption to Cu^{2+} is small. Biosorption to Cu^{2+} increases with the increase of pH from 2.5 to 4.5. The highest biosorption efficiency is observed in the pH range of 4.5–5.0. At low pH, the surface of biosorbent would also be surrounded by hydronium ions which decrease the copper interaction with binding sites of the aminated ephedra waste by greater repulsive forces. When the pH increases, fixation capacities are improved due to the lower competition between the protons and copper (II) ions (Gerente et al. 2000). As a result the working pH value for Cu^{2+} biosorption was chosen as 4.7 and the other biosorption experiments were performed at this pH value.

3.3 Biosorption kinetics of Cu^{2+}

The effect of contact time on adsorption was given in Figure 3. The adsorptive quantity of Cu^{2+} increased with increase in contact time and reached equilibrium after 180 min. So keeping these observations in view, 180 min contact time was opted for further experiments.

The kinetics of Cu^{2+} ion biosorption on aminated ephedra waste biosorbent were analyzed using pseudo-second order (Ho & Mckay 1998):

$$\frac{t}{q_t} = \frac{1}{k_2 q_e^2} + \frac{t}{q_e} \tag{3}$$

where k_2 (g/(mg · min)) is the constant of pseudo-second-order rate; q_e (mg/g) and q_t (mg/g) are the biosorption capacity at equilibrium and time t, respectively.

The values of pseudo-second-order equation parameters together with correlation coefficients are listed in Table 1. The correlation coefficients for the pseudo-second-order equation were 0.999. The calculated q_e values also agree very well with the experimental data. This strongly suggests that the biosorption of Cu^{2+} onto aminated ephedra waste is most appropriately represented by a pseudo-second-order rate process and the biosorption rate is controlled by chemical biosorption.

3.4 Biosorption isotherms

The adsorption isotherm of Cu^{2+} on aminated ephedra waste was given in Figure 4. The equilibrium adsorption q_e increases with the increase in Cu^{2+} concentrations. Langmuir

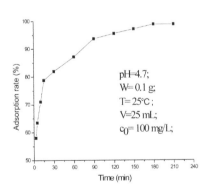

Figure 3. Biosorption kinetics of Cu^{2+} ions.

Table 1. Kinetic parameters of pseudo-second-order equation for Cu^{2+} adsorption.

Metal ions	Experimental value q_e (mg/g)	Kinetic parameters of pseudo-second-order		
		k_2 (g/(mg · min))	q_e (mg/g)	R^2
Cu^{2+}	24.77	0.033	25.38	0.9992

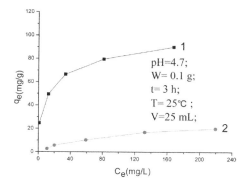

Figure 4. Adsorptions isotherms curves of modified (1) and unmodified (2) ephedra waste.

Table 2. The conform parameters of Langmuir equation.

Adsorbents	q_{max} (mg/g)	b (L/mg)	R^2
Modified	93.11	0.155	0.9855
Unmodified	17.61	0.062	0.9259

isotherm was employed to calculate adsorption capacity. Langmuir isotherm model is as follows:

$$\frac{c_e}{q_e} = \frac{1}{q_{max}b} + \frac{c_e}{q_{max}}$$ (4)

where q_m is the maximum monolayer capacity of the biosorbent, and b is the biosorption constant.

The parameters resulted from the Langmuir plot for Cu^{2+} are presented in Table 2. As seen from Table 2, very high regression correlation coefficients (>0.98) were found for Langmuir isotherm model. In accordance with the result, Langmuir model is suitable for describing the biosorption equilibrium of Cu^{2+} by the aminated ephedra waste in the studied concentration range. According to the Langmuir equation, the maximum adsorption capacities of modified adsorbent for Cu^{2+} are 93.11 mg/g, which are higher than untreated adsorbent (17.61 mg/g). Amino was introduced by amination reaction, and amino can chelate with copper ions, thus amination modification can improve the adsorption quantity of the biosorbent.

4 CONCLUSIONS

This preliminary study has shown that aminated ephedra waste biosorbent could be an interesting low-cost biosorbent for copper removal from aqueous solutions. The optimum absorption conditions of aminated ephedra waste: pH is 4.7; contact time is 3 h; Pseudo-second-order model is more applicable for the adsorption process. The biosorption of Cu^{2+} on aminated ephedra waste biosorbent obeys the Langmuir isotherm. According to the Langmuir equation, the maximum adsorption capacities of modified adsorbent for Cu^{2+} are 93.11 mg/g. The results obtained with aminated ephedra waste may be tested using metal-industry wastewater containing Cu^{2+}, since ephedra waste is an inexpensive source and there-fore may have the advantage of economic viability.

ACKNOWLEDGEMENTS

This work was supported by the National Natural Science Foundation of China (No. 21266026).

REFERENCES

Feng, D. & Aldrich, C. 2004. Adsorption of heavy metals by biomaterials derived from the marine alga Ecklonia maxima [J]. *Hydrometallurgy* 73: 1–10.

Feng, N., Guo, X., Liang, S. 2009. Kinetic and thermodynamic studies on biosorption of Cu (II) by chemically modified orange peel [J]. *Transactions of Nonferrous Metals Society of China* 19: 1365–1370.

Feng, N. & Zhang, F. 2013. Untreated Chinese ephedra residue as biosorbents for the removal of Pb^{2+} ions from aqueous solutions [J]. *Procedia Environmental Sciences* 18: 794–799.

Gnanasambandam, R. & Protor, A. 2000. Determination of pectin degree of esterification by difuse reflectance Fourier Transform Infrared Spectroscopy [J]. *Food Chemistry* 68(3): 327–332.

Gerente, C., Couespel du Mesnil, P., Andres, Y., Thibault, J.F., Le Cloirec P. 2000. Removal of metal ions from aqueous solution on low cost natural polysaccharides: sorption mechanism approach [J]. *Reactive and Functional Polymers* 46: 135–44.

Ho, Y.S. & Mckay, G. 1998. The kinetics f sorptions of basic dyes from aqueous solutions by sphagnum moss peat [J]. *Canadian Journal of Chemical Engineering* 76: 822–827.

Pan, H., Sun, G., Zhao T. 2013. Synthesis and characterization of aminated lignin [J]. *International Journal of Biological Macromolecules* 59: 221–226.

Pehlivan, E., Yanik, B.H., Ahmetli, G., Pehlivan, M. 2008. Equilibrium isotherm studies for the uptake of cadmium and lead ions onto sugar beet pulp [J]. *Bioresource Technology* 99: 520–3527.

Saygideger, S., Gulnaz, O., Istifli, E.S., Yucel, N. 2005. Adsorption of Cd (II), Cu (II) and Ni (II) ions by Lemna minor L.: Effect of physicochemical environment [J]. *Journal of Hazardous Materials* B126: 96–104.

Seco, A., Marzal, P., Gabaldon, C., Ferrer J. 1999. Study of the adsorption of Cd and Zn onto an activated carbon: Influence of pH, cation concentration, and adsorbent concentration [J]. *Separation Science and Technology* 34(8): 1577–1593.

Yu, R., Wu, J., Huang, H. 2010. Adsorption of Three Wastewater Dyes by Modified Pineapple Peel Fiber [J]. *Modern Food Science and Technology* 26(7): 674–675.

Hydraulic Engineering III – Xie (Ed)

Diatomaceous Earth precoat filtration for drinking water treatment

Wenqi Zhou, Hui Ye & Dong Zhang
National Engineering Research Center of Urban Water Resources, Shanghai, China

Chunlin Zhang
Shanghai Safbon Water Service Co. Ltd., Shanghai, China

ABSTRACT: Taking the raw water of Yangtze Chenhang Reservoir as the research object, the characterization of diatomaceous earth, and the removal efficiency of COD_{Mn}, turbidity by precoat filtration process were investigated by pilot experiment in this paper. Diatomaceous earth patterns having porous structure and the diatoms range in size from under 5 to more than 100 micrometers. Precoat filtration has a good stability to remove solid suspended particle, effluent turbidity was remained about 0.05 NTU. The running time is linear with the operating pressure, $R^2 = 0.942$, which indicate a good correlation. The precoat layer has little effect on operating pressure. The operating pressure (total head loss) is proportional to square of filtration rate. Precoat filtration can be defined as a filtration technique for relatively low turbidity raw waters.

1 INTRODUCTION

Precoat filtration is a US Environmental Protection Agency (USEPA) accepted filtration technique for potable water treatment (George P, 2000). During World War II, the US Army developed the new filter, which successfully removed the Entamoeba histolytica, a protozoan parasite prevalent in the Pacific war zone, from drinking water. This filter commonly called Diatomaceous Earth (DE) filtration, or precoat filtration. After the WWII, precoat filtration has been used in the filtration of sugar syrups, fruit juices, wine, beer, and water. More than 170 potable water treatment plants using precoat filtration have been constructed. As a new water treatment technology, diatomaceous earth precoat filtration could remove suspended particles without adding chemicals (AWWA, 1995). The main objectives of this study were to evaluate the feasibility of the application of precoat filtration enhance the turbidity removal in Yangtze Chenhang Reservoir raw water.

2 MATERIAL AND METHOD

2.1 Raw water

The main feature parameters of experimental raw water were shown in Table 1.

Table 1. Water quality of raw water.

Parameter	COD_{Mn} (mg/L)	Turbidity (NTU)	UV_{254} (cm⁻¹)	Conductivity (μS/cm)	Fe (mg/L)	Mn (mg/L)
Value	1.4–2.7	5–45	0.027–0.103	280–860	0.33–1.04	0.002–0.58

2.2 *Pilot process parameters*

Pilot-scale was 6 m³/h. Coagulant used PAC (alumina content 10%) and the dosage was about 20 mg/L. After coagulation and sedimentation, raw water were flowing into sand filtration or ultrafiltration (3 m³/h) and precoat filtration (3 m³/h). The filter area was 1.5 m². The dosage of precoat was 0.9 kg, turn on the precoat pump to bring the diatomaceous earth to the filter. When all the filter media is on the filters, change the valving to bring raw water and body-feed to the filter. Dosage of body-feed was 4.32 mg/L.

2.3 *Materials and physical characterizations*

The diatomaceous earth sample was provided by Lin Jiang Imerys Diatomite Co, Ltd. The surface textures of the samples were observed using Scanning Electron Microscopy (SEM) with a JEOL JSM-6700F (JEOL Co., Japan) apparatus.

3 RESULTS AND DISCUSSION

3.1 *Characterization of diatomaceous earth*

SEM micrographs of Diatomaceous Earth (DE) are shown in Figure 1 (Angela F, 2012 and Wen-Tien Tsai, 2006). The porous structure of DE can be clearly seen in Figure 1(b). The filtering medium used in precoat filtration is composed of fossil-like skeletons of microscopic

Figure 1(a).   SEM photographs (×400) of diatomaceous earth.

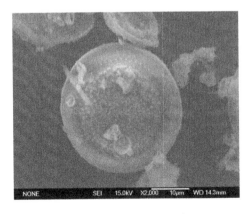

Figure 1(b).   SEM photographs (×2000) of diatomaceous earth.

water plans called diatoms, which range in size from under 5 to more than 100 micrometers. These algae have the capability of extracting silica from water to produce their skeletal structure. Diatomaceous earth is distinguished not only by the Diatom species, but also by its purity of SiO_2. Most of the constituents of diatomaceous earth are practically inert. The soluble portion of diatomite is very low (less than 1%). The chemically inert characteristics mean that diatomaceous earth could safely be used for filtration of water, beverage, or other liquids for human consumption.

3.2 COD_{Mn} removal

As shown in Figure 2, under different operation conditions, the effluent COD_{Mn} of precoat filtration was 1.38 mg/L, which was closer to the effluent of sedimentation (1.36 mg/L) and sand filtration (1.32 mg/L). It was showed that precoat filtration had no effect on removing COD_{Mn}. The removal of organics combined process mainly depends on the process of coagulation and sedimentation. Therefore, in order to make the effluent quality meet the national drinking water standard (GB5749-2006, less than 3 mg/L), the production units should use coagulation and sedimentation, or other pre-oxidation process to ensure the effective removal of organic matter in raw water.

3.3 Turbidity removal

The turbidity of different process was shown in Figure 3, within 40 hours of running time, effluent turbidity of ultrafiltration was remained about 0.05 NTU, the removal rate of

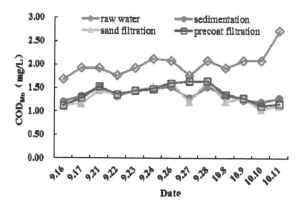

Figure 2. The effluent COD_{Mn} of different process.

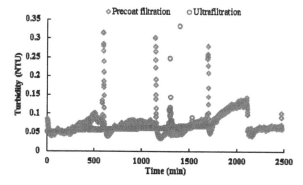

Figure 3. The effluent turbidity of ultrafiltration and precoat filtration.

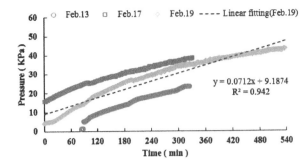

Figure 4.   The hydraulics of precoat filtration process.

turbidity was more than 90%. Compared with the ultrafiltration process, the effluent turbidity of precoat filtration was affected by the operation pressure, influent turbidity, and other factors. The effluent turbidity of precoat filtration was changed form 0.05 NTU to 0.1 NTU. The removal rate of turbidity in precoat filtration was more than 80%, which indicate that precoat filtration has a good effect on remove solid suspended particles. Diatomaceous Earth (DE or diatomite) filtration can be defined as a filtration technique for relatively low turbidity waters, generally below 10 turbidity units (NTU). A survey of direct filtration practice, including precoat filtration plants, indicated that 80 percent of the plants had an average source water turbidity of 5 NTU or less. The maximum turbidity at 80 percent of the plants was 40 NTU or less (AWWA, 1995).

3.4   Hydraulics

The hydraulics of precoat filtration process was showed in Figure 4. By analyzing operation pressure and running time during one cycle, the hydraulics of the device was investigated. After opening the valve, the operation pressure rise slowly increased with time. By curve fitting, the result showed that running time is linear with the operating pressure. Fitting equation is $y = 0.0712\,x + 9.1874$ ($R^2 = 0.942$), which indicate that a good correlation. The result of other research demonstrated that $H = h_1$ (precoat) + h_2 (body feed layer) or:

$$H = K_3 Q W_1 + K_4 C_D Q^2 t \times (8.33 \div 10^6) \tag{1}$$

where, H = operating pressure, Q = filtration rate, W_1 = precoat weight, K_3 = permeability of precoat layer, C_D = body feed weight, t = running time, K_4 = permeability of body feed layer (George P, 2000). Equation 1 is the empirical equation of operating pressure and other parameter. The precoat layer has little effect on operating pressure. Thus, the operating pressure (total head loss) is proportional to Q^2 (square of filtration rate), i.e., the greater the filtration rate, the faster the operating pressure rise. Otherwise, the operating pressure rises more slowly.

4   CONCLUSIONS

Diatomaceous earth patterns having porous structure and the diatoms range in size from under 5 to more than 100 micrometers. Precoat filtration had no effect on removing COD_{Mn}. Precoat filtration has a good stability to remove solid suspended particle, effluent turbidity was remained about 0.05 NTU. The precoat layer has little effect on operating pressure. The operating pressure (total head loss) is proportional to square of filtration rate. Precoat filtration can be defined as a filtration technique for relatively low turbidity raw waters.

ACKNOWLEDGEMENTS

This work was financially supported by the Shanghai Natural Science Foundation (12ZR1410000), National science and technology major project of water pollution control and treatment (2012ZX07414-001). We also appreciate the assistance of Shanghai Safbon water service Co. Ltd. in the experimental setup operation.

REFERENCES

American Water Works Association (AWWA). Precoat Filtration Manual M30, 2nd edition. Denver, Colorado: American Water Works Association (1995).

Angela F. Danil de Namor, Abdelaziz El Gamouz, Sofia Frangie, et al. Turning the volume down on heavy metals using tuned diatomite. A review of diatomite and modified diatomite for the extraction of heavy metals from water diatomite and modified diatomite for the extraction of heavy metals from water, Journal of Hazardous Materials, Vol. 241 (2012), 14–31.

George P. Fulton. Diatomaceous Earth Filtration for Safe Drinking Water, Reston, Virginia: American Society of Civil Engineers (2000).

Wen-Tien Tsai, Chi-Wei Lai, Kuo-Jong Hsien. Characterization and adsorption properties of diatomaceous earth modified by hydrofluoric acid etching, Journal of Colloid and Interface Science, Vol. 297 (2006) 749–754.

Hydraulic Engineering III – Xie (Ed)
© *2015 Taylor & Francis Group, London, ISBN 978-1-138-02743-5*

Lake sediment records on climate change of the Qinghai Lake catchment in Southwest China based on wavelet analysis

Xian'e Long
Key Laboratory of Plateau Lake Ecology and Global Change, Kunming, China
College of Tourism and Geography Science, Yunnan Normal University, Kunming, China

Wenxiang Zhang, Hucai Zhang, Qingzhong Ming, Zhengtao Shi & Jie Niu
Key Laboratory of Plateau Lake Ecology and Global Change, Kunming, China

Guoliang Lei
Key Laboratory of Humid Subtropical Eco-Geographical Process, Ministry of Education, Fuzhou, China

ABSTRACT: Qinghai Lake is located in the Tengchong County, Yunnan Province. The climate is affected by the Asian southwest monsoon. Based on the wavelet analysis, the climate change of Qinghai Lake are studied by the environmental proxies, such as grain-size, Total Organic Carbon (TOC), and carbonate content ($CaCO_3$), magnetic susceptibility. Combine with the AMS14C datings, we use the wavelet analysis method to explore the climate periodic signal of Qinghai lake during the Holocene. The results show that the environmental proxies are exist ~1000a, ~500a, ~200a cycles, and the ~1000a, ~500a cycles are obvious during the whole Holocene. It is reflects the characteristics of Asian southwest monsoon during the Holocene, which indicates that the monsoon from strong to weak, the mechanism of Southwest monsoon is solar activity.

1 INTRODUCTION

To study the past climate, we usually use the tree rings, coral, ice cores, stalagmite, lakes, wetlands and marine sediment records to achieve the information of climate and environment change. Lake sediments is a good record, which have characteristics of continuity, high resolution and abundant information (Wang, 1999). Holocene climate was unstable and shift with cyclicities, which found it exist centennial or millennial scale cycles. The North Atlantic region has a cyclicity close to 1470 ± 500 years during the Holocene (Bond, 2001); The speleothem records from south Oman has 200a, 80–120a cycles (Burns, 2001). The lake sediments of central Europe exists Gleissberg cycle (88a), Vries cycle (210a), 500-year- and 1000-year-cycles (Kern, 2012). The peat deposits record of Tibetan Plateau exists series cycles of ~1.428 ka, ~0.512 ka, ~0.255 ka and ~0.217 ka (Xian, 2006).

Wavelet analysis is a method for time-frequency analysis, which can obtain weak information from climate records. Wavelet transform can translation and stretch the original climate record, then transform with multiple scales to get the cycles. Qinghai Lake is located at the western side of Gaoligong Mountain in Yunnan Plateau, which mainly influenced by the Asian southwest monsoon. In this paper, we investigating the Holocene climate variance by analysis the grain-size, Total Organic Carbon content (TOC), carbonate content ($CaCO_3$) and Magnetic susceptibility records, then use the wavelet analysis to get the Holocene climate cycles for further study the characteristics and mechanism of Asian southwest monsoon.

2 STUDY AREA AND METHODS

2.1 Study area

Qinghai Lake is an acid lake, it is located in the Tengchong County, Yunnan province (Fig. 1). It is situated at 1950 m.s.l and the catchment area is 0.4 km². The deepest depth of the lake water is 27 m. Qinghai Lake is supply by the deep phreatic water, where the hydrothermal activities is frequently (Wang, 1998). The climate in the catchment area is dominate by the Asian southwest monsoon, Annual mean temperature is 4.8 °C and the annual mean precipitation is 1463.8 mm, more than 84.3% falling between May and October (Wang, 2002). The vegetations are mainly compose of broadleaved forests and Pinus yunnanenisis forests, the vertical distribution characteristics is significantly.

In 2004, we were collected a 5.0 m core from the southeast of Qinghai Lake, the core was sampled at 1 cm intervals in the fields. Then put the samples into the plastic bags and transported to the laboratory.

2.2 Experimental analysis

Grain size was determined by using a Mastersizer-2000 laser diffraction particle size analyzer (Malvern Instruments, Britain), the samples was processed by 10 ml of 10% H_2O_2 and 10 ml of 10% HCL to remove organic matter and carbonate, then added sodium hexametaphosphate to disaggregate the sediments. We used the hydrated heat potassium dichromate oxidation-colorimetry method to measured Total Organic Carbon (TOC) content. Carbonate content ($CaCO_3$) was determined by the calcimeter method. Magnetic susceptibility was measured with a Bartington MS2 magnetic susceptibility meter. All these proxies was analyzed in the key laboratory of the plateau lake ecology and global change of Yunnan province.

2.3 Wavelet analysis

Wavelet transform is a tool which can transform the space (time) and frequency. It attempts to decompose signals in multiple-scales, and get extract informations from signal. The essential is to make the signal in low frequency part with high frequency resolution and low time resolution, the high frequency part with high temporal resolution and low frequency resolution (Guan, 2007; Zhang, 2007). In this study, we use the morlet wavelet function to discuss climate cycles. In order to get equally spaced series, we need to use the cubic spline function

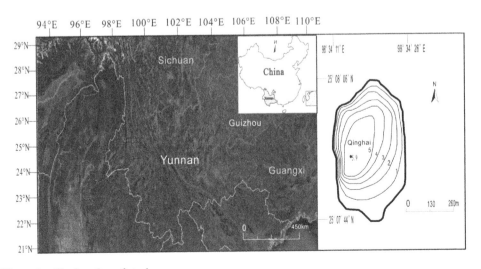

Figure 1. The location of study area.

192

to interpolate all the time series and proxies, the resolution is 30 years. Then transforming the data with continuous wavelet (Torrence, 1998; Martin, 2007).

3   RESULTS AND DISCUSSIONS

3.1   *Age model and chronology*

Thirteen AMS^{14}C dating samples were obtained from Qinghai sediment core, dates were converted to calendar years before present using IntCal013 (Table 1 and Fig. 2). Then used the linear interpolating method to get the age of each sample, the top age of sediments core was 0, we were used 471 samples to analysis the Holocene climate change of Qinghai Lake.

3.2   *The significance of climate proxies*

Lake sediments grain size is an important proxy in paleoclimate reconstruction, it can be used to indicate the changes of lake water level. When the grain size is large, it indicates that the lake hydrodynamic conditions is strong, lake water level is rising, the climate becomes humid; On the contrary, the climate becomes dry. The organic matter is divide into endogenous and terrigenous, it reflects the changes of lake productivity. Magnetic susceptibility is an indicator

Table 1.   Radiocarbon dates for Qinghai Lake sediments.

Laboratory number	Depth (cm)	AMS ^{14}C age (a BP)	Calibrated ^{14}C age (cal.BC/AD)	Age (a cal.BP)
QH-A98C	98	350 ± 40	1546.5 ± 91.5	404.25 ± 0.75
QH-E125C	125	965 ± 40	1079.5 ± 82.5	870.25 ± 0.25
QH-E177C	177	2505 ± 45	615 ± 181	2565
QH-A207C	207	3305 ± 45	1576.5 ± 114.5	3525.75 ± 0.75
QH-A223C	223	3830 ± 45	2303 ± 157	4254 ± 1
QH-E265C	265	4965 ± 45	3791.5 ± 140.5	5740.75 ± 0.75
QH-E295C	295	5535 ± 50	4377 ± 109	6326 ± 1
QH-F12C	312	6180 ± 45	5145 ± 4145	7095
QH-F35C	335	6815 ± 50	5712 ± 85	7661 ± 1
QH-F63C	363	8040 ± 65	6936.5 ± 235.5	8885.75 ± 0.75
QH-F98C	398	9195 ± 50	8421 ± 126	10370.5 ± 0.5
QH-H34C	450	10835 ± 55	10794 ± 68	12744.5 ± 0.5
QH-H55C	471	11210 ± 65	11117.5 ± 159.5	13066.25 ± 1.25

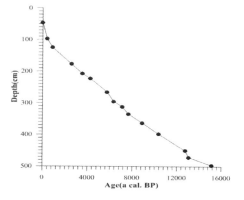

Figure 2.   OxCal constructed age model of Qinghai Lake sediments.

193

of the ferromagnetic material content in the sediments, it is relate to the lakes hydrodynamic and sediments source. The higher the magnetic susceptibility is, the weaker hydrodynamic becomes; On the contrary, the hydrodynamic becomes stronger (Wu, 2009). The carbonate of lake sediments is compose of authigenic and exogenous carbonates. Authigenic carbonate is mainly affect by chemical and biological effects, the carbonate content reflects climate change. And exogenous carbonate is affect by the surface runoffs (Shen, 2010). Qinghai lake is a volcanic lake, the carbonate content is influence by the water level and water Ph value.

3.3 Wavelets analysis results

In order to study the paleoclimate variance information, we use the wavelet analysis to analyze the climate proxies of Qinghai Lake (Fig. 3). Standardized all the dates before wavelet transform, then get the wavelet coefficients and variances. The wavelet spectrum show clearly periodicity for the grain size, TOC, magnetic susceptibility and $CaCO_3$ (Fig. 4). For the grain size, the ~1000a, ~500a cycles are obvious, the ~200a, ~100a cycle is only appearance during the late Holocene (0~5000 a cal.BP). According to the variance spectrum, we can see the peak is ~1000a cycle. For the TOC also show ~1000a, ~500a cycles, it also shows ~1000a cycle is the peak in the variance spectrum. For the magnetic susceptibility show obvious ~1000a, ~500a, ~200a cycles, and the peak is also ~1000a cycle. The $CaCO_3$ content show obvious ~1000a, ~500a, ~200a, ~100a cycles, which are obvious during all the Holocene, the peak is ~100a cycle in the variance spectrum.

The ~500a, ~200a cycle is obvious in the wavelet spectrum. The Qinghai-Tibet plateau record has 500a, 250a, 200a cycles (Hong, 2004; Duan, 2012); Dongge cave stalagmite records found the 208a, 88a cycle (Dykoski, 2006). The sediment records of Arabian sea found 470a, 320a, 220a cycle (Thamban, 2007); ~500a was relate to the instability of the north Atlantic thermohaline circulation, ~200a cycle was similar to the Suess cycle (210 a) of solar activity (Zhu, 2009). ~1000 a cycle was also related to solar activity, M. Debret found 1000a cycle in the North Atlantic climate records during the Holocene, and the cycle disappeared in the period of late Holocene (Debret, 2007). The TOC content from lake sediment in the northwest arid areas has found 1190a cycle (Jin, 2004). All these records indicate climate cyclicity was influenced by solar activity. Solar activity would increase the convective activity in the equatorial region, it would bring much precipitation and the cyclicity information was recorded in the sediments.

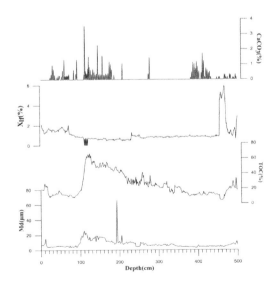

Figure 3.   Multi-proxies results from Qinghai Lake.

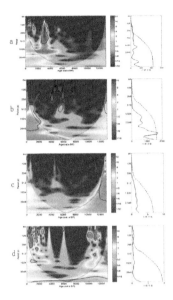

Figure 4.   The wavelet analysis results (a: grainsize, b: TOC, c: magnetic susceptibility, d: CaCO₃).

The Qinghai Lake climate cycles is mainly record the characters of Asian southwest monsoon variation. During the period of 12000–9000a cal.BP, it is only obvious ~500a, ~200a cycle. Asian southwest monsoon was became strong, it brings many precipitation. In the middle of Holocene (9000–4000a cal.BP), all the cycles signal are significance, especially the ~1000a cycle. Asian southwest monsoon was stronger, the grain size is large, the region climate was warm and wet. It experienced abundant rainfall and high effective moisture, and vegetation in the lake catchment was flourishing, which caused by the strongly Asian southwest monsoon. This coincided with the global Holocene Optimum. From 4000a cal.BP to present, ~1000a cycle is get together with ~500a cycle, it means the Asian southwest monsoon was weakened. The climate of the region is becoming cold and wet. In a conclusion, we can know the evolutionary phase of Asian southwest monsoon from the periodicity.

4   CONCLUSION

Based on the environmental proxies grain-size, TOC, CaCO₃ and magnetic susceptibility with wavelet analysis and AMS¹⁴C datings, we have identified the climate cycle of Qinghai Lake catchment during the Holocene. The results are follows:

Qinhai Lake sediment records are exist ~1000a, ~500a and ~200a cycles, And the ~1000a, ~500a cycles are obvious during the Holocene, it is reflects the characteristics of Asian southwest monsoon during the Holocene. Asian monsoon climate cyclicity was mainly influenced by solar activity. During 9–4 ka cal.BP, it experienced abundant rainfall and high effective moisture of the lake catchment, and vegetation in the lake catchment was flourishing, which caused by the strongly Asian southwest monsoon. This coincided with the global Holocene Optimum.

ACKNOWLEDGEMENTS

Special thanks to Dr. Guangjie Chen provided valuable comments and suggestions for the article, and Dr. Lunqing Yang, Kunwu Yang, Qinglei Li, and Jie Chen for the field work and their assistance during the laboratory work. This study was supported by the Science and

Technology Planning Project of Yunnan Province (Grant No. 2012FB143), the Research Fund for the Doctoral Program of Higher Education of China (Grant No. 20125303120004), and Geographical environment and region development on low-latitude plateau of superior and special group of Yunnan Province.

REFERENCES

Bond, G., Kromer, B., Beer, J., et al. 2001. Persistent solar influence on North Atlantic climate during the Holocene. *Science*, 294: 2130–2136.

Burns, S.J., Fleitmann, D., Mudelsee, M., et al. 2002. A 780-year annually resolved record of Indian Ocean monsoon precipitation from a speleothem from south Oman. *Journal of geophysical research*, 107(D20): 4434.

Debret, M., Bout-Roumazeilles, V., Grousset, F., et al. 2007. Origin of 1500-year cycles in Holocene North-Atlantic records. *Clim.past*, 3: 569–579.

Duan, K.Q., Yao, T.D., Wang, N.L., et al. 2012. The unstable holocene climatic change recorded in an ice core from the central Tibetan Plateau (in Chinese). *Sci Sin Terrae*, 42: 1441–1449.

Dykoski, C.A., Lawrence Edwardsa, R., Cheng, Hai., et al. 2005. A high-resolution, absolute-date Holocene and deglacial Asian monsoon record from Dongge Cave. *China Earth and planetary science letters*, 233: 71–86.

Guan Lvtai. 2007. *The wavelet analysis method and application*. Beijing: Higher education press.

Hong Bing, Lin Qinghua, Hong Yetang, et al. 2004. Evolution of southwest monsoon in the eastern of Qinghai-Tibet Plateau during Holocene. *Earth and environment*, 32(01):42–48.

Jin Linya, Chen Fahu, ZhuYan. 2014. Holocene climatic periodicities recorded from lake sediments in the arid-semiarid areas of Northwestern China. *Marine geology and Quaternary geology*, 24(02):10.

Kern, A.K., Harzhauser, M., Piller, W.E., et al. 2012. Strong evidence for the influence of solar cycles on a Late Miocene lake system revealed by biotic and abiotic proxies. *Palaeogeography, Palaeoclimatology, Palaeoecology*, 329–330: 124–136.

Martin, H., Trauth. 2007. *MATLAB recipies for earth sciences*. Springer.

Shen Ji, XueBin, Wu Jinglu, et al. 2010. *Lake sedimentary and environment evolution*. Beijing: Science Press.

Thamban, M., Kawahata, H., Rao, V.P. 2007. Indian summer monsoon variability during the Holocene as recoded in sediments of Arabian sea: timing and implications. *Journal of oceanography*, 63: 1009–1020.

Torrence, C., Compo, G.P. 1998. *A practical guide to wavelet analysis*. Atmospheric and oceanic sciences, 79(1): 61–78.

Wang Suming, Zhang Zhenke. 1999. New progress of study in environmental evolution and lacustrine sedimentation, China. *Chinese science bulletin*, 44(6): 579–587.

Wang Suming, Dou Hongshen. 1998. *The preface of memoirs of Chinese lakes*. Beijing: Science Press.

Wang Yunfei, Zhu Yuxin, Pan Hongxi, et al. 2002. Environmental characteristics of an acid Qinghai Lake in Tengchong, Yunnan Province. *Jurnal of lake sciences*, 14(02): 117–124.

Wu Jian, Shen Ji. 2009. Paleoenvironmental and paleoclimatic changes reflected by diffuse reflectance spectroscopy and magetic susceptibility from Xingkai lake sediments. *Marine geology and quaternary geology*, 29(3): 123–131.

Xian Feng, Zhou Weijian, Yu Huagui. 2006. The abrupt changes and periodicities of climate during Holocene. *Marine geology and quaternary geolog*, 26(05): 109–115.

Zhang Shanwen, Lei Yingjie, Feng Youqian, et al. 2007. *The application of Matlab in time series analysis*. Xi an: Xidian University.

Zhu,Y., Chen,Y., Zhao, Z.J, et al. 2009. Record of environmental change by α-cellulose $\delta^{13}C$ of sphagnum peat at Shennongjia, 4000–1000 a BP. *Chinese Sci Bull*, 54:3731–3738.

Hydraulic Engineering III – Xie (Ed)
© 2015 Taylor & Francis Group, London, ISBN 978-1-138-02743-5

The sedimentary records and climate environment evolution of Tengchong Qinghai lake since the 15.0 cal. ka BP

Guoqi Gan
Key Laboratory of Plateau Lake Ecology and Global Change, Kunming, China
College of Tourism and Geography Science, Yunnan Normal University, Kunming, China

Wenxiang Zhang, Hucai Zhang, Qingzhong Ming, Zhengtao Shi & Jie Niu
Key Laboratory of Plateau Lake Ecology and Global Change, Kunming, China

Guoliang Lei
Key Laboratory of Humid Subtropical Eco-Geographical Process, Ministry of Education, Fuzhou, China

ABSTRACT: According to analyze environmental proxies of Qinghai Lake in southwest China, e.g., the grain-size, low frequency magnetic susceptibility (x_{lf}), $CaCO_3$ and Total Organic Carbon (TOC), with the AMS ^{14}C datings and Principal Component Analysis (PCA), the evolution processes of climate environment was reconstructed since 15.0 cal. ka BP. The results show that during 12.7 to 9.6 cal. ka BP, the southwest monsoon was stronger. During 9.6 to 5.0 cal. ka BP, the lake water level was continued to rise, the southwest monsoon was further enhanced. During 5.0 to 0.26 cal. ka BP, the climate was arid gradually. The intensifying human activity and land-use in the lake catchment since 0.26 cal. ka BP were associated with the ancient culture within Qinghai's catchment.

1 INTRODUCTION

Lakes are widely distributed in the whole world, and it is the natural archives of environment change on the continent (Liu, 2012). At the same time, because lakes have the characteristics of rich information, good continuity, fast deposition rate, high resolution and wide geographical coverage, etc (Wang, 2011), recorded the process of lake basin ecological environment evolution and information to human activity. So it is a good indicator of research on global climate change (Jiang, 2012).

Qinghai lake is located in the southwest monsoon area, it is less influenced by the east Asian monsoon system (Chen, 2004). The water supply mainly from groundwater and precipitation, less influenced by the outside (Tong etc., 1989), so the Qinghai lake's sedimentary record has good continuity and integrity. It is one of the ideal areas to study southwest monsoon evolution. This paper according to analyze the Qinghai lake 500 cm sedimentary records of median grain-size, χ_{lf}, TOC and $CaCO_3$ content, to explore the environment evolution mechanism of Qinghai lake since 15.0 cal. ka BP. To provide important basis information to predict Qinghai lake and even the entire southwest monsoon climate and environment change in the future.

2 STUDY AREA

Qinghai lake is located in the southwest Yunnan Province, China. It border on Eurasian plate and Indian plate (Yun, 2013). Earthquake activity is frequent in this area, geothermal resources is rich, magmatic activity is stronger, and is a semi-enclosed lake, less influenced by external conditions, it is also one of the only three acid lakes in China (Wang etc., 2002),and it is the lowest salinity lake discovered in China so far (Wang etc., 1998). Qinghai Lake is

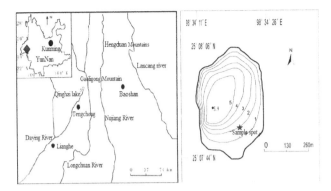

Figure 1.   Qinghai lake location and sampling points.

influenced by southwest monsoon all the year round for many years, the average temperature is 14.7 °C, and annual average rainfall is 1425 mm. Among the rainy season form May to October precipitation accounts for 84.3% of total rainfall (Wen, 2007). Regional vegetation type is given priority to with semi-humid evergreen broad-leaved forest (Editors Committee of Yunnan Vegetation, 1987).

3   SAMPLING AND EXPERIMENTAL

3.1   *Sampling*

In June 2007, using the floating platform and the piston sampling equipment, to picked at continuous columnar lacustrine sediment core 500 cm in Qinghai of Yunnan province, lake water depth of 5.5 meters (98°34′19.7′E, 25°7′56.4′N, Fig. 1). In order to ensure the core has high resolution and good continuity, and we had drilled two parallel cores to ensure its accuracy. we have be cutting the core 1 cm intervals in wild, sealed saved back to the lab for testing of each environmental indicators.

3.2   *Experimental*

The experimental analysis of sedimentary proxies was carried out at the Key Laboratory of Plateau Surface Process and Environment Changes of Yunnan Province. Grain size was measured with a Mastersizer-2000 laser diffraction particle size analyzer (Malvern Instruments Ltd., UK) after treatment with H_2O_2 and HCl to remove organic matter and carbonate. From this analysis, the relative standard deviation of parallel analyses for individual samples obtained was <1%. Magnetic susceptibility was measured with a Bartington MS2 magnetic susceptibility meter and mass-specific magnetic susceptibility (χ_{lf}) was also calculated. Total Organic Carbon (TOC) content was measured by titration with potassium dichromate. The analysis precision was better than ±0.8%. The $CaCO_3$ content of the samples was measured using the calcimeter method of Bascomb. This involved measuring the amount of CO_2 produced after adding the HCl, and stoichiometrically calculated this into $CaCO_3$ content, and error of <1% was achieved during this analysis. In the interest of obtaining accurate data, all of above proxies were measured three times, under the same conditions, treatment and analysis methods.

4   RESULTS AND DISCUSSIONS

4.1   *Age model and chronology*

Fourteen plant macrofossil samples were collected from the core, and Accelerated Mass Spectrometry (AMS) ^{14}C dates were measured at the AMS Laboratory of Beijing University,

China. The conventional ages were converted to calibrate with Intcal13 calibration data. Based on linear-fitting analysis, the age of the profile is in 15.0 cal. ka BP (Fig. 2).

4.2 The characteristics of environmental proxies

500–450 cm: Md value was stay below average, $CaCO_3$ content is low, and the content of TOC was continue decline to the historically low in about 460 cm, but χ_{lf} value was suddenly rise, its record high in 460 cm. 450–380 cm: Md and χ_{lf} was still keep steady, the content of TOC was greater volatility, $CaCO_3$ content was suddenly increased. 380–235 cm: Md and χ_{lf} value was keep steady, $CaCO_3$ content was almost zero, the TOC content was increased volatility. 235–80 cm: Md, TOC and $CaCO_3$ content was drastically growth, and in 110 cm, Md, TOC and $CaCO_3$ content was increased maximum value, but χ_{lf} value was decline to minimum. 90–0 cm, Md value was decline, χ_{lf}, TOC and $CaCO_3$ content was fluctuation rise within a narrow range (Fig. 3).

4.3 Environmental significance of the proxies

Lake sediment grain-size parameters was directly reflected the sedimentary hydrodynamic conditions, thus speculated that the changes of sedimentary environment, sediment grain-size and there is close relationship between climate (Yang etc., 1999). Magnetic susceptibility is more like a reflection of lake basin land environment information. That are

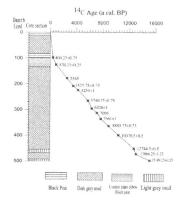

Figure 2. The chronology mode of core.

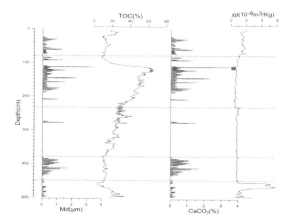

Figure 3. Characteristics change analysis results.

specific to analyze water loss and soil erosion, and the influence of human activity (Hu etc, 2001). When the temperature and precipitation was increased, a large number of creatures grown on land, a large number of biology had brought in lakes, so terrigenous organic matter deposited a huge increase in lakes. Besides, in condition of the humid and rainy climate, the runoff into the lake is big, bring abundant terrestrial plants and nutrient, also make the aquatic plankton prosper and lead the organic matter productivity. On the other hand, climate and rainfall, TOC content in the lake is reduced. Lake self-generating carbonate is an important part of the lake sediments, its formation by the regional geological background, climate environment, the lake water chemistry properties and biological activity of many factors such as control, it can well reflect the depositional environment (Chen etc., 2002).

4.4 PCA analysis

The article is an application of environmental proxies $CaCO_3$ content, TOC content, χ_{lf} value and median grain-size, and Tengchong Qinghai lake environmental changes impact factors of PCA, to explore the $CaCO_3$ content, TOC content, median grain-size and χ_{lf} was respond to the changes of precipitation and temperature. Further explore the influence of precipitation and temperature to the environmental change in Tengchong Qinghai lake.

The article combined with Tengchong Qinghai lake basin natural environment condition and development history to comprehensively analyze. Achieved a conclusion that RC1 representative environmental impact factor is precipitation and RC2 representative environment impact factor is temperature (Fig. 4).

4.5 Asia Southwest monsoon evolution stages and human activities since 15.0 cal. ka BP

From Lake sedimentary records and compared with history literature, in this region. The southwest monsoon evolution and human activities history can be divided into five stages (Fig. 5), as follows:

Stage 1: 15.1–12.7 cal. ka BP

Lake sediments of media grain-size was below of average value, the content of $CaCO_3$ was high, χ_{lf} was bigger also. That shows in this period, the precipitation of Qinghai lake basin was declined, the climate was relatively dry and low lake water level. The content of TOC was decline trend, means that the creatures were underdeveloped on land, the organic was decline what was bring into the lake by runoff. The Asia southwest monsoon activity was weaker, the small dry-cold and wet. In general, the period of Qinghai lake was in low water level and rainfall was decreased, the climate was dry.

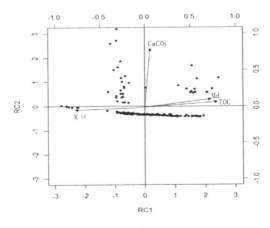

Figure 4.   PCA analyses of the environmental proxies.

200

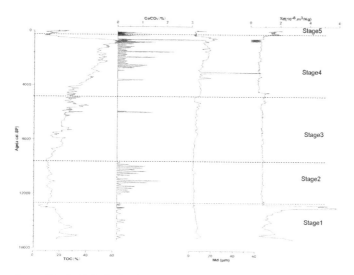

Figure 5.   Tengchong Qinghai lake Md, TOC, CaCO$_3$ and χ_{lf} analysis results.

Stage 2: 12.7–9.6 cal. ka BP

Media grain-size of lake sediments was below of average size, the content of TOC and χ_{lf} content have no obvious change.but in a low. Indicating in this period, Qinghai lake basin's land creatures had no without too big change. The content of CaCO$_3$ was growing in this period, it shows that Qinghai lake basin was warm and wet, the precipitation was increased, lead to hydrodynamic force was stronger, lake level was expanded. Therefore,the content of CaCO$_3$ was higher, most of them were above 0.45%. It shows that the Asia southwest monsoon activity was stronger and climate had changed from warm and warm to wet and warm.

Stage 3: 9.6–5.0 cal. ka BP

Median grain-size of sediment was low in the average value and has a tendency of rise. Indicating in this period, rainfall was high, hydrodynamic conditions has increased, the surrounding hillside particles larger debris washed into the lake and lead to grain-size of lake sediment become larger. The content of TOC was between 9.99% ~ 41.37%, and has a tendency of increasing. It shows that the lake water level has rise. The content of CaCO$_3$ and χ_{lf} content were low, even has kept steady, due to precipitation was increased, lead to CaCO$_3$ of flow into the lake and ferromagnetic minerals can be fully decomposition, resulted in CaCO$_3$ and χ_{lf} content was low. It shows that the Asia southwest monsoon activities are stronger.

Stage 4: 5.0–0.26 cal. ka BP

Median grain-size of sediment was grow above the average, the content of TOC and CaCO$_3$ was dramatically increased, x_{lf} value was fall at the same time, but overall was stable. These indicators reflect that the climate was arid gradually, lake water level was fall, lake shrinking and the Asia southwest monsoon activity was weaker. The climate had changed from cold and wet to warm and dry. And in 1.2 cal. ka BP came around median grain-size of sediment was suddenly increased, more than 21.26 μm, the content of TOC and CaCO$_3$ was largely increasing, TOC content over 60.64%,and χ_{lf} value had appeared suddenly decline. It suggests that the Qinghai lake's drying events at this time.

Stage 5: 0.26–0 cal. ka BP

Media grain-size of lake sediments was low average size. The content of TOC was small amplitude fluctuation declined. The content of CaCO$_3$ had increased. X$_{lf}$ value had rise, it's that human activities was aggravate fluence on Qinghai lake basin environment change. The proxies of lake sediment fluctuated intensively because of the frequent human agricultural activities and cutting down of forest vegetation.

5 CONCLUSION

1. The environmental proxies of Qinghai lake show that the median grain-size, χ_{lf} and TOC content change was mainly affected by precipitation.
2. During 12.7 to 9.6 cal. ka BP, the southwest monsoon was stronger. During 9.6 to 5.0 cal. ka BP, the lake water level was continued to rise, the southwest monsoon was further enhanced. During 5.0 to 0.26 cal. ka BP, the climate was arid gradually. The intensifying human activity and land-use in the lake catchment since 0.26 cal. ka BP were associated with the ancient culture within Qinghai's catchment.

ACKNOWLEDGMENTS

This study gets supports from Science and Technology Plan Projects of Yunnan Province (2012FB143), the Ministry of Education of Doctoral Fund (20125303120004) and Geographical environment and region development on low-latitude plateau of superior and special subject group of Yunnan Province. We thank L.Q. Yang, K.W. Yang and J. Chen for their assistance during the laboratory work.

REFERENCES

Chen, J.A., Wan, G.J., Wang, F.S., et al., 2002. Research of the Carbon Environment Records in the Lake Modern Sediments. Science in China (Series D), 32: 73–80.

Ditors Committee of Yunnan Vegetation. 1987. Vegetation in Yunnan [M]. Science Press in Beijing.

Jiang Shan. 2012. Over the past 3000 years of the typical north and south poles in the ecological environment change of sedimentary records and contrast [D]. University of Science and Technology of China.

Longxun Chen et al. 2004. East Asia and tropical Indian summer monsoon circulation convection disturbance of zonal propagation characteristics [J]. Science in China. Edit(D): Earth Science, 11:171–179.

Shan Yun. 2013. Yunnan tengchong volcanic geology park [J]. North China land and Resources, 11:42.

Shouyun Hu, Chenglong Deng, 2001. Appel E etc. Lake sediment magnetic properties of environmental significance [J]. Science Newspaper, 46–48 (17): 1491–1494.

Sunmin Wang, Hong Shen Dou. 1998. Lakes in China. Science Press in Beijing.

Wei Tong, Ming Tao Zhang. 1989. Tengchong geothermal. Science Press in Beijing.

Xiaomin Liu. 2012. The zoig plateau peat mirobial GDGTs distribution and climate environment [D]. Northwest University.

Xiaolei Wang. 2011. Yunnan plateau lake modern sedimentary environment change research [D]. Nanjing Normal University.

Xiaoqiang Yang, Huamei Li. 1999. Mud river basin sediment magnetic susceptibility and grain size parameters on the response of the sedimentary environment [J]. Journal of Sedimentology. (S1): 763–767.

Yunfei Wang, Yuxin Zhu et al. 2002. Yunnan Tengchong in Qinghai—acid lake environment [J]. Journal of Lake Science, 11: 117–124.

Yanxing Wen. 2007. Tengchong beihai wetland sedimentary records and south Asian summer monsoon evolution since 150 ka. [D]. China University of Geosciences (Beijing).

Hydraulic Engineering III – Xie (Ed)
© 2015 Taylor & Francis Group, London, ISBN 978-1-138-02743-5

Experimental analysis of low air pressure influences on the fire detection

S.L. Yang, S. Lu, J. Wang & C.C. Liu
State Key Laboratory of Fire Science, University of Science and Technology of China, Hefei, China

K.K. Yuen
Department of Civil and Architectural Engineering, City University of Hong Kong, Hong Kong, China

ABSTRACT: Most of cargo fire detectors are based on light scattering of soot particles. The air pressure in many aircraft class C cargo compartment is about 80 kPa which may affect the performance of smoke detectors in case of fire. The main purpose of the work is to investigate the influence of air pressure on the parameters of combustion (temperature, CO, smoke extinction value) generated by N-heptane pool fires experimentally. The overall test has been conducted in an altitude chamber. The result has shown that low air pressure obviously decreases the mass loss rate and increases the concentration of CO and the temperature above the fire. It is noteworthy to mention that the extinction coefficient of smoke is less under lower air pressure than that under higher pressure. Multi-criteria fire detectors may have a better performance in the low air pressure environment.

1 INTRODUCTION

Temperature, concentration of carbon monoxide (CO) and smoke extinction value are very important information implying fire occurring. Most of the fire detectors in the aircraft cargo compartment are based on smoke particle optical properties such as light scattering or light extinction. As for aircraft cargo compartment, requirement of the fire detectors, which must alarm within one minute after the fire occurring, is very rigid. So the sensitivity of cargo fire detectors is very sensitive to smoke particles. However, the environment in the cargo is complex where the dust, vapour, feathers of animals etc. could disturb the performance of fire detectors. According to FAA report (Blake, B. 2000), the ratio of false cargo compartment alarms to actual fire or smoke events was increasing and was at 200 to 1 from 1995 to 1999.

According to previous work, fire Behaviours under low air pressure are different from that under the ordinary condition (Blake, B. 2000; Zhou, Z.H. et al. 2014; Yin, J. et al. 2013; Yin, J.S. et al. 2013). It may be a challenge for the fire detector to work well under the low air pressure. However, research about the parameters of detection is not enough and little work has been done to study the properties of smoke under the low air pressure.

The purpose of this work is to find out the differences of the detection parameters (temperature, CO, smoke extinction value) between under the low air pressure and under the ordinal condition.

2 EXPERIMENTAL METHODS

In order to simulate the environment in the cargo compartment in flight, experiments have been conducted in a low air pressure chamber (3 m*2 m*2 m) which has a powerful pressure controlling system to simulate both static and dynamic low pressure environments. In our study, N-heptane pool fires were tests under static pressures including 101 Kpa.A (equivalent altitude: 0 m) 90.9 Kpa.A (905 m), 80.8 Kpa.A (1869 m), 70.7 Kpa.A (2935 m), 60,6 Kpa.A (4157 m), 50.5 Kpa.A (5535 m). The fire pan is 10 cm in diameter, filled with 50 ml 99% pure liquid n-heptane.

High precision electronic balance with precision of 0.01 g was in charge of the mass loss rate of fuel. Sony HDR-CX360 took charge of record the experimental process and the flume of fire. Smoke meter AML was employed as the extinction measuring instrument to measure the smoke's extinction value (%/m). AML was mounted on the top ceiling above the fire flume. Delta-65 and K-type thermocouple have took records of concentrations of CO and temperatures near the ceiling above the fire respectively.

During the overall tests, the environment temperature was between 20°C–27°C, and the relative humidity was between 40–70%RH. The fire was lighted by a hot electrically heated wire above the liquid fuel level but not touching the liquid fuel.

3 RESULTS AND DISCUSSION

3.1 *Mass loss rate*

According to Figure 1, the mass loss rate is increasing to the first maximum at beginning of fire because of the high concentration of O_2 and then slowing down while the oxygen is consumed. The temperature inside the chamber is rising up owing to heat energy generated by combustion. Volatilization rate of fuel is growing up with the temperature. Therefore, the mass loss rate is growing up again faster and faster during the next long period of time till the fuel runs out.

Overall the whole process of combustion, mass loss rates are less under lower air pressures than those under higher air pressures. Especially, the first maximum is 0.064 g/s under 101 Kpa pressure which is 1.68 times larger than that under 50 Kpa pressure (0.038 g/s) at beginning of fire (within 60 s of fire occurring). For the early detection of fire, it may be more difficult for the detector to detect fire under lower pressure whose burning rate is much lower.

3.2 *Concentration of carbon monoxide*

The concentration of CO is important information to analyze the risk of fire for it's harmful for people's health. Figure 2 shows that fires have generated nearly the same amount of CO whatever the air pressure is during a long period of time after fire occurring. It is noteworthy

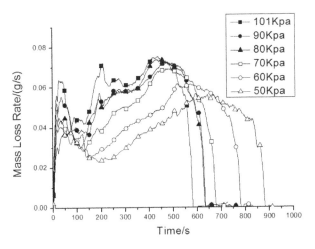

Figure 1.   Mass loss rate with time under different air pressures.

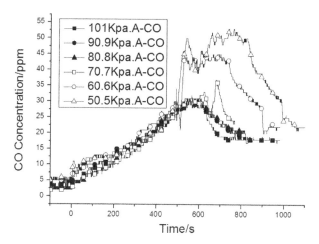

Figure 2. Concentration of carbon monoxide.

that volume fraction of CO is increasing sharply under lower pressures (especially under 50.5 Kpa.A and 60 Kpa.A) when the fuel is going to run out. It has turned out that CO sensors, which will work regularly at the beginning of fire and perform excellently during the later stage of combustion, could be very useful to detect the fire under low air pressure conditions.

3.3 *Temperature*

It's very interesting that temperatures near the ceiling of the altitude chamber under lower air pressures are not lower and even higher than those under higher air pressures as shown in Figure 3, even though their burning rates are much less than those under higher pressures obviously. The reason for it is that the average height of flame increases under lower pressure as validated by the averaged visible flame heights obtained from the flame images in Hefei and Lhasa (Zhou, Z.H. et al. 2014). The distance between the ceiling and the top of flame is shorter and the heat plume is easy to reach the ceiling under low pressures. Most of temperature fire sensors which are mounted near the ceiling would work well even better under lower pressures than under higher pressures.

3.4 *Intensity of light scattering*

The voltage of the light sensor is nearly proportional to light scattering intensity. Voltages of the scattering light sensors, which could detect the forward scattering light intensity scattered by smoke particles, are illustrated in Figure 4 along with different air pressures. It has revealed that the light intensity of forward scattering by soot is obviously related to the air pressure when the amount of fuel is the same. Most of present smoke detectors confirm fire occurring based on the value of the photoelectric sensor which can represent the light scattering intensity and the smoke particles' optical density. The voltage, subtracted the initial value, under 80 Kpa.A (the pressure in class C cargo compartment) is about half of that under 101 Kpa.A. Photoelectric smoke detector may not work very well under the lower pressure.

3.5 *Extinction coefficients of smoke*

In order to study the smoke optical properties further, extinction coefficient m (dB/m) obtained by smoke meter AML has been shown in Figure 5. The value of m under 80 Kpa.A is about one half of that under 100 Kpa.A. From the movie of the n-heptane fire process recorded by the Sony camera, we can easily see the frame of film (not illustrated in this paper

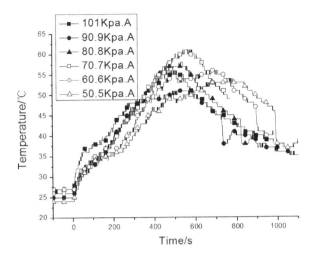

Figure 3. Temperature near the ceiling above the fire.

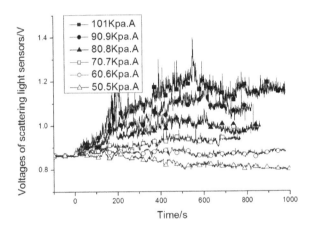

Figure 4. Voltages of forward scattering light sensors.

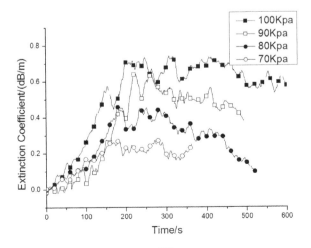

Figure 5. Extinction coefficients of smoke under different pressures.

because of space limitation) at the time when fire is going to burn out is more clarity under lower air pressure than that under higher air pressure qualitatively. It could be concluded that lower air pressure could cause n-heptane fire to generate a smaller number of particles when the amount of fuel is the same. It implies that it may be more difficult for the optical smoke detector to detect fire occurring under lower air pressure timely.

4 CONCLUSIONS

A series of experiments has been conducted in an altitude chamber to record burning rates/ mass loss rates, concentrates of CO, temperatures, light intensities of forward scattering and smoke extinction values using oil pan of 10 cm in diameter, filled with 50 ml 99% pure liquid n-heptane. The major conclusions are as follows:

The low air pressure obviously decreases the mass loss rate, but increases the concentration of CO and the temperature above the flame.

Extinction coefficients of smoke are much less under lower pressures and light intensities of forward scattering of soot are less than those under higher pressures though amounts of fuel are the same.

As for most of smoke detectors (using in cargo compartments or other low air pressure regions), it may be difficult for them to work well as under ordinary air pressure. Multi-criteria fire detectors using temperature, CO sensors may have a better performance in low air pressure environments to enhance the safety.

REFERENCES

Blake, B. 2000. Aircraft Cargo Compartment Smoke Detector Alarm Incidents on U.S.-Registered Aircraft. *FAA, DOT/FAA/AR-TN*00/29, 1974–1999.
Niu, Y. et al. 2013. Experimental study of burning rates of cardboard box fires near sea level and at high altitude. *Proceedings of the Combustion Institute* 34(2):2565–2573.
Yin, J. et al. 2013. Experimental study of n-Heptane pool fire behavior in an altitude chamber. *International Journal of Heat and Mass Transfer* 62(0):543–552.
Yin, J.S. et al. 2013. Experimental Study of N-Heptane Pool Fire Behaviors under Dynamic Pressures in an Altitude Chamber. *Procedia Engineering* 52(0):548–556.
Zhou, Z.H. et al. 2014. Experimental analysis of low air pressure influences on fire plumes. *International Journal of Heat and Mass Transfer* 70(0):578–585.

Hydraulic Engineering III – Xie (Ed)
© 2015 Taylor & Francis Group, London, ISBN 978-1-138-02743-5

The summary and prospect of environment risk research on Yongjiang River basin

Jiaoyan Ai, Zongming Wei, Zongshu Wu & Yajuan Cai
Electrical Engineering College of Guangxi University, Nanning, Guangxi, China

Chaobing Deng & Bei Chen
Guangxi Autonomous Region Environmental Monitoring Central Station, Nanning, Guangxi, China

ABSTRACT: Yongjiang River is the main water resource of live and industry of Nanning city, and the environmental quality of its basin relates to the success of ASEAN Expo, the economic development and social stability of Nanning city. In this article, we summarize the research methods and contents about the environmental risk of Yongjiang River basin and then we put forward some opinions for the environmental quality control.

1 INTRODUCTION

Nanning city locates in southern Guangxi, it is an economic, politics, culture, education, financial and trade centre of Guangxi province in China. As the core city of The Beibu Gulf Economic Development Area and trade centre in southeast of Asia, its environmental quality is a crucial factor for regional development and ASEAN-China Exposition. Yongjiang River is the main water resource of live and industry of Nanning city with total length of 133.8 km, basin area 6120 km² which includes all tributaries and regional lakes, reservoirs. Figure 1 shows Yongjiang River and its main tributaries and Figure 2 shows its basin. In recent years, regional development cause more environmental problem, such as soil desertification, disappearance of forest vegetation, reduction of aquatic organisms, basin residents' health and ecological stability under serious threat. Therefore, the study of environmental risk of Yong River basin has become particularly important, especially on human health risk and ecological environment risks.

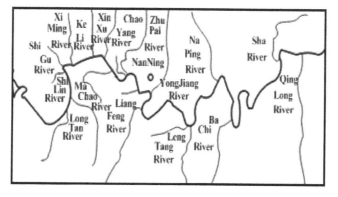

Figure 1. Yongjiang River and its tributaries.

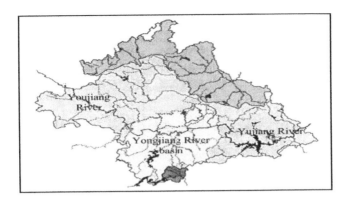

Figure 2.   Yongjiang River basin.

2   RESEARCH STATUS OF YONGJIANG RIVER BASIN ENVIRONMENTAL RISK

Current Yongjiang River Basin Environmental Risk Research focused on three aspects: the detection of pollutants, water pollution accident emergency and quality management.

2.1   *The status quo of contaminants detection research*

In recent years, with the development of Nanning agriculture, the use of pesticides and fertilizers were significant increased that lead to heavy soil contaminant in agricultural land. It will become biological potential health hazards. Through stormwater runoff pesticide residues will make more environmental problems. Paper mills, sugar mills, chemical plants, electroplating factory increased the amount of industrial waste water year by year. In contrast, Yongjiang River release capability doesn't increase, the result is that its water and tributaries pollutant content increased and the probability of the risk of accidents environment increases.

Some researchers have study the environmental risk of Yongjiang River basin: Tang Yan-hong (2006) analyzed pollutants data and the monitored data in recent years of Yongjiang River, the results showed that the main pollutants of Yongjiang River are organic pollutants. And then made several pollution prevention, protection and improvement views for water quality; Su Xiao-Chuan (2008) used liquid-liquid extraction-gas chromatography-mass spectrometry to detected Semi-Volatile Organic Compounds of Yongjiang River water (SVOC), the results showed that Yongjiang River water has been pollute by SVOC, the main pollutants comes from the paper mills, pesticide plants, sugar mills; Wu Lie-Shan (2009) used solid phase extraction and GC-ECD to analyzed organochlorine pesticides content on Yongjiang River, the results showed that organochlorine pesticides in Yongjiang River did not exceed the relevant standards; Deng Chao-Bing (2010) used Purge and Trap-GC-MS section to detected the content of Volatile Organic Compounds (VOCs) in Yongjiang River, the results show the content of VOCs less than other rivers in China; Liang Liu-Lin (2010) used an improved solid phase extraction, accelerated solvent extraction and gas chromatography-electron capture to detected the content of organochlorine pesticides (OCPs) in the Yongjiang River sediment, the results show that the pollutions in urban water and sediment in the river is OCPs, DDE and DDD. The content of Pesticide residues of surface water are not exceeded quality standard limits; Tian Yan (2012) used solid phase extraction-the content gas chromatography/pulsed flame to analyzed water organophosphate pesticides photometric in the wet season and dry season Yongjiang River; Bai Bao-Qiang (2012) assessed quality of water eutrophication.

The result shows that higher organic content on lakes and reservoirs. Local residents living waste is mainly reason for water eutrophication; Liang Liu-Lin (2012) used an improved solid phase extraction, accelerated solvent extraction and high performance liquid chromatography to analyzed the residue content of polycyclic aromatic hydrocarbons in Yongjiang River water and sediment. The results showed that the main polycyclic aromatic hydrocarbons in Yongjiang River is fluoranthene-based, the content of pollutant limit on the environmental quality standards, inland polycyclic aromatic hydrocarbons has a impact to water quality of Yongjiang River; Li Li-He (2012) used Nemerow pollution index and trophic state index method analyzed the change of water quality about Nanhu lake, the results showed TN, TP content in Nanhu lake is excess, some water are in moderately polluted and some are heavily polluted; Tian Yan (2013) used accelerated solvent extraction-gas chromatography-mass spectrometry to analyzed the content of Semi-Volatile Organic Compounds (SVOCs) in Yongjiang River sediments, the result showed that there are 12 kinds of Semi-Volatile Organic Compounds (SVOCs) in Yongjiang River sediments; WU Ting-Ting (2013) used ultrasonic extraction, solid phase extraction and HPLC-MS/MS to analyzed the content of antibiotic contaminated in Yongjiang River sediments, the results showed that there are macrolides and sulfonamides antibiotics in sediments, antibiotic resistance gene pollution can't be ignored.

2.2 *Research status of Yongjiang River watershed pollution accident emergency and water management*

Some researchers have study for Yongjiang River basin environmental risk evaluation: Huang Hui-Jing (2004) analyzed the sewage into Youjiang River and Yongjiang River and the causes and characteristics of 14 cases water pollution accidents, proposed water pollution accident prevention and treatment measures; Lv Jun (2006) used WebGIS, SQL databases, Visual Basic6.0 and other computer technology to design a Yujiang River water quality forecast system. The system can used in Yujiang River water quality and water pollution accident emergency warning; Tao Zeng-cai (2006) based on the status of water resources pollution of Yongjiang river and Visual Basic 6.0 to build a water quality prediction system. This system can provide some basis suggests for Yongjiang River water conservation and management; Tao Zeng-Cai (2006) used Geographic Information Systems (GIS), Visual Basic6.0 to develop a Yongjiang River water information management system, this system efficient management those information of Yongjiang water; Huang Zhen-Xian (2007) used two-dimensional water quality model and capacity analysis in the formula and water environmental capacity of the added value to analyzed the function of Nanning Yongjiang River water sewage treatment plant. The results showed that the Yongjiang River water quality will be improved if sewage treatment plant into operation[16]; Peng Bin (2007) analyzed those into the river pollution sources which and based on the corresponding water protection project implementation impact on pollution emissions predict the future trend of pollutants of Nanning mainly drinking water sources; Yu Ge (2010) used network technology, database technology, geographic information software established an environmental pollution emergency integrated platform, it main function is the pollution accident case analysis, pollutant monitoring, pollute emergency early warning and incident response disposal; Huang Tie-Ming (2011) analyzed two drinking water resource points monitoring data of Yongjiang River derived natural water features and the water pollutants character of Yongjiang River; Guangxi University, Environmental Monitoring Centre, etc. (2012) base on 863 items of "significant environmental pollution of sudden emergency response technology development and demonstration in the Beibu Gulf" for a field exercise. At this exercise by assumed a chemical burst environmental pollution accidents to drill the background, the drill 863 research projects has been used for sudden environmental pollution accident emergency response, emergency monitoring, emergency treatment and disposal, emergency resource coordination configuration, emergency command in the exercise.

3 INADEQUATE AND PROSPECT

3.1 *Inadequate*

To sum up, we conduct the conclusion.

1. In contents, there are few research on non-sudden environmental risks and potential watershed secondary pollution. Research on environmental risk factors mainly are single factor, few studies on the effects of interactions of multiple factors. Current work mainly study chemical environmental risk in basin water body, but few research on soil environmental risks, air environmental risks and ecological environmental risks.
2. There is no work on basin risk assessment and regionalization because of the lack of effective risk system model for basin region that make it is impossible to undertake reasonable and feasible environmental quality control and management.
3. Practical applications are not enough to verify reasonableness, practicality and reliability on real nature. The prediction of water quality and pollutant simulation lack real-time online function.

3.2 *Prospect*

According to these inadequate of the research status, we can propose some views for Yongjiang River basin environmental risk research.

1. We should establish a relationship model of pollutants, human health and ecological environmental stability. Based on the analysis results we should find out some appropriate measures to reduce the basin environmental risks.
2. The research trend is to combine human health risks and ecological environmental risks to establish comprehensive assessment for basin risk, including soil, air, and water quality, human and ecological health.
3. Due to the complexity, uncertainties and development of basin environmental risk system, we think the best way is to build a new risk system model by combining macro and micro method, spatiotemporal dynamic and intelligent method, so that we can undertake reasonable comprehensive analysis as the basis of regional environmental quality control.

4 CONCLUSION

The risk research on Yongjiang River basin is important for regional development of Nanning city and The Beibu Gulf Economic Development Area. We summarize the current research methods and contents and then we point out the deficiency and put forward some opinions to push up study breadth and depth for more scientific and reasonable environmental risk assessment and quality control.

ACKNOWLEDGEMENTS

The work is supported both by Guangxi Province Natural Science Foundation Key Project No. 2013 jjEA20003 and Guangxi Province Distinguish Experts Research Fund in Sudden Pollution Accident Emergency Technologies.

REFERENCES

Bai Bao-Qiang & Lv Bao-Yu. 2012. Eutrophication Assessment of River-type Lakes and Reservoirs in Guangxi. Journal of Guangxi Academy of Science, 04:330–332.

Cao Chang-Chun, Liu You-Liang & Zou Jin-Xing. 2012. Guangxi heavy metal pollution and water environmental behavior analysis. Collected papers of Industry Branch of China Civil Engineering Society water drainage committee. National Drainage Committee 2012 Annual Conference Proceedings. Collected papers of China Civil Engineering Institute of water Industry Branch drainage Committee:5.

Deng Chao-Bing, Tian Yan, Li Xin-Ping, Jiang Jian-Hong, Liao Ping-De & Liangliu-Ling. 2010. Volatile Organic Compounds in Yongjiang River and Inland Surface Waters of Nanning. Journal of Environmental Science & Technology, 01:119–123.

Deng Chao-Bing, Liang Liu-Ling, Liao Ping-De, Wu Lie-Shan, Tian Yan & Xu Gui-Ping. 2010. Distribution Characteristics of Organochlorine Pesticides in water and surface sediments from the rivers of Nanning city. Journal of Research of Environmental Sciences, 12:1506–1510.

Huang Zhen-Xian & Hu Qing-Feng. 2007. Improvement of water quality in Yongjiang River by construction of wastewater treatment plant in Nanning City. Journal of Water resources protection, 05: 38–40 +54.

Huang Hui-Jin. 2004. Analyzed of Zuojiang, Youjiang and Yongjiang water pollution accidents and countermeasures. Journal of Resources protection, 04:48–51.

Huang Tie-Ming, Weng Wei-Man & Bei De-Guang. 2011. Features and variation tendency of source water quality in Yongjiang River. Journal of water technology, 02:32–35 +41.

Liang Liu-Ling, Deng Chao-Bing, Liao Ping-De & Yang An-Ping. 2012. Distribution Characteristics of PAHs in Water and Surface Sediments from the Rivers of Nanning City. Journal of Environmental Monitoring in China, 05:67–70.

Li Li-He, Zhou Xiao-Ning, Wei Li-Qun, Huang Yu-Qin & Liao Hai-Ying. 2012. Eutroprophication evaluation of Water Quality in Nanning Nanhu Lake. Journal of Guangxi Academy of Sciences, 03:219–223.

Lv Jun, Peng Bin & Tang Qi-Shan. 2006. Research on the construction mode of water quality forecast system of Yujiang River. Journal of Water Resources Protection, 05:81–83.

Peng Bin, Lv Jun & Huang Zhen-Xian. 2007. The main source of pollution which into the river impact for drinking water sources analysis. Journal of Peoples Yangtze, 04:125–126.

Su Xiao-chuan, Tang Zhen-zhu, Qin Zhi-ying & Zhang Rui. 2008. Recent situation of semi-volatile organic compounds in fountainhead water of Yongjiang River passes through Nanning.Chinese Journal of Health Laboratory Technology, 10:1986–1991.

Tang Yan-hong, Tong Zhang-fa, Tang Yan-kui & Wei yuan-an. 2006. Current situation and protection measures of water quality in the NanNing section of YongJiang river. Geological Journal of Hazards and Environment, 02:33–36.

TianYan, Liao Ping-De, Deng Chao-Bing, Wu Lie-Shan, Liang Liu-Ling & Jiang Yun-Yun. 2012. Investigation of Organic Phosphorus Pesticides in Yongjiang River and Nanning Urban River. Journal of Guangxi Academy of Sciences, 03:216–218 +223.

Tian Yan, Deng Chao-Bing, Liao Ping-De, Liang Liu-Ling, Yang An-Ping & Wu Lie-shan. 2013. Distribution and Risk Assessment of the Semi-Volatile Organic Compounds(SVOCs) in Sediments from Urban Rivers of Nanning. Journal of Environment of Monitoring in China, 04:11–14.

Tao Zeng-Cai, Wei Yan-Song, Zhang Tong-Fa & Wu Lie-Shan. 2006. The Design and Realization of Prediction System for Water Quality Evaluation of Yongjiang River. Journal of HeChi university, 02:34–37.

Tao Zeng-Cai, Wang Shu-Xian, Tong Zhang-Fa, Wu Lie-Shan & Yu Ge. 2006. The water information management system developed with Mapobjects2.3 fro Yongjiang River in Nanning. Journal of Guangxi Academy of Sciences, 04:330–332 +336.

Wu Ting-Ting, Zhang Rui-Jie, Wang Ying-Hui, Leng Bing, Xue Bao-Ming, Liu Xiang & Lin Wei-dong. 2013. Investigation of the typical antibiotics in the sediments of the Yongjiang River, Nanning City, South China. Journal of China Environmental Science, 02:336–344.

Wu Lie-shan, Liang liu-Ling, Deng Chao-Bing, Liao Ping-De, Xu Gui-Ping & Tian Yan. 2009. Determining the Residues of Organochlorinated Pesticides in YongJiang River by SPE-GCPECD. Journal of Environment of Monitoring in China, 06:58–62.

Yu Ge, WuXiao-Yin, Fan Yu-Hang, Mo Rong-Xu, Yin Qi-Ming & Liu Qian. 2010. Study of Nanning Environmental Pollution Emergency Management Integrated Platform (volume II). Collected papers of Chinese Society for Environmental Sciences. 2010 China Environmental Science Institute Conference Proceedings. Collected papers of Chinese Society for Environmental Sciences:4.

Hydraulic Engineering III – Xie (Ed)
© 2015 Taylor & Francis Group, London, ISBN 978-1-138-02743-5

Analysis of the dynamic and vapor characteristics of continuous heavy rain during 6–8 June 2013 in Guilin China

Xianghong Li, Junjun Wang & Jin Hu
Guilin Meteorological Bureau, Guangxi, China

Weiliang Liang
Guangxi Meteorological Observatory, Guangxi, China

Rongqun Jiang
Menshan County Weather Station, Guangxi, China

ABSTRACT: By using conventional data and numerical models, radar, satellite data, the continuous heavy rain were analyzed in Guilin which occurred during 8–10 June 2013. The results showed that: a significant upper rise force of the trough was the cause of heavy rain. The low-level cyclonic curvature was the maintaining mechanism of the continuous rainstorm. Radar wind profilers showed that: the enhancement of the southwest low level jet formed the bow echo along the valley between Hunan to Guangxi, and the corresponding convergence line led to the convective precipitation. Convection over the Bay of Bengal developed 3 days ahead of the convection in Guangxi, indicated that the wave motion broadcast eastward from the Bay of Bengal to Guangxi and produced the heavy rainstorm.

1 INTRODUCTION

Guilin locates in the Northeast Guangxi. Guangxi locates in the southwest of China. It is belong to the semi-tropical monsoon region. At the same time the front and troughs often move from plateau affecting Guilin very year. So in every June, the continuous heavy rain often occur in Guilin, lead to flood and geological disasters.

Lots of researches have been done for the initiation, features, cause of the rainstorm. The low jet can lead to strong updraft to the rainstorm (Wilson J. W. & Mueller C. 1993). The thunderstorms often occur near the large-scale boundary layer. The strong atmospheric forcing can lead updraft and form quasi-linear rainstorm convection (Matthew J. B. & Mark R. H. 2006). Ming X. & William J. M (2006) studied the storm convergence by using high-resolution model. Juan Z. S. & Ying Z. (2008) stated that the warm shear and the vortex system were the main reason of convective rainstorm events.

Many Chinese meteorologists study the heavy rain and found many mechanism. For example the stable low level jet were important to rainfall (Ding et al. 2008). The stable shear line and vortex could provide moisture condition and uplift condition for continuous rainfall (Huang et al. 2008). Some researchers studied on continuous rainstorms in single station and got common results, such as a stable circulation was benefit to the generation, retention, and regeneration of rainstorm systems (Hu et al. 2007 Zhang et al. 2004).

2 RAINFALL DISTRIBUTION

2.1 *Continuity heavy rain daily distribution*

The 8–10 June 2013 continuity heavy rain spanned more than 60 hours. The 9 June 2013 heavy rain (Fig. 1) affected widespread all Guilin area. There are more than 10 stations

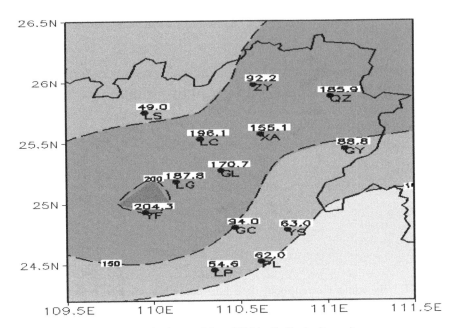

Figure 1.   The daily rainfall distribution on 9 June 2013 in Guilin (unit: mm).

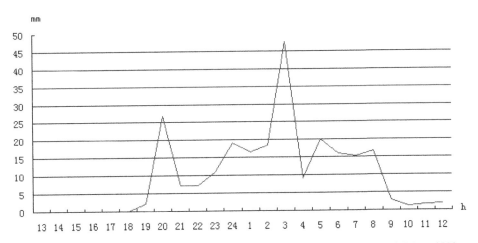

Figure 2.   The hourly rainfall of Dutou station from 1200 UTC 8 June to 1200 UTC 9 June 2013.

which precipitation exceed 100 mm. The center of the precipitation spread along the railway area. The precipitation of the Linchuan, Guilin, Lingui, Yongfu counties also exceeded more than 100 mm. The precipitation of the Ziyuan and Xingan counties exceeded 50 mm.

2.2   *The features of convective rainfall*

The heavy rainfall mainly occurred in the daytime of 9 June with the obvious convectional features. On 9 June, the rainfall mostly occurred between 2300UTC 8 June and 0800 UTC 9 June. For example, the rainfall of Dutou town which locates along the railway reached 70–90 mm in only 10 hours (Fig. 2).

3 THE FEATURE OF HEAVY RAIN SUPERCELL

At 0300 UTC 9 Jun, the rainfall in Lingui and Dutou ware more than 30 mm, with obvious convectional features. The reflectivity distribution (Fig. 3a) on 9 June 2013 showed that the heavy rain convective echo in Du tou and Lujiao villages was a bow echo which moved from southwest to northeast. It finally evolved to a heavy precipitation super cell.

At 0300 UTC 9, a mount of floccose echo gradually formed a large bow echo, and moved to the Dutou and Lojiao villages. The reflectivity intensified up to 53dBz. At the same time, the radial velocity reflected obvious convergence line characteristics in the Lingui county.

From the radial velocity (Fig. 3b), it could be found that convergence line ran from southwest to northeast, along the railway. In the radar wind profile, a deep southerly controlled the area from low to high level (1–9 km), at 850 hPa and 700 hPa the wind ware strong enough to be a low level jet. The deep low level jet provided moisture and heat to the rainstorms.

4 DYNAMIC AND VAPOR

4.1 *500 hPa vertical velocity*

The feature of physical parameters was analyzed by using FNL reanalysis data. 500 hPa vertical velocity indicated that the middle level in south of Guilin was controlled by the uplift at 0000 UTC 9 June 2013. The vertical velocity of Yongfu county was 0.6 Pa · s^{-1}, corresponding to the location of rainstorms exactly (Fig. 4).

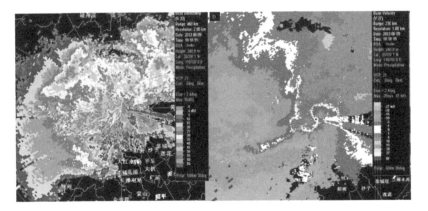

Figure 3.   The reflectivity (a) and the radial velocity (b) at 0300 UTC 9 June 2013 of Guilin radar.

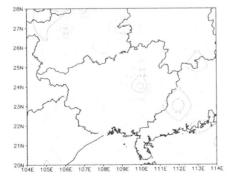

Figure 4.   500 hPa vertical velocity at 0000 UTC 9 Jun 2013 (unit: Pa · s^{-1}).

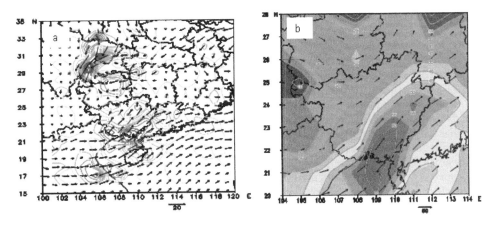

Figure 5.   The vapor situation at 0000 UTC 9 Jun 2013 (a: 850 hPa vapor flux divergence and wind barb, unit: 10^{-8} g.hPa^{-1}, b: Vapor flux vertical integral from 1000 hPa to 100 hPa, unit: g.s^{-1}.cm^{-1}).

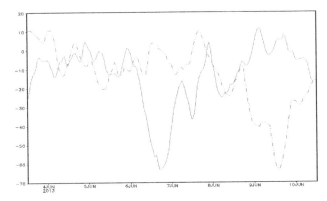

Figure 6.   The regional average TBB over Guangxi (dashed line) and Bay of Bengal (solid line) (the region of Bay of Bengal: 80–100°E, 5–23°N).

4.2   *The vapor flux divergence and the vapor flux vertical integral*

At 0000 UTC 9 June 2013 850 hPa vapor flux divergence reached -3×10^{-8} g.hPa^{-1}.cm^{-2}.s^{-1}, as showed in the distribution of vapor flux divergence and wind (Fig. 5). Convergence center was in Guilin, corresponding to the rainstorms. Vapor converged intensively in this region. At 0000 UTC 9 June 2013 the vapor flux vertical integral from 1000 hPa to 100 hPa got to 35 g.s^{-1}.cm^{-1} in the heavy rain region, so led the heavy rain storm.

5   PRECURSORY SIGNAL OF HEAVY RAIN

The evolution (Fig. 6) of the TBB over Guangxi (dashed line) and Bay of Bengal (solid line) indicated that the convection over Bengal Bay developed 3 days in advance than that of Guangxi. The regional average TBB over the Bay of Bengal reached the minimum value (−65 °C) at 0600 UTC 6 June, while TBB over Guangxi reached the minimum value (−60 °C) at about 0600 UTC 9 June. The convection over the Bay of Bengal increased to a maximum at about 3 days in advance the heavy rain began in Guangxi. So the convection over Bengal Bay have precursory signal meaning to Guangxi heavy rain.

6 HEAVY RAIN FORECAST

The medium-range forecasting of this case accurately predicted that the rainstorms would happen in the period of 9–10 June. But in the short-range forecasting, it predicted that the rainstorms would happen on 10 June. In fact, the heavy rain occurred at the daytime on 9 June. The error of the numerical model product of the 500 hPa weather systems led the forecasting error.

The numerical model products showed that Guangxi located at the uplift region in front of the 500 hPa trough on 9 June. On 10 June the Guilin still located at the ahead of the 500 hPa through. The through did not pass the 110°E at 1200 UTC 10 June. So the heavy rain maybe occurred between the night of 9 and the daytime of 10 June. But the 850 hPa system passed faster than the 500 hPa system. So the heavy rain occurred in the daytime of 9 June.

7 SUMMARY

The rainstorms happened in front of 500 hPa trough and nearby the cyclonic curve region at low level. The trough forced the uplift while the cyclonic curve led to the vapor convergence in Guilin.

EC flow field showed a weak wind environment. The low level jet did not exist. But the radar wind profile indicated the southwesterly jet increased rapidly and supported vapor for the rainstorms.

The rainfall had convectional feature. There was bow echo in radar reflectivity and convergence line in radial velocity in the Hunan to Guangxi corridor region.

The convection over Bengal Bay developed 3 days in advance than that of Guangxi. When the convection over Bengal Bay weakened, the convection strengthened in Guangxi. So the convection oover Bengal Bay have precursory signal meaning to Guangxi heavy rain.

ACKNOWLEDGEMENTS

Project funding: 2013 Guangxi Natural Science Fund Project (2013GXNSFAA019288. Southern China Regional Meteorological Center Project (CRMC2012M07).

REFERENCES

Ding Zhiying, Chang Yue & Zhu Li. et al. 2008. Research on the Reason of the Double Rain-Bands' Forming in A Sustaining Storm Rainfall of South China. Journal of Tropical Meteorology 24(2): 117–126.

Hu Liang, He Jinhai & Gao Shouting. 2007. An Analysis of Large-Scale Condition for Persistent Heavy Rain in South China. Journal of Nanjing Institute of Meteorology 30(3): 345–351.

Huang Zhong, Wu Naigeng & Feng Yerong. et al. Huang Z. 2008. Causality Analysis of the Continuous Heavy Rain in Eastern Guangdong in June 2007. Meteorological Monthly 34(4): 53–60.

Juan Z.S. & Ying Z. 2008. Analysis and Prediction of a Squall Line Observed during IHOP Using Multiple WSR-88D Observations. Mon Wea Rev, 136:2364–2388.

Matthew J.B. & Mark R.H. 2006. Smith. An Observational Examination of Long-Lived Supercells. Part I: Characteristics, Evolution, and Demise. Wea Foreca, 21: 673–688.

Ming X. & William J.M. 2006. High-Resolution Modeling Study of the 24 May 2002 Dryline Case during IHOP. Part I: Numerical Simulation and General Evolution of the Dryline and Convention. Mon Wea Rev, 134:149–171.

Wilson J.W. & Mueller C. 1993. Nowcast of thunderstorm initiation and evolution. WeaForcast, 8:113–131.

Zhang Xiaoling, Tao Shiyan & Zhang Shunli. et al. 2004. A Case Study of Persistent Heavy Rainfall over Hunan Province in July 1996. Journal of Applied Meteorological Science 15(1): 21–30.

Hydraulic Engineering III – Xie (Ed)

A model-based diagnostic study of "99.6" Meiyu front rainstorm

Wen Wang & Xiaojun Cai
Key Laboratory of Meteorological Disaster of Ministry of Education, NUIST, Jiangsu, Nanjing, China

Xiao Long
College of Atmospheric Sciences, Lanzhou University, Gansu, Lanzhou, China

ABSTRACT: The long-lasting rainy season over East Asia in early summer is called Meiyu in China and Baiu in Japan. In June 1999, the precipitation in middle and low basin of Yangzi River is twice the average in normal years. A model simulation of "99.6" Meiyu Front rainstorm by nonhydrostatic mesoscale model MM5V3 is analyzed in an effort to study the mechanism of the Meiyu front. The diagnosis of Convective Momentum Transport (CMT) shows that, the budget residual X of the horizontal momentum has different effects in different periods of the low vortex with the shear line: On low levels, X strengthened the southwest flow to north on the occurring stage, on middle levels, X accelerated northwest flow behind East Asia Trough to north and also accelerated northwest flow ahead of East Asia Trough to the south. All these were in favor of deepening the East Asia Trough and mingling cold and warm air, and gave favorable conditions to produce rainstorm.

1 INTRODUCTION

The early summer long-lasting rainy season over East Asia is called Meiyu in China and Baiu in Japan. In this season, The Meiyu front extends eastward from the lee of Tibet Plateau, over the Yangzi-Huai River basin with its eastern edge extending to Japan. Heavy rainstorms developed along the Meiyu front often cause severe flooding over those areas, and Meiyu front is one of the most significant circulation systems for the hydrological cycle in the east Asia monsoon region.

The synoptic structure of the Meiyu front, a moist, quasi-stationary front characterized by weak temperature gradient but strong equivalent potential temperature (θ_e) gradient is different from that of typical mid-latitude fronts (Ding, 1992). In the lower troposphere (850–700 hPa), a low-level cyclonic wind shear line is located about 200–300 km north of the surface front. On the southern side of the shear line there often occurs a well-defined synoptic-scale Low-Level Jet (LLJ), which transports warm, moist air northward at lower levels and creats a convectively unstable layer. At the high-level, Upper-Level Jet (ULJ) appears on the northern side of Meiyu front (Zhang et al., 2000).

The meso- and micro-scale synoptic systems which form rainstorm occur and develop along the large-scale circumfluence. The post studies were mainly describing the large-scale and meso-scale systems (Tao et al., 1998). Therefore, it is still difficult to explain the heavy rain event with meso-β scale weather systems with the existing mesoscale dynamical theories.

2 THE NUMERICAL MODEL AND INITIAL CONDITIONS

The numerical model used in this study is a nonhydrostatic, primitive equation with terrain-following sigma(σ) vertical coordinate MM5 version 3. The simulation performed for this study was run on two-way interactive nesting model. They consist of a 6-km grid (Do$_3$, mesh size of 181×181), a 18-km grid (Do$_2$, mesh size of 121×121) and a 54-km grid (Do$_1$, mesh size of 91×91). All the grids have 23 levels in the vertical. The top of model was set to 100 hPa.

The model physics used in this experiment includes the Reiner Graupel scheme for the resolvable-scale motion, in which six prognostic equations are used for water vapor, cloud water, rain water, cloud ice, snow and graupel, the subgrid-scale convection of Grell cumulus parameterization scheme and Blackdar's high-resolution Planetary Boundary Layer (PBL) scheme which is implemented to calculate the vertical fluxes of sensible heat, latent heat and momentum and turbulent mixing in the PBL. The radiation scheme was CCM_2.

The global analyses data with resolution of $1.875° \times 1.875°$ (T_{106}) were used as the boundary conditions and the first guess field, and the initial field obtained by successfully correcting the first guess field with twice rawinsonde and four times surface observations a day. The simulation was started at 00UTC/23 June 1999 for Do_1, 06UTC/23 for Do_2 and 12UTC/23 for Do_3, time step was 90 s for Do_1.

3 DIAGNOSTIC OF CONVECTIVE MOMENTUM TRANSPORT

In order to study the dynamic mechanism of "99.6" Meiyu Front rainstorm, the momentum budget residual $\vec{X}$ is taken into account for the simulation output with the following formula of Tung and Yania (Tung et al., 2002a, b):

$$\vec{X} = (X, Y)$$

$$\equiv \frac{\partial \bar{\vec{V}}}{\partial t} + \bar{\vec{V}} \bullet \nabla_p \bar{\vec{V}} + \bar{\omega} \frac{\partial \bar{\vec{V}}}{\partial p} + \nabla_p \bar{\Phi} + f\bar{k} \wedge \bar{\vec{V}} \tag{1a}$$

$$= -\nabla_p \bullet \overline{\vec{V}'\vec{V}'} - \frac{\overline{\partial \vec{V}'\omega'}}{\partial p} \tag{1b}$$

where $\vec{V}$ is the horizontal velocity, $\omega(= dp/dt)$ the vertical p velocity, Φ the geopotential height, f the Coriolis parameter, ∇ the isobaric del operator. The overbar denotes the running average with respect to a large-scale area, and the prime denotes the deviation from the average. Formula (1) indicates that the horizontal momentum budget residual is the contribution of subgrid scale convection and the other eddy momentum flux convergence to the acceleration of large scale wind field.

Taking the dot product of formula (1) with $\bar{\vec{V}}$, the dynamic equation of large scale system and the dynamic transfer between large and subgrid scale system can obtain:

$$\bar{\vec{V}} \cdot \vec{X} \equiv \frac{\partial \bar{K}}{\partial t} + \bar{\vec{V}} \cdot \nabla_h \bar{K} + \bar{\omega} \frac{\partial \bar{K}}{\partial p} + \bar{\vec{V}} \cdot \nabla_h \bar{\Phi} + \bar{\vec{V}} \cdot f\bar{k} \wedge \bar{\vec{V}} \tag{2}$$

where $\bar{K} = (\bar{\vec{V}} \cdot \bar{\vec{V}})/2$ is the large scale kinetic energy. Define the quantity

$$E = -\bar{\vec{V}} \cdot \vec{X} = -(\bar{u}X + \bar{v}Y)$$

to measure the kinetic energy loss from the mean flow to subgrid-scale eddies. $E > 0$ means the kinetic energy being transferred from the large scale motion to the subgrid-scale convection (Downscale transfer) and $E < 0$ means upscale transfer.

When the mesoscale vortex with shear line initially created on 01UTC/23 June, there were three vortices at the level of 850 hPa, the strongest $\vec{X}$ vector located on (30°N, 110°E) where is in the southeast of Qinghai Xizang Plateau, that indicated momentum budget residual $\vec{X}$ strengthened the southwest flow to move northward. East Asia Trough was very strong at the levels of 700 and 500 hPa, the angle between $\vec{X}$ and $\vec{V}$ was smaller than 90° both behind and ahead of the Trough, that means $\vec{X}$ accelerated northwest warm and moist air behind East Asia Trough to the north and also accelerated northwest cold air ahead of East Asia Trough

222

to the south. These conditions were propitious to the development of East Asia Trough. At the level of 300 hPa, kinetic energy transfer E had a belt distribution, E < 0 between 27°N–39°N latitudes and E > 0 in the other areas, which means kinetic energy are transferred from the southward cold air and northward southwest flow. And the negative and positive maximum centers of E were mainly located on the each side of upper Jet, that showed the upper Jet played a very important role in kinetic energy transfer.

On 06UTC/24, the developing period of low vortex with shear line, strong $\vec{X}$ had a reverse direction with southwest flow at level of 850 hPa located on (25°N, 105°E), which prevented southwest flow to the north. At the same time, the momentum residual (the angle between $\vec{X}$ and $\vec{V}$ was larger than 90°) decelerated the cold (warm) flow behind (ahead of) the East Asia Trough at the level of 500 hPa.

When the low vortex was being to launch out on 18UTC/24, the distribution of $\vec{X}$ and E were the same as that in developing period except intensity became infirm.

The simulated result shows that the low vortex with shear line developed to the strongest on 06UTC/24, there were four β-mesoscale systems A, B, C and D on the head of the low vortex at the level of 850 hPa (Fig. 1). The distribution of $\vec{X}$ shows that intense $\vec{X}$ almost coupled with intense ascend motion, β-mesoscale system D has hardly been accelerated, so the development level is lowest. The distribution of E suggested that the area of E > 0 and E < 0 were alternating, it indicated the energy transfer was very complicated between large

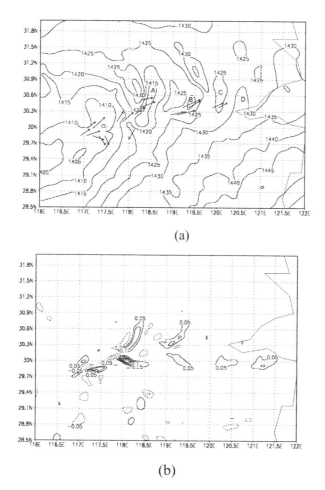

(a)

(b)

Figure 1. Mesoscale analyses of CMT. (a) meso-β systems with $\vec{X}$ ($\times 10^{-2}$ ms^{-2}), (b) E($\times$ m^2 s^{-3}) at 06UTC/24 at the level of 850 hPa.

scale and subgrid scale systems, rather than described as a energy cascade process. That positive value of E was much greater than the negative suggested the energy was mainly transferred from large scale to subgrid scale.

On 18UTC/24, The low vortex was being to launch out, distribution of $\bar{X}$ showed southwest flow in warm sector was decelerated strongly except for the head of the low vortex was accelerated that indicated the low vortex would dissipate. The calculated energy transfer at low level of troposphere mainly occurred in warm sector of low vortex, which in cold sector was small.

4   SUMMARY AND CONCLUSIONS

A diagnostic analysis based on the nonhydrostatic mesoscale model MM5V3 simulation of "99.6" Meiyu Front rainstorm on 00UTC/23-00UTC/25 is presented. On low levels, the southwest Jet and the northeast flow increased convergence on low levels as the low meso-α vortex with shear line developing, and the westly Jet and the eastly Jet increased divergence on high levels. It is this collocation from low levels to high levels that urged meso-α low vortex to develop. Analyses of trajectory also illuminate those characters.

The diagnosis of CMT shows that, the budget residual $\bar{X}$ of the horizontal momentum has different effects in different periods of the low vortex with the shear line: On low levels, $\bar{X}$ strengthened the southwest flow to north on the occurring stage, on middle levels, $\bar{X}$ accelerated northwest flow behind East Asia Trough to north and also accelerated northwest flow ahead of East Asia Trough to the south. All these were in favor of deepening the East Asia Trough and mingling cold and warm air, and gave favorable conditions to produce rainstorm. On the developing period, $\bar{X}$ has the reverse direction with southwest flow, which prevented southwest flow to the north. At the same time, it decelerated the cold (warm) flow behind (ahead of) the East Asia Trough. on the level of 500 hPa. The diagnosis of CMT on the violently developing period of the mesoscale system shows that strong $\bar{X}$ always related to the meso-β systems which brought intense ascending motions. While the shear line launched out, $\bar{X}$ has strong decelerating effects for southwest flow. The result of diagnosis also revealed that large-scale environment field has overt energy interconversion with subgrid scale systems on the intense developing period of the mesoscale system, and energy primarily was transported to microscale systems. Furthermore, energy exchanges on high levels were closely with the location of Upper Jet.

REFERENCES

Ding, Y.-H. (1992), Summer monsoon rainfalls in China. J. Meteor. Soc. Japan, 70:337–396.

Tao Shiyan, Ni Yongqi, Zhao Sixiong etc, (2001) Studies on mechanism and prediction for China summer Rainstorm in 1998, Meteorology Press (in Chinese).

Tao ShiYan (1980), Rainstorm In China, Science Press (in Chinese).

Tung, Wen-Wen and M. Yanai, Convective Momentum Transport Observed during the TOGA COARE IOP. Part I: General Features, J. Atmos. Sci., 2002, 59:1857–1871.

Tung, Wen-Wen and M. Yanai, Convective Momentum Transport Observed during the TOGA COARE IOP. Part II: Case Studies, J. Atmos. Sci., 2002, 59:2535–2549.

Zhang Qinghong, Liu, Q.-H., Wang, H.-Q., Chen, S.-J., Numerical simulation for mesoscale system on Meiyu Front in South China. 2000, 45(18):1988–1992.

Hydraulic Engineering III – Xie (Ed)
© 2015 Taylor & Francis Group, London, ISBN 978-1-138-02743-5

Scale invariance analysis of the decadal wind velocity in Henan Province

Ji-Jun Wang
Henan Provincial Climate Center, Zhengzhou, P.R. China

Cai-Hong Hu
School of Water Conservancy and Environment, Zhengzhou University, Zhengzhou, P.R. China

Dan-Dan Zhou
Xuchang Meteorological Office, Xuchang, P.R. China

Lei-Lei Zhu
Henan Provincial Climate Center, Zhengzhou, P.R. China

ABSTRACT: Using the decadal wind velocity data of 107 meteorological stations in Henan Province from 1961 to 2012, variations on Spatial-temporal distributional characteristics were analyzed, and the scale invariance analysis of decadal wind velocity was analyzed. The results showed that: the mean decadal rainfall and its standardized errors among the meteorological stations were 2.3 m/s and 0.68 m/s, respectively, which probability distributions were not subject to normal distribution. The cumulative difference of decadal wind velocity and its standard error among the meteorological stations had the same extremely linear upward trends, which could pass the significance test at the 0.001 significant level. The scale indexes of decadal wind velocity were all above 0.5, indicating that time series of decadal wind velocity of all meteorological stations were persistent. Spatial variability on scale index was small, which showed that the scale index of decadal wind velocity obeyed normal distribution among the meteorological stations.

1 INTRODUCTION

The prediction of wind velocity is one of the most important aspects when variation of wind velocity when dealing with renewable energy. The variation of wind velocity, in a certain site, was strictly related to the economic aspects of a wind farm, such as maintenance operations and evaluation of a new site. However, the forecast of wind velocity depended on the research of discipline in variation on wind velocity. Many studies on characteristics of wind velocity were carried out. Si Cheng et al (2013) analyzed the temporal and spatial distribution of the wind velocity and its monthly cycle, inter-annual, inter-decadal variations in the last 50 years in Jiangsu Province by using the regular statistical methods. Ya-Wei Ma et al (2013) presented the variation periods and abrupt phenomenon of Tsingtao surface wind velocity by using the observed data. Jing Pan et al (2014) studied the change trends of global sea surface wind velocity from 1958 to 2001 based on the ERA-40 sea surface 10 m wind field data of the European Center for Medium-Range Weather Forecasts (ECMWF), including the overall change trend, seasonal differences, regional differences and change cycles. Xiao-Yong Geng and Guang-Chao Cao analyzed the temporal and spatial characteristics of wind velocity in Nanjing and its possible causes. Yan-Qing Li et al proposed a hybrid prediction method in order to improve the prediction performance of wind velocity series with chaotic characteristics. Tie-Jun Liu et al analyzed the long-term linear trend of sea surface wind velocity from 1958 to 2001 in the Northern Indian Ocean based on the ERA-40 wind field from ECMWF.

All of the current papers were study on wind velocity of climate change and prediction, the research on detrended fluctuation analysis of wind velocity were rare. This paper was organized as follows. First of all, in section 2, we describe the database and methodologies used for the study. In section 3, the characteristics of temporal and spatial variation on decadal mean wind velocity were analyzed, at the same time, characteristics of temporal and spatial variation on cumulative difference of decadal mean wind velocity were also studied, in the end, the spatial distribution of scale index of decadal wind velocity were studied. Finally in section 4 some including remarks were presented.

2 DATA AND METHODOLOGIES

2.1 Summary of study area

Henan is located in the middle east of China and in the middle and lower reaches of Yellow River, bounded on the north latitude 31° 23′–36° 22′, longitude 110° 21′–116° 39′. The mean annual temperature was generally between 12–16°C, which extreme maximum temperature was 44.2°C appeared in Luoyang on 20th, June, 1966.

2.2 Data

The database used for the analysis in this work were freely provided by Henan Provincial meteorological information and technological supporting center and were composed of 107 meteorological stations in Henan Province of daily wind velocity from 1961 to 2012. The decadal mean wind velocity at every station per year were calculated, then, decadal mean wind velocity were systematically analyzed.

2.3 Methodologies

Detrended Fluctuation Analysis (DFA)

The DFA method, proposed by Peng et al., was based on random walk theory, allowed the calculation of a signal quantitative parameter—the scaling exponent to quantify the correlation properties of a signal. The DFA method was briefly introduced.

We consider a fluctuating time series $X_i(i = 1, 2, 3, ..., N – 1, N)$ sampled at equidistant times $i\Delta t$. We assume that $\bar{X}$ was the global mean of the time series, then the new time series were integrated.

$$Y_i = \sum_{j=1}^{i} X_j - \bar{X} \qquad (1)$$

We divide the integrate time series into N_S non-overlapping segments of equal length S, where $N_S = N/S$. In order to accommodate the fact that some of the data points may be left out, the procedure was repeated from the other end of the data set and $2 N_S$ segments were obtained.

We detrend the integrated time series Y_i by subtracting the local trend in each segment and we calculate the detrended fluctuation function $\tilde{Y}_i^j$.

$$\tilde{Y}_i = Y_i - P_v^i \qquad (2)$$

where $P_v^i(i = 1, 2, 3, ..., S-1, S; V = 1, 2, 3, ..., N_S - 1, N_S)$ was the local trend of the segment of V.

For a given segment size S, the variance $F^2(V, S)$ were calculated from

$$F^2(V,S) = \frac{1}{S}\sum_{i=1}^{S} \tilde{Y}_v^2[(V-1)S+i], \ (V = 1, 2, ..., N_S - 1 N_S) \qquad (3)$$

226

$$F^2(V,S) = \frac{1}{S} \sum_{i=1}^{S} \tilde{Y}_V^2 [N - (V - N_S)S + i], \quad (V = N_S + 1, N_S + 2, ..., 2N_S - 1, 2N_S) \tag{4}$$

the root mean square function of all the trended fluctuation function was

$$F(S) = \sqrt{\frac{1}{2N_S} \sum_{S=1}^{2N_S} F^2(V,S)} \tag{5}$$

The above function was calculated by changing S value and then the relationship between $F(S)$ and S could be obtained. A power-law relation between $F(S)$ and the segment size S indicated the following the scaling:

$$F(S) \sim S^\alpha \tag{6}$$

The α is called correlation exponent or scaling exponent, representing the correlation characteristics of the signal:

$\alpha < 0.5$ means that it is anti-correlated signal; $\alpha = 0.5$ means that it is uncorrelated signal; $\alpha > 0.5$ means that it is a long-range power-law correlated signal.

Normal distribution test

The skewness coefficient C_{sk} and sharpness coefficient C_{sh} were calculated from

$$C_{sk} = \frac{1}{N} \sum_{i=1}^{N} \left(\frac{\chi_i - \bar{\chi}}{s} \right)^3, \quad C_{sh} = \frac{1}{N} \sum_{i=1}^{N} \left(\frac{\chi_i - \bar{\chi}}{s} \right)^4 - 3 \tag{7}$$

where $\bar{\chi} = 1/N \sum_{i=1}^{N} \chi_i$, $s = \sqrt{1/N \sum_{i=1}^{N} (\chi_i - \bar{\chi})^2}$.

If the sample is large enough and the sample obey the normal distribution, then the skewness coefficient and sharpness coefficient all obey the normal distribution, and both of the standardized error could be calculated from

$$S_{sk} = \sqrt{\frac{6(n-2)}{(n+1)(n+3)}}, \quad S_{sh} = \sqrt{\frac{24(n-2)(n-3)}{(n+1)^2(n+3)(n+5)}}$$

Hypothesis H0: the sample obeys normal distribution. At the confidence degree $\alpha = 0.05$, if $|C_{sk}/s_{sk}| < 1.96$ and $|C_{sh}/s_{sh}| < 1.96$, then we could accept the null hypothesis, that is, the sample obeys normal distribution; otherwise, reject the null hypothesis, i.e. sequence does not obey the normal distribution.

3 RESULTS AND DISCUSIONS

3.1 Characteristics of temporal and spatial variation on decadal mean wind velocity

Annual and inter-annual variability on decadal mean wind velocity in Henan Province were showed in Figure 1. The mean decadal wind velocities were 2.3 m/s on average from 1961 to 2012, which standard error was 0.15 m/s, which median were 2.2 m/s. The maximum and minimum of wind velocity were 5.4 m/s (occurred on the 3rd decads of 1963), 1.1 m/s (occurred on the 29th of 1990 and 2004, 26th of 2006, and 22nd of 2011), respectively. The Mean Standard Errors (MSE) of the decadal wind velocity were 0.66 m/s. The mean decadal wind velocity had extremely significant linear downward trend, which explained variance could reach 45.2%. The wind velocity decreased 0.008 m/s every 10 decads, and the trend could pass the significance test at the 0.001 confidence level.

227

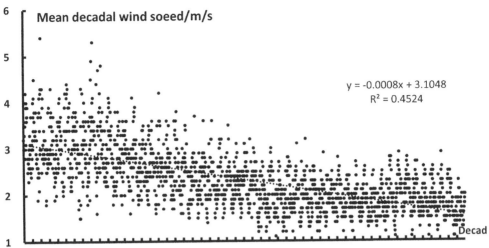

Figure 1.   Annual and inter-annual variability on decadal wind velocity in Henan Province from 1961 to 2012.

The frequencies of decadal wind velocity between 1.7 m/s and 2.8 m/s were 60%, and the frequencies of wind velocity less than 1.7 m/s or more than 2.8 m/s both accounted 20%. The wind velocity less than 1.6 m/s occurred 178 times, concentrated from October to April, which could account for 69.7%, appeared from May to September was only 30.3%. The frequencies occurred before May or after September had the equal opportunity.

The wind velocity more than 3.6 m/s occurred 82 times, concentrated from May to September, which could account for 69.5%, appeared from October to April was only 30.5%. The frequencies occurred before May generally equaled the frequency occurred after September.

The skewness coefficient and sharpness coefficient of mean decadal wind velocity were 0.82 and 0.60, respectively, which standard error was 0.06 and 0.11, indicated that the time series is not subject to normal distribution. There was an extreme significant logarithmic relationship between the decadal wind velocity and its frequency, which explained variance could reach 91.6%, and the fitting equation:

$$P = 38.601 \ln \mu - 37.16$$

where P was the cumulated frequency, and μ was the decadal wind velocity.

Annual and inter-annual variability on MSE of decadal mean wind among meteorological stations in Henan Province were showed in Figure 2. The MSE of decadal wind velocity were 0.68 m/s on average from 1961 to 2012, which standard error was 0.004 m/s, which median were 0.65 m/s. The maximum and minimum of MSE were 1.92 m/s (occurred on the 3rd decad of 1963), 0.37 m/s (occurred on the 7th of 2001), respectively. The maximum of MSE appeared in the decad of maximum wind velocity. The MSE of MSE were 0.16 m/s. As with decadal wind velocity, the MSE also had extremely significant linear downward trend, which explained variance could reach 25.3%. The wind velocity decreased 0.001 m/s every 10 decads, and the trend could pass the significance test at the 0.001 confidence level.

The frequencies of MSE of decadal wind velocity among meteorological stations between 0.56 m/s and 0.78 m/s were 60%, and the frequencies of MSE of wind velocity less than 0.56 m/s or more than 0.78 m/s both accounted 20%. The wind velocity less than 0.5 m/s occurred 154 times, concentrated from October to April, which could account for 68.8%, appeared from May to September was only 31.2%. The MSE of wind velocity among stations more than 1.0 m/s occurred 65 times, concentrated from May to September, which could account for 70.8%, appeared from October to April was only 29.2%. The frequencies

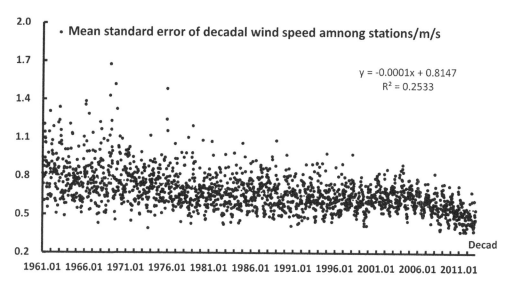

Figure 2. Annual and inter-annual variability on MSE of decadal wind velocity among the meteorological stations from 1961 to 2012.

occurred before May were 12 times, While the frequency occurred after September were 7 times.

The skewness coefficient and sharpness coefficient of MSE of decadal wind velocity among stations were 1.47 and 5.12, respectively, which standard error was 0.06 and 0.11, indicated that the time series is not subject to normal distribution. There was an extreme significant linear relationship between the decadal wind velocity and its frequency, which explained variance could reach 99.9%, and the fitting equation:

$$P = 0.0671\sigma - 1.1615$$

where P was the cumulated frequency, and σ was the MSE of decadal wind velocity among meteorological stations.

3.2 Characteristics of temporal and spatial variation on cumulative difference of decadal mean wind velocity

Annual and inter-annual variability on cumulative difference of decadal wind velocity in Henan Province were showed in Figure 3. The cumulative difference of decadal wind velocity had extremely significant linear upward trend, which explained variance could reach 98.2%. The cumulative difference of wind velocity increased 291.9 m/s every decads, and the trend could pass the significance test at the 0.001 confidence level.

Annual and inter-annual variability on cumulative difference of decadal wind velocity in Henan Province were showed in Figure 4. The cumulative difference of MSE of decadal wind velocity among the meteorological had extremely significant linear upward trend, which explained variance could reach 99.1%. The cumulative difference MSE of wind velocity increased 49.4 m/s every decad, and the trend could pass the significance test at the 0.001 confidence level.

3.3 Scale index of decadal wind velocity

Spatial distribution of scale index of mean decadal wind velocity in Henan Province were showed in Figure 5. The mean of scale index of decadal wind velocity was 0.90, which standard error was 0.004, and the median was close to the mean. Mean standard error of scale

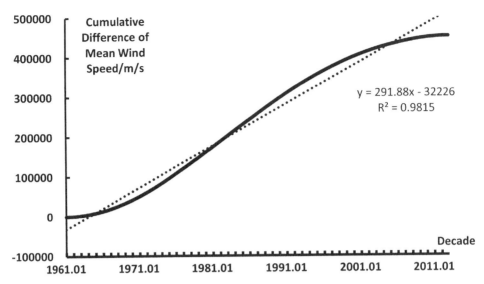

Figure 3.　Cumulative difference curve of decadal wind velocity in Henn Province from 1961 to 2012.

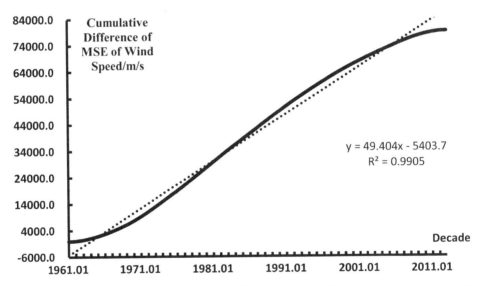

Figure 4.　Cumulative difference curve of MSE of decadal wind velocity among the meteorological stations in Henan Province from 1961 to 2012.

index among the meteorological stations was 0.004, indicated that the scale index had little spatial variability. Among 107 meteorological stations the maximum scale index was 0.97, appeared in Xiayi, Shangqiu. There were 2 stations' scale index more than 0.95, appeared in Songxian station of Luoyang and Xinxian station of Xinyang. The minimum scale index was 0.72, appeared in Yiyang station of Luoyang, as well as there was a station's scale index less than 0.77, occurred in Linzhou of Anyang, which value was 0.77.

The skewness coefficient and sharpness coefficient of scale index of decadal wind velocity were −1.19 and 2.99, respectively, which standard error was 0.23 and 0.46, indicated that the scale index among meteorological station in Henan Province were not subject to normal distribution. All of the scale index of decadal wind velocity of all meteorological stations were

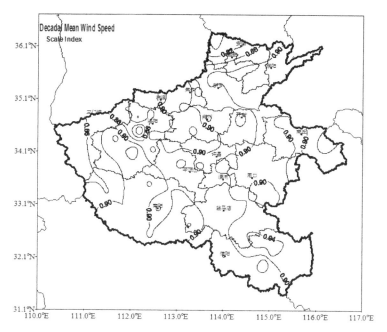

Figure 5. Spatial distribution of scale index of mean decadal wind velocity.

above 0.5, indicated that the sequence of decadal wind velocity was persistent, implied trends of a period of past was as same as that of the coming.

4 CONCLUSION

Using mean decadal wind velocity of 107 meteorological stations in Henan Province from 1961 to 2012, characteristics of temporal and spatial distribution of decadal wind velocity were studied, at the same time, the scale index of wind velocity were also analyzed, the results showed that:

The mean decadal wind velocity was 0.23 m/s during 51 years, and the mean MSE of decadal wind velocity among meteorological stations were 0.68 m/s. Whether a mean decadal wind velocity or MSE of decadal wind velocity among meteorological stations were both not subject to normal distribution, which skewness coefficients were 0.82 and 1.47, respectively.

Cumulative difference of decadal wind velocity and cumulative difference of MSE of decadal wind velocity among meteorological stations both had an extreme significant linear upward trends, which trend rates were 291.8 m/s and 49.4 m/s every decad, and the trends can pass the significance test at the 0.001 significant level.

All of the scale index of decadal wind velocity of 107 meteorological stations in Henan Province were above 0.5, indicated that the sequence of decadal wind velocity was persistent. There were smaller spatial variability among the meteorological stations, which obeyed the normal distribution.

ACKNOWLEDGEMENT

The study is supported by the Chinese Natural Research Fund (51079131), Climate Change Science Foundation (CCSF201312), and CMA/Henan Key Laboratory of Agrometeorological Support and Applied Technique (AMF201304).

REFERENCES

Jing Pan, Zhu-Piao Liu, Chong-Wei Zheng, et al. Analysis of global ocean surface wind velocity over recent 44 years. Meteorological Science and Technology, 2014, 42(1): 104–109 (In Chinese).

Peng C.K, Buldyrev S.V, Havlin S, et al. Mosaic organization of DNA nucleotides [J]. Physical Review E, 1994, 49(2):1685–1689.

Si Cheng, Peng Chen, Junfeng Miao. Spatial and temporal characteristics of surface wind velocity in the last 50 years over Jiangsu Province. Meteorological, Hydrological and Marine Instruments, 2013, 3: 121–124 (In Chinese).

Xiao-Yong Geng, Guang-Chao Cao. Changes in wind velocity and its causes in Nanjing during the period of 1960–2004. Science and Technology, 2013,13:122–125 (In Chinese).

Ya-Wei Ma, Ai-Li Cong, Chong-Wei Zheng, et al. Periods and abrupt analysis of temperature and surface wind velocity in Tsingtao during the last 50 year. Science and Technology Information, 2013, 17: 119–124 (In Chinese).

Yan-Qing Li, Yi Cheng, Xin-Ting Liu. Study on hybrid prediction method of wind velocity series. Journal of Tianji Polytechnic University, 2013, 32(5):47–50,56 (In Chinese).

Hydraulic Engineering III – Xie (Ed)
© 2015 Taylor & Francis Group, London, ISBN 978-1-138-02743-5

Research on causes and stability of sand ridges in Liaodong Bay—take Majia shoal for example

Zhiyuan Han

Tianjin Research Institute for Water Transport Engineering, Key Laboratory of Engineering Sediment of Ministry of Communications, Tianjin, China

ABSTRACT: Based on the analysis of underwater topographic evolution, sedimentary characteristics, and the sediment depositional age, the causes and stability of Majia shoal are studied. The conclusions are shown as follows: (1) Majia shoal is a linear sand ridge sheltered by Majia Cape, and its underwater topography can keep stable; (2) Majia shoal consists of median or coarse sand, and formed during geological processes in Holocene, existing as thousands of years; (3) Wave at local sea area is not strong and current velocity is not large; (4) Suspended sediment concentration is low and sediment sources are limit; (5) Seabed sediment it is not active and cannot transport under wave and current action, so Majia shoal can maintain a stable state for many years.

1 INTRODUCTION

Liaodong Bay is a semi-closed bay in northern Bohai Sea. There are many underwater linear sand ridges in Liaodong Bay, which can be seen in the western bay near Liugu River Estuary, in the eastern bay near Changxing Island, and the southern bay near Bohai Straight. Majia Shoal, a longshore sand ridge, lies in the northwest of Changxing Island (See Fig. 1).

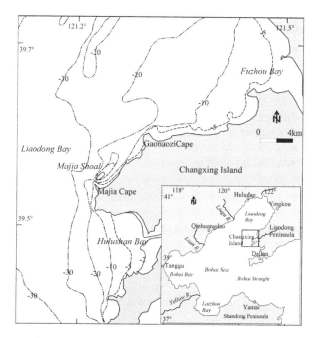

Figure 1. Sketch map of the research zone.

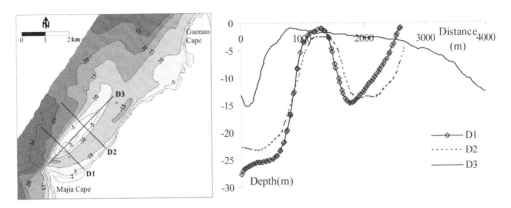

Figure 2.   Skep of underwater terrain and underwater profiles of D1~D3 section.

The sand ridge is sheltered by Majia Cape and about 1.4 km offshore. The length of Majia shoal is about 4 km with width of 1 km. At the top of Majia Shoal, water depth is only about −1.5 m (Chart Datum). A trough lies between Majia Shoal and Changxing Island, with water depth as −10 m~−15 m. Water depth at the west of Majia Shoal is open sea with water depth greater than −20 m, and seabed slope as 1/40~1/30 (see Fig. 2).

Some researchers have conducted a lot of research works on the linear sand ridges near Bohai Strait (Xia Dongxing & Liu Zhengxia, 1983 & 1984; Li Fan, etc., 1984; Tang Yuxiang, etc., 1993). They usually focused on the underwater terrain, depositional characteristics, sedimentary sources, hydrodynamic conditions, and migration and formation mechanisms of the sand ridges (Liu Zhengxia, etc., 1993; Liu Zhengxia, etc., 1996; Dong Tailu, 1996; Yin Ping, 2003). There are very few research works on linear sand ridges in eastern Liaodong Bay. This paper take Majia Shoal as an example, based on the analysis of underwater topography evolution, sedimentary characteristics, hydrodynamic conditions, and study the causes and stability of Majia Shoal.

2   RESEARCH DATA AND METHODS

1. 35 seabed sediment samples were sampled using grab sampler in Majia Shoal sea area in March 2009. All the samples were sent to laboratory for grain size analysis.
2. 10 sediment cores were drilled on Majia Shoal in March 2008. All the core samples were sent to laboratory for analysis and vertical deposition profiles were drawn. 7 core sediment samples of different water depth with shell fragments were chosen and sent to AMS Center of Peking University for ^{14}C dating.
3. 3 charting maps were measured in years of 1938, 1998 and 2008 respectively, with sale of 1:200 000, 1:150 000, and 1:150 000.

3   RESULTS

3.1   *Underwater topography evolution*

According to chart maps in Majia Shoal sea area measured in years of 1935, 1998 and 2008, underwater topography evolution are shown as the following characteristics (see Fig. 3):

1. From 1935 to 1998, isobaths of −5 m, −10 m and −20 m at both north and south sides of Changxing Island changed slightly and underwater topography could keep a stable state; isobaths of −2 m, −5 m, −10 m and −20 m at Majia Shoal, the northwest side of Changxing Island, changed slightly, so Majia Shoal could keep a stable state also; isobaths of −30 m at open sea changed slightly and could keep stable.

2. From 1998 to 2008, isobaths of −5 m, −10 m and −20 m at both north and south side of Changxing Island changed slightly and underwater topography could keep a stable state; isobaths of −2 m, −5 m, −10 m and −20 m at Majia Shoal changed slightly, so Majia Shoal could keep a stable state also; isobaths of −30 m at outer sea changed slightly and could keep stable.

Overall, underwater topography around Majia Shoal changed slightly and had no significant trend of erosion or siltation. So the shape and underwater topography can maintain a stable sate for more than 70 years.

3.2   *Sedimentary characteristics*

1. Vertical deposition characteristics
   Majia Shoal mainly consists of median-coarse sand and median-fine sand. Median sand is deposited from top to −15 m~−20 m depth, with layer thickness of 10 m~15 m (see Fig. 4). d_{50} of the sandy layer is between 0.20 mm and 0.40 mm in vertical, with an average of 0.30 mm. Sorting coefficient of sand is between 0.23 and 0.64, and the sand is sorted very well.

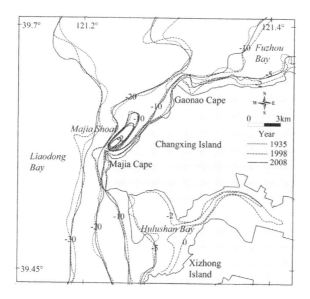

Figure 3.   Underwater topography evolution from 1935 to 2008.

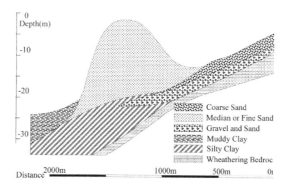

Figure 4.   Deposition profile of D1 section.

235

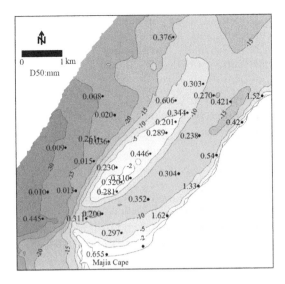

Figure 5.   Distribution of d50 of seabed sediment.

Sediments deposited between sandy layer and bedrock are mainly silty clay, with layer thickness of 5 m~7 m and d50 less than 0.05 mm. Sediments deposited at the west of Majia shoal are mainly muddy clay, with layer thickness of about 5 m and d50 less than 0.02 mm. Sediments deposited at between Majia shoal and Changxing Island are mainly coarse sand and gravel-sand, with layer thickness of about 3 m~5 m.

According to ^{14}C dating results, dating age of shell samples with water depth of −5 m~−18 m on Majia Shoal is from 2135 ± 35 a to 4515 ± 35a. So deposition age of sediment on Majia shoal is 2100~4550 yr. B.P. and marine sediments on Majia Shoal are mainly deposited during Holocene. Majia Shoal has been formed and existed for thousands of years.

2. Seabed sediment characteristics

Seabed sediment types on Majia Shoal are mainly median-fine sand and median-coarse sand (see Fig. 5), with d50 of 0.20 mm~0.61 mm and the sediments are sorted very well. Sediment types along the coastline are mainly coarse sand and gravel-sand, with d50 greater than 0.45 mm; sediment types in deep water at the west side of Majia Shoal are mainly clayey silt, with d50 less than 0.02 mm.

Overall, seabed sediments on Majia Shoal are mainly sandy sediments with sandy content greater than 90%, and clay content less than 5%.

4   DISCUSSION

4.1   *Wave*

1. According to wave data of Changxing Station observed during from December 2004 to November 2005, the prevailing wave direction of Majia Shoal sea area is NE, with frequency of 22.3%, and the second prevailing wave direction is SW, with frequency of 13.5%. The strong wave direction is NE and NW, with maximum $H_{1/10}$ of 5.4 m. Wave frequency with $H_{1/10}$ less than 1.5 m is 92.0%, and wave frequency with $H_{1/10}$ greater than 2.0 m is only 2.5%.
2. Underwater topography of Majia Shoal can keep sable, so sand on not Majia Shoal cannot move by wave condition and wave action in Majia Shoal sea area is not strong.

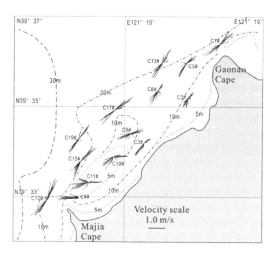

Figure 6.   Current ellipses during spring tide.

4.2   *Current*

According to tidal current data observed during August 2008, local current flowed as the following characteristics:

1. At deep water area in the west side of Majia Shoal, tidal current cannot be affected by Majia Cape, and flow reciprocatingly along the shoreline (see Fig. 6). Flow direction during flood tide is NE and is SW during ebb tide. During spring tide, mean flood velocity is 0.72 ~ 0.85 m/s and mean ebb tide is 0.55 ~ 0.75 m/s. Maximum flow velocity in full tide flood is 0.95 ~ 1.58 m/s.
2. At shallow area on Majia Shoal, tidal current is affected by Majia Cape and nearshore shallows, and flow reciprocatingly, with flood direction as E and ebb direction as W. During spring tide, mean flood velocity is 0.42 ~ 0.49 m/s, and mean ebb velocity is 0.52 ~ 0.57 m/s. Maximum velocity in full tide flood is 0.88 ~ 1.16 m/s. Current velocity on Majia shoal is not large.
3. At the trough between Majia Shoal and Changxing Island, tidal current flow reciprocatingly along the trough. During spring tide, mean flood velocity is 0.25 ~ 0.29 m/s, and mean ebb velocity is 0.58 ~ 0.70 m/s; maximum flood velocity is 0.48~0.66 m/s and maximum ebb velocity is 1.08~1.35 m/s. The ebb velocity is 2 times than flood velocity.

4.3   *Suspended sediment concentration*

1. Suspended Sediment Concentration (SSC) in Majia Shoal sea area is low, and the differences between spring tide and neap tide are small. In winter, mean SSC of flood and ebb tide are 0.099 kg/m³, 0.093 kg/m³ respectively during spring tide, and are 0.076 kg/m³, 0.077 kg/m³ respectively during neap tide; maximum SSC of flood and ebb tide are about 0.169 kg/m³, 0.179 kg/m³ respectively during spring tide. In summer, mean SSC of flood and ebb tide are 0.025 kg/m³, 0.021 kg/m³ respectively during spring tide, and are 0.012 kg/m³ and 0.012 kg/m³ respectively during neap tide; maximum SSC of flood and ebb tide are about 0.047 kg/m³, 0.058 kg/m³ respectively during spring tide.
2. Mean d50 of suspended sediment in Majia Shoal sea area is about 0.013 mm. Suspended sediment types are mainly clayey silt, with clay content greater than 30% and sandy content less than 10%. Suspended sediment is different to seabed sediment, so exchange between seabed sediment and suspended sediment is very limit and seabed sediment on Majia Shoal is not active.

4.4 *Sediment sources*

Changxing Island is a hilly bedrock island, so sediment supply from the island is very limit. Fuzhou River is the longest river nearby and flow into Fuzhou Bay with annual runoff of 263 million m³. The river discharge sediment mainly concentrated at river mouth in the north of Changxing Island and cannot transport to Majia Shoal sea area.

Northwest coast of Changxing Island are mainly belong to bedrock coast, so its erosion rate is very slow and sediment supply is small amount each year. Local coastline is controlled by Majia Cape and Gaonao Cape, so longshore sand from both north and south sides cannot transport into Majia shoal sea area. Overall, local sediment sources are very limit.

5 CONCLUSION

Based on the analysis of underwater topographic evolution, depositional environment, sediment sources, and the sediment depositional age, this paper discuss the causes and stability of Majia shoal. The conclusions are shown as follows:

1. Majia shoal is a linear sand ridge sheltered by Majia Cape, and its underwater topography can keep stable;
2. Majia shoal consists of median or coarse sand, and formed during geological processes in Holocene, which has existed for thousands of years;
3. Wave at Majia Shoal sea area is not strong and current velocity is not large;
4. Suspended sediment concentration is low and sediment sources are limit;
5. Seabed sediment at Majia Shoal is not active, and cannot transport under wave and current action, so Majia shoal can maintain a stable state for many years.

REFERENCES

Dong Tailu. Modern deposition models in Bohai Sea [J] Marine Geology and Quaternary Geology, 1996, 16 (4): 43–53.

Li Fan, Lin Meihua. Liaodong Bay seabed topography and residual deposition [J]. Bulletin of Marine Science, Episode 23, Beijing: Science Press, 1984: 56–67.

Liu Zhenxia, Xia Dongxing, Tang Yuxiang, etc. Holocene sedimentary system in eastern Bohai Sea [J] Science in China, 1994, 24 (12): 133–138.

Liu Zhenxia, Tang Yuxiang, Wang Kuiyang, etc. Tidal dynamic geomorphic system in the east part of the Bohai Sea [J]. Journal of Oceanography of Huanghai & Bohai Seas, 1996, 14 (1): 7–21.

Tang Yuxiang, Yao Lanfang, Liu Zhengxia. Characteristics of the tidal current movement and its relationship with the development of the sand ridges in Liaodong Bay [J]. Journal of Oceanography of Huanghai & Bohai Seas, 1993, 11 (4): 9–18.

Xia Dongxing, Liu Zhengxia. The submarine sand ridges in China Waters [J]. Yellow Sea and Bohai seas, 1983, 1 (1): 45–49.

Xia Dongxing, Liu Zhengxia. Formation mechanism and developmental conditions of sand ridges [J]. Acta Oceanologic Sinica 1984, 16(3): 361–367.

Yin Ping. Topography and internal structural characteristics of sand ridges in shelf of East China Sea during late Ice Age [J]. Advances in Marine Science, 2003, 21 (2): 181–187.

Hydraulic Engineering III – Xie (Ed)
© 2015 Taylor & Francis Group, London, ISBN 978-1-138-02743-5

Research on performance evaluation index system of the large-exhibition project risk management

Qiang Zhang
Shanghai Jianke Engineering Consulting Co. Ltd., Shanghai, P.R. China

ABSTRACT: The boom of large-scale complex projects over the world leads the risk management more and more imperative, which achieves good economic and social benefits. The risks of safety, quality, progress and investment of the large complex project construction were discussed, and a large complex project performance evaluation model of risk control was established, which can accomplish the fuzzy evaluation of risk control performance. This method makes the project managers to control the risk of project construction quantitatively and qualitatively.

1   INTRODUCTION

The current research has already covered the performance evaluation of the index system of engineering project. Huang Yu-kun (Huang Yu-kun, 2005) advanced the engineering project risk evaluation index was set up two layers of index factors, the first layer is the main factor, the evaluation indicators include: natural risk, design risk, construction risk, financial risk, economic risk, contract default risk and environment risk, political risk. Wu Jian-nan (Wu Jian-nan, 2007) divided the index into two dimensions: one is to form the good management of large public works projects in the hierarchical framework for effective set, which can be divided into four levels: project investment, project processes, project, project output influence. The second is the large public engineering project management of the whole process of modernization. The application of project management ideas were divided into three stages of project decision-making, project implementation and project operation. Man Yi (Man Yi, 2007) proposed performance evaluation through eight aspects which includes the quality management, economic management, schedule management, security management, organization management, project activity, social environment and the owners satisfaction. Wu Shao-li (Wu Shao-li, 2007) considers the engineering project risk index of the fuzzy comprehensive evaluation model based on cost, schedule, quality, safety, personnel organization and engineering efficiency as the main monitoring objects. Chen Min (Chen Min, 2012) proposed that the influence factors of engineering project risk could be divided into human factors, external factors, material purchasing and management factors, capital input factors.

Combining with China's expo exhibition complex projects's construction objectives and requirements, "Three guarantees a control" (Keep safety, Ensure quality, Ensure progress, control investment), the evaluation method was discussed focus on safety, quality, progress and investment, which provide an evaluation method of risk control.

2   THE PRINCIPLE OF BUILDING THE INDEX SYSTEM

In engineering projects, the evaluated objects were in large quantity, thus, the indexes can be dozens, hundreds, but the contribution degree of each index is distinct. When choosing the indexes, the indexes should be serious mixed with irrelevant indicators, which were not only increase the difficulty of the evaluation index system, but also easy to dilute the main indicators. When select the indexes, it should follow the principle of objective science, representative, commonality, independence, convenient operation, combining of quantitative and qualitative, etc. In this article, through expert interviews, as well as referring to a large number of literature, the construction phase of the project were divided into construction preparation stage, construction stage, completion inspection and acceptance stage. The elements of performance evaluation index system should include: 1) Criteria: the performance evaluation index system usually contains a number of levels, and each level contains a number of elements, which usually called the elements for the guidelines. 2) Goals: the level of the last time is often referred to as target layer, and the target is a kind of expectations, which shows the organization level of expectation result finally. 3) Indicators: the performance evaluation index system of the bottom of the generally called index layer. Indicators for the specific content of evaluation and general items were used to measure the realization of project objectives. Index layer is controlled by the criterion layer.

3 THE FIRST SELECT OF RISK INDICATORS

According to the results of the expert consultation, preliminary selection of general indexes was selected. As to the evaluation index's of schedule risk, in the construction preparation stage, according to the expert interview and literature research, preliminary selected in the index layer of schedule management planning stage evaluation index have a study and analysis of the total package contract, conformance with the construction procedure, construction general contract signing, project starting time effectiveness, construction progress plan, construction schedule planning conditions. In the construction phase, the evaluation index includes planning management, mechanical equipment to ensure continuous improvement and progress management process inspection, etc. In the completion acceptance stage, the evaluation index includes contract time limit for a project implementation situation, schedule management, customer complaints and completed summary writing situation and other factors.

In terms of investment, the construction preparation stage index were market price stability, the stability of laws and regulations policy, site conditions, owners, designers and management factors such as the bidding process. In the construction phase, the indicators of risk factors were contract risk, complex technology, site conditions, etc.

As to the safety, in the preparation stage, the indicators have project safety management organization intends to set up and all kinds of personnel related certificates of the project, project staff security responsibility intends to decomposition, intends to employees to be awwere of the "contract" and so on. In the construction phase, the subcontracting, specialized contracting (collectively referred to as subcontractors) resources such as security, construction machinery and equipment safety management, environmental protection material safety environmental management, safety education, safety acceptance, dismantle management and hidden dangers rectification of evaluation index. In the completion acceptance stage, safety evaluation index includes civilization construction effect, safety accidents and bad social influence and other factors.

In terms of quality, in the construction preparation stage, the evaluation index were design disclosure and construction organization design, the construction unit and the overall quality, engineering measurement. During the construction stage, the evaluation index includes the technical clarification situation, ISO 9000 certification, mechanical equipment operation situation, the three indexes of inspection system enforced. In the final acceptance stage, the acceptance of a unit, finished product protection appropriately and other factors. According to the above analysis of various risk factors, the importance of the relationship between a comprehensive consider index, based on the idea of multidimensional performance evaluation, combining with the characteristics of large conference and exhibition complex construction project management, safety, quality, progress, investment risk of key indicators summarized as follows, constructing evaluation index system were shown in Table 1.

Table 1. Safety, quality, progress, investment risk index system.

Target layer	Criterion layer	Index (indicators)
Schedule management performance evaluation system	Construction preparation stage	Study and analysis of the general contracting
		Construction procedures conform to the situation the situation of construction general contract signed
		Construction must be approved by branch
		The preparation of construction management planning report material
		The time when the project start is effective
		Total construction schedule planning
		Mechanical equipment planning management
		Labor and the subcontract plan management
		The project intends to on schedule performance index decomposition
		The civilization construction planning
	The construction phase	Planning and management of continuous improvement
		Mechanical equipment guarantee
		Human resources guarantee
		The security situation
		Check schedule management process
		Delays single sign
		Shutdown compliance situation
		Project department self-evaluation
	The completion acceptance stage	The contract time limit for a project performance
		Schedule management customer complaints and completed summary writing
Investment management performance evaluation system	Construction preparation stage	The market price stability
		Laws and regulations policy stability
		Site conditions
		The owner management ability
		Stylist ability
		The bidding process
	The construction phase	The contract risks
		Technical complexity
		Site conditions
		Contractor's ability to
		Stylist ability
		The owner management ability
The safety management performance evaluation system	Construction preparation stage	Project safety management organization intends to set up related certificates and all kinds of personnel
		Project staff security responsibility intends to break down
		Project staff intends to the understanding of the "contract"
		Construction organization design and approval
		Risk factors for identification
		Production safety accident emergency rescue plan
	The construction phase	The subcontracting, specialized contracting (collectively known as subcontractors), and other resources
		Construction machinery safety environmental management
		Supplies ehs management
		Safety education to carry out
		Technical measures to guarantee
		The implementation of safety technical measures, construction management
		Security special acceptance, remove the management
		Dangerous operations management
		Fire management

(Continued)

241

Table 1. *Continued*

Target layer	Criterion layer	Index (indicators)
Quality management performance evaluation system		Health and epidemic prevention and occupational disease prevention and control of management and effect
		Safety check
		Hidden dangers rectification of
		Management improvement
	The completion acceptance stage	Civilization construction effect
		Safety accidents and bad social influence
	Construction preparation stage	Design drawings and/or disclosure came
		Construction organization design (quality)
		Construction units in qualification and comprehensive quality
		Engineering measurement is accurate
	The construction phase	Technical clarification
		ISO9000 certification
		Construction materials inspection rate
		The workings of the mechanical equipment
		Process quality acceptance rate at a time
		Three inspection system were enforced
		Among the professional collaboration
	The completion acceptance stage	The qualification rate of acceptance of a unit
		Complete and accurate timeliness of completion data
		Finished products protection appropriately

4 INDICATORS OF RISK SCREENING

In order to enhance the scientific nature, the rationality of the evaluation index, the theory of the evaluation indicators for membership degree analysis, correlation analysis, discrimination analysis and reliability test were selected.

We sent through the design of 57 experts consulting tables and received 55 pieces, including 50 effective feedback reports. From the initial screening of the evaluation index which they think the most important evaluation indexes were selected. Based on the effective feedback reports, membership degree of evaluation index is analyzed. Assumptions of the No. i evaluation expert selected a total number of M_i, that is to say a total of M_i experts think No. i is the quality, safety, investment, schedule management performance evaluation index, the important then the membership degree of evaluation index were as follows:

$$R_i = \frac{M_i}{50}$$

If R_i value is large, the evaluation indexes in the evaluation system is very important, which could be used as a formal evaluation index. On the other hand, the evaluation index is necessary to be deleted. Through the statistical analysis of 50 effective feedback reports, the membership degree was lower than 0.3 of 15 evaluation indexes were deleted, and the 41 indicators were retained, which formed the second round safety, quality, progress, investment risk index system.

To eliminate or reduce the assessment indexes repeated reflection evaluation object information and great influence on the evaluation results, it is necessary for the second round of assessment system of evaluation indicators for correlation analysis. In this article, the method of SPSS 16.0 was used through in-depth field investigation and collected eight construction unit safety, quality, progress, investment risk evaluation system of performance evaluation of the second round. A given threshold M 0.9, the last 10 indicators of the membership degree were lowest are removed, and the rest of the evaluation index of 31 were retained, which formed the third round of safety, quality, progress, investment risk index system.

Table 2. The significance of a factor.

a	0.00–0.30	0.30–0.50	0.50–0.70	0.70–0.90	0.90–1.00
Significance	Not credible	A little confidence	Credible	Very reliable	Highly trusted

Through study of the discrimination of evaluation index analysis, the safety, quality, progress, investment risk management performance evaluation index were the ability assessment of different construction enterprises or the same enterprise in different construction engineering. In constructing evaluation system, if all the comments in a construction project evaluation index on present were almost uniformly high (or low) score, then this evaluation index was almost no discrimination, and it is difficult to identify the differences between the project's quality performances. On the contrary, if the evaluation of the project in score was significantly different on one of the indicators, which suggests that the evaluation index has high resolution, then it is able to identify differences between the engineering quality performances. In practice, coefficient of variation to describe evaluation index of discrimination was usually used, namely:

$$V_i = \frac{S_i}{\overline{X}}$$

In this formula, $\overline{X} = 1/n \sum_{i-1}^{n} X_i$ is the average value, $S_i = \sqrt{1/n - 1 \sum (X_i - \overline{X})^2}$ is the standard deviation. The greater of the coefficient of variation, the index's identification ability is stronger. According to actual needs, if the coefficient of variation is relatively smaller, the evaluation index could be deleted. According to the above principles, using method of SPSS 16.0 to variance analysis, the evaluation index on the basis of analysis of variance computation in the evaluation system of 31 evaluation index variation analysis, the coefficient of variation of six evaluation indexes is relatively smaller, then they were deleted, and the other 25 indicators were retained, which constitutes the fourth round of safety, quality, progress, investment risk index system.

According to the requirements of performance evaluation theory, the construction stage of construction quality management performance evaluation system of index content and the structure is reasonable, good results can be trusted, eventually the need for reliability test. This article adopts the method of internal consistency reliability to verify the reliability of the evaluation system. The meaning of the coefficient is shown in Table 2.

This article collected eight engineering construction progress, quality, safety, investment performance evaluation index system of data, and the method of SPSS 16.0 was used in the fourth round of the evaluation index of reliability analysis. Then the evaluation system of overall and various indicators of a coefficient, namely the "a" is 0.842 in the overall project, the "a" is 0.828 before construction, the "a" is 0.833 in the construction period, and the "a" is 0.834 after the construction. Therefore, the evaluation system of the internal structure is the basic consistent, which could achieve the basic requirements of evaluation theory.

The evaluation indexes were mutual connection, restriction and influence. With the change of an indicator, the related indicators will be changed, which affects the whole index system changes, thus to influence the outcome of the performance evaluation. For example, if the engineering quality was improved, the lower profit margins and the time may be delay; If the time is shorter, it may cause a drop in the quality of the rising in cost. Therefore, in view of the performance evaluation of the project should give full consideration to the relations among the evaluation index, optimize the various indicators of decision making, in order to achieve the project's overall goal optimal. In addition, this paper argues that in considering the risk indicators, management based indicators were also considered, as shown in Table 3.

Table 3. The index after screening.

Target layer	Criterion layer	Index (indicators)
Schedule management performance evaluation	Construction preparation stage	Construction general contract signed Total construction schedule planning Mechanical equipment planning management Labor and the subcontract plan management Project staff security responsibility intends to break down
	The construction phase	Planning and management of continuous improvement Delays single sign Shutdown compliance situation
Investment management performance evaluation system	Construction preparation stage	The market price stability Laws and regulations policy stability The bidding process
	The construction phase	The contract risks Technical complexity Site conditions
Safety management performance evaluation system	Construction preparation stage	Risk factors for identification Production safety accident emergency rescue plan
	The construction phase	Construction machinery safety environmental management Safety education to carry out Security special acceptance, remove the management Dangerous operations management Safety check Hidden dangers rectification of
Quality management performance evaluation system	Construction preparation stage	Construction units in qualification and comprehensive quality
	The construction phase	Technical clarification Construction materials inspection rate Process quality acceptance rate at a time Among the professional collaboration
	Foundations of risk management	Management system of rules, organization management, human resources, quality, project culture

5 PERFORMANCE EVALUATION OF PROJECT RISK CONTROL TECHNOLOGY RESEARCH

5.1 *The weight of evaluation index set*

When dealing with multi-objective and decision making, many factors should be considered to have more or less, greatly or small, light or weight, and so on. However, they have a common characteristic, the importance, influence or priority level of these factors were often hard to quantify when analysis, comparison, judgment, evaluation, decision-making. It make the person's subjective choice will play a very important role. When determining the weight of each index in the index system, the need for the project of industry characteristics, the project's life cycle, project specific operating environment objective comprehensive consideration of various factors on the assigned and appropriate weights were different, thus ensure the truthful and accurate evaluation results.

The analytic hierarchy process (SHP) to determine the index weight was used. By establishing the hierarchy model, to construct judgment matrix, hierarchical sorting and consistency check to confirm the weight. By designing the questionnaire, more than 50 invited experts to participate in the questionnaire, and calculate the survey results, the weights were obtained as show in Table 4.

5.2 *The project performance evaluation based on fuzzy membership degree evaluation analysis*

After confirmed the project performance evaluation index weight, the project's overall performance could be evaluated with the further quantitative evaluation. The quantitative evaluation

Table 4. Construction project schedule, safety, quality, investment performance index weight.

Target layer	Criterion layer	Index (indicators)	Ratio
Schedule management performance evaluation system	Construction preparation stage	Construction general contract signed	6%
		Total construction schedule planning	20%
		Labor and the subcontract plan management	9%
		The project intends to on schedule performance index decomposition	13%
	The construction phase	Planning and management of continuous improvement	23%
		Delays single sign	5%
		Shutdown compliance situation	11%
	Foundations of risk management	Management system of rules, organization management, human resources, quality, project culture	12%
Investment management performance evaluation system	Construction preparation stage	The market price stability	5%
		Laws and regulations policy stability	2%
		The bidding process	33%
	The construction phase	The contract risks	15%
		Technical complexity	14%
		Site conditions	13%
	Foundations of risk management	Management system of rules, organization management, human resources, quality, project culture	12%
The safety management performance evaluation system	Construction preparation stage	Risk factors for identification	5%
		Production safety accident emergency rescue plan	7%
	The construction phase	Construction machinery safety environmental management	5%
		Safety education to carry out	1%
		Security special acceptance, remove the management	6%
		Dangerous operations management	17%
		Safety check	22%
		Hidden dangers rectification of	31%
	Foundations of risk management	Management system of rules, organization management, human resources, quality, project culture	12%
Quality management performance evaluation system	Construction preparation stage	Construction units in qualification and comprehensive quality	12%
	The construction phase	Technical clarification	16%
		Construction materials inspection rate	31%
		Process quality acceptance rate at a time	21%
		Among the professional collaboration	8%
	Foundations of risk management	Management system of rules, organization management, human resources, quality, project culture	12%

indexes for processing are divided into two steps: first, the evaluation index of qualitative indexes and quantitative values, so as to the final evaluation result is obtained by quantitative calculation. Secondly, evaluation index dimensionless processing should be carried out, because each evaluation index may have a different dimension unit. Only after the dimensionless processing was carried out, operations between different indicators could be merged. In this article, the quantitative indicators can be mainly comes from the relevant department of financial statements, management, log files, such as for the other to acquire data, you can use the questionnaire survey, field investigation, query and other means to get online.

First set index of fuzzy comprehensive evaluation sets U = {excellent, good, pass, fail}, corresponding values {4, 3, 2, 1}. Design and sent a total of 57 experts reports, and receive 50 effective reports, M_i shows a total number of expert choices,

Table 5. Quality management performance indicators valuations.

Target layer	Criterion layer	Index (indicators)	Ratio	Value of assessment
Quality management performance evaluation system	Construction preparation stage	Construction units in qualification and comprehensive quality	12%	4
	The construction phase	Technical clarification	16%	4
		Construction materials inspection rate	31%	3
		Process quality acceptance rate at a time	21%	3
		Among the professional collaboration	8%	2
	Foundations of risk management	Management system of rules, organization management, human resources, quality, project culture	12%	3

$$R_i = \frac{M_i}{50}$$

Among them, the maximum value of membership degree for the index is selected as the judge value. If the same value of R_i appears, the lower values is selected.

We take the quality management performance evaluation system as an example, the indexes of valuation are as shown in Table 5.

So, the value of the quality management system assessment is

$$= 4 * 12\% + 4 * 16\% + 3 * 31\% + 3 * 21\% + 2 * 8\% + 2 * 12\% = 0.48 + 0.64 + 0.93 + 0.63$$
$$+ 0.16 + 0.36 = 3.2$$

The evaluation result is passing the exam, and then the project manager could control the basic information of project quality management and could also strengthen the weak links of quality management. The other three risk evaluation could be assessed in the same way.

6   CONCLUSION

In the exhibition project risk management performance evaluation, the risk index system is established based on the principle of selecting important factors and indexes, and the safety, quality, progress and investment risk are selected as the performance evaluation, which could be the guide of risk management for the manager. However, it still has some deficiency, such as in adopting the combination of quantitative and qualitative methods to analyze the factors affecting the performance of large conference and exhibition complex projects. So, the factors and indexes need to be constantly improved, and the index system of multiple objective risk management should be established.

REFERENCES

Chen Min. Performance evaluation research of Engineering project risk management [J] science and Technology Innovation Herald. 2012(18): 193.
Huang Yu kun. Fuzzy comprehensive evaluation on project risk [J]. Journal of Zhejiang Institute of Communications, 2005(6): 1–4.
Hu Fang. Research on performance evaluation of large public engineering project [D]. Hunan: Hunan University, 2012.
Hu Fang. Performance evaluation research on large public engineering project [D]. Hunan: Hunan University, 2012.
Man Yi. Study on performance evaluation of city infrastructure project [D]. Hunan: Central South University, 2007.
Wu Shao li. Risk identificate and evaluation method Design of the project [D]. Sichuan: Southwest Jiao Tong University, 2007.

Hydraulic Engineering III – Xie (Ed)

Distribution pattern and potential ecological risk assessment of heavy metals in the sediments of Jiulong River estuary

Xiaofeng Guo, Zhouhua Guo, Cui Wang & Keliang Chen
Third Institute of Oceanography, SOA, Xiamen, Fujian, China

ABSTRACT: An analysis and evaluation of the monitoring results on seven heavy metals in 18 sample points of Jiulong River estuary was conducted in May 2011. Heavy metal pollution was assessed based on the geo-accumulation index and the potential ecological risk index. The result of geo-accumulation index indicates the pollution degree of the seven heavy metals is Cd > Zn > Pb > Cu > As > Hg > Cr; and the result of potential ecological risk index indicates that the ecological hazard degree of the seven heavy metals is: Cd > Hg > As > Pb > Cu > Zn > Cr. The potential ecological risk of heavy metals in the surface sediments is of intermediate level. Results of the two assessment methods both show that the pollution degree of Cd is the severest, which indicates that the heavy metal pollutants at Jiulong river estuary mainly come from the drainage basin.

1 INTRODUCTION

Estuary waters are the main passage that carries terrigenous matters to the seas. Masses of pollutants produced during human activities go to sea through estuaries and then spread across the open sea. Heavy metal pollutants from the land are very complex in the marine biogeochemical process in the offshore area, which settle into the sediments with the fine particle matters in the sea, partly return to the sea under the influence of human activities, physical, chemical and biological factors and result in secondary pollution of sea water quality. Heavy metal is a kind of accumulative and toxic pollutant that is not decomposable by microorganisms. It can accumulate in biological bodies. And some heavy metals can convert into metal-organic compounds of severer toxicity, posing potential threats to marine ecology and human health. Thus research on heavy metal environmental quality in sea sediments has become a hot topic for domestic and international marine environmental protection.

Jiulong River is the second largest river in Fujian Province. In recent years, with the rapid development of the social economy in Jiulong River drainage basin, there has been an increasing amount of pollutants flowing to the sea, leading to frequent red tides and degrading ecosystem in Jiulong River estuary sea area. The research on heavy metals source in the sediments of Jiulong River estuary and its environmental quality is an important aspect for overall evaluation of the marine environmental quality of this area. The present paper, with the surface sediments in Jiulong River estuary sea area as the research object, carries out a comprehensive analysis of the heavy metal concentration and distribution in the surface sediments, probes into the source of the pollutants, and gives an environmental quality assessment, and consequently, provids a scientific basis for environmental protection and ecological restoration in the sea area.

2 MATERIALS AND METHODS

2.1 Sampling station and sample collection

Based on the characteristics of Jiulong River estuary sea area, altogether 18 sampling stations are set up and the samples are collected in May 2011. The investigated zone and

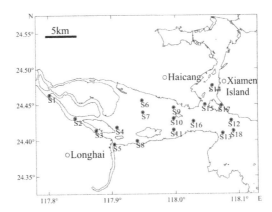

Figure 1.   Sampling Stations in Jiulong River estuary.

sampling stations are illustrated in Figure 1. Sample collection conforms to Part 5 of *The Specification for Marine Monitoring—Sediments Analysis (GB17378.5-2007)*. The bottom mud is obtained with the gravity bottom sampler; then the sediments within 0~2 cm on the mud surface are taken with a wood spoon and put into the sample bag for laboratory analysis.

2.2   *Monitoring items and analysis method*

Monitoring items include the measurement of the 7 heavy metal indexes: Cu, Pb, Zn, Cd, Cr, Hg and As. Sediments decomposition and monitoring methods can be found in Part 5 of *The Specification for Marine Monitoring—Sediments Analysis (GB17378.5-2007)*. For Cu and Zn, the flame atomic absorption spectrophotometry is adopted; for Pb, Cd and Cr, the non-flame atomic absorption spectrophotometry is used; For Hg and As, the atomic fluorescence spectroscopy is applied. The accuracy of standard substances is also measured according to national offshore sediment analysis. The analysis results meet the requirement. During the analysis, reagent blank and parallel samples are casually measured, the results of which show that the analysis process is not polluted and the relative standard deviation for the parallel samples are all lower than 10%.

3   HEAVY METAL CONCENTRATION AND DISTRIBUTION IN SEDIMENTS

The concentration of the 7 heavy metals in Jiulong River estuary surface sediments can be seen in Table 1, which shows that the average concentration of Cu, Pb, Zn, Cd, Cr, Hg and As is respectively 16.5, 46.4, 122.7, 0.31, 0.05 and 3.41 mg·kg^{-1}. Compared with heavy metal concentration in the surface sediments of other sea areas of Xiamen in the year 2011 (refer to Table 2), the heavy metal concentration in Jiulong River estuary is of intermediate level, lower than that of Xiamen western sea area and Maluan Bay but higher than that of southern sea area, Tong'an Bay and Dadeng sea area. Refer to Figure 2 for the division of Xiamen sea area.

The distribution of heavy metal concentration illustrated in Figure 3 shows a relatively consistent distribution of Cu, Pb, Zn, Cd and Cr in the sediments of Jiulong River estuary sea area. High concentrations appear in the northeast of Zini Island. The maximum concentration of Cu, Pb, Zn and Cd appears at Station 6 while that of Cr appears at Station 7.

Table 1. Heavy metal concentration in surface sediments (mg·kg⁻¹).

Heavy metal	Maximum	Minimum	Average
Cu	37.9 (S6)	1.93 (S2)	16.5
Pb	128.0 (S6)	2.91 (S13)	46.4
Zn	407.0 (S6)	52.1 (S14)	122.7
Cd	2.51 (S6)	0.04 (S18)	0.31
Cr	35.0 (S7)	0.38 (S13)	14.0
Hg	0.154 (S1)	0.014 (S2)	0.05
As	6.0 (S1)	1.0 (S2)	3.41

Table 2. Comparison of heavy metal concentration in surface sediments in different sea areas of Xiamen Bay (mg·kg⁻¹).

Area	Cu	Pb	Zn	Cd	Cr	Hg	As
Maluan Bay	58.3	40.1	156.8	1.16	54.6	0.29	7.0
Western sea area	42.0	54.8	143.2	0.12	46.7	0.12	5.9
Southern sea area	9.7	27.5	66.7	0.05	11.7	0.07	3.1
Eastern sea area	11.7	27.1	68.4	0.05	18.1	0.09	3.3
Tong'an Bay	12.1	25.1	60.1	0.04	23.4	0.05	4.7
Dadeng sea area	9.4	32.8	60.4	0.03	12.1	0.06	6.5
Jiulong River estuary	16.5	46.4	122.7	0.31	14.0	0.05	3.41

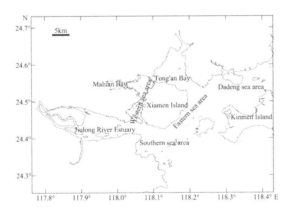

Figure 2. The division of Xiamen sea area.

4 ASSESSMENT METHOD OF HEAVY METAL POLLUTION IN SURFACE SEDIMENTS

4.1 Geo-accumulation index

Geo-accumulation index is to determine the pollution level of heavy metal by means of checking the relationship between the total concentration of heavy metals and the background value. As a quantitative index developed in Europe in late 1960s and widely used in researches on heavy metal pollution level in sediments and other matters, it is extensively applied to the evaluation of heavy metal pollution in modern sediments. The following is its calculation formula:

$$I_{geo} = \log_2\left[C_n/(k \times B_n)\right] \tag{1}$$

249

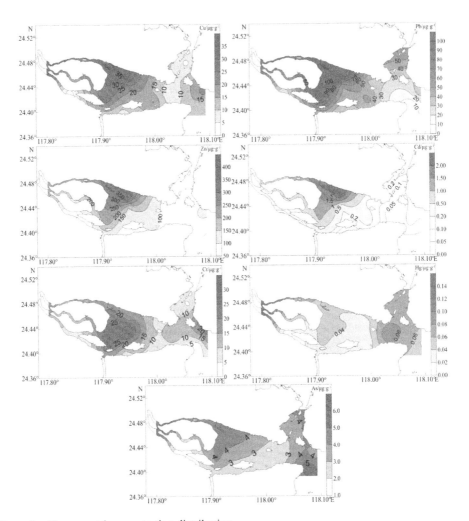

Figure 3. Heavy metal concentration distribution.

In the formula above, I_{geo} means geo-accumulation index, C_n stands for the measured heavy metal concentration in the sediments, and B_n represents the geochemical background value of the measured element in sedimentary rock, which in this study adopts the heavy metal background value in the soil of the coastal areas of Fujian Province: the value for Cu, Zn, Hg, Cr, Pb, As and Cd is respectively 18.0, 69.9, 0.112, 32.1, 37.6, 5.52 and 0.046 mg/kg. k is the constant set for the background value change possibly arising in diagenesis. Generally, $k = 1.5$.

4.2 Potential ecological risk index

Potential ecological risk index is a method of assessing heavy metal pollution in sediments or soil proposed by Swedish scholar Hakanson from the perspective of sedimentology according to the property of heavy metal and the environmental behavior features. This method, which is widely used, can not only reflect the influence of one pollutant in the environment but also the integrated influence of multiple pollutants. The risk index is assessed by quantitative means. The calculation formula is like this:

$$C_f^i = C^i / C_n^i; \quad C_d = \sum C_f^i; \quad E_r^i = T_r^i \times C_f^i; \quad RI = \sum E_r^i \qquad (2)$$

Table 3. Geo-accumulation index and pollution level.

Igeo	<0	0~1	1~2	2~3	3~4	4~5	>5
Grades	0	1	2	3	4	5	6
Pollution levels	None	Slight	Moderate low	Moderate	Moderate high	Bad	Severe

Table 4. Potential ecological hazard index of heavy metal and grading.

The potential ecological risk of a single element		The potential ecological risk of multiple elements	
$E_r^i < 40$	Low	RI<150	Low
$40 \leq E_r^i < 80$	Moderate	$150 \leq RI < 300$	Moderate
$80 \leq E_r^i < 160$	Considerable		
$160 \leq E_r^i < 320$	High	$300 \leq RI < 600$	High
$E_r^i \geq 320$	Very high	$RI \geq 600$	Very high

In the formula above, C_f^i refers to the pollution index of a single metal; C^i is the measured value of heavy metal in surface sediments; and C_n^i is the reference value for the sediments in heavy metal. C_d means the integrated pollution level of surface sediments; E_r^i represents the potential ecological risk factor of one element; and T_r^i stands for the toxic coefficient of each heavy metal in the sediments. The toxic coefficient of standardized heavy metal according to Hankanson is: Zn(1) < Cr(2) < Cu(5) = Pb(5) < As(10) < Cd(30) < Hg (40). RI is the potential ecological risk index of surface sediments. Table 4 demonstrates the relationship between the potential ecological hazard index of heavy metal and the grading.

5 ASSESSMENT RESULTS FOR HEAVY METAL POLLUTION IN SURFACE SEDIMENTS

5.1 Geo-accumulation index

Using geo-accumulation index (I_{geo}) to assess the heavy metal pollution in surface sediments of Jiulong River estuary sea area, the geo-accumulation indices I_{geo} of heavy metal measured at all sampling stations are listed in Table 5. Generally, I_{geo} of Cu lies between –3.81 and 0.49; I_{geo} of Pb lies between –4.28 and 1.18; I_{geo} of Zn lies between –0.96 and 1.96; I_{geo} of Cd lies between –0.79 and 5.18; I_{geo} of Cr lies between –6.99 and –0.46; I_{geo} of Hg lies between –3.58 and –0.13; I_{geo} of As lies between –3.05 and –0.46. Based on the classification of pollution grade and pollution level, the pollution of the 7 heavy metals goes in the following order: Cd > Zn > Pb > Cu > As > Hg > Cr. The average pollution level of Cd is the severest, which falls in the "Moderate" pollution level (Grade 3). Zn follows and belongs to the "Slight" pollution level (Grade 1).

5.2 Potential ecological risk index

The pollution coefficient (E_r^i) of single heavy metal and the potential ecological risk index (RI) calculated with potential ecological risk index for the surface sediments in Jiulong River estuary are shown in Table 6. In terms of the pollution coefficient (E_r^i) of single heavy metal, E_r^i value of Cu lies between 0.54 and 10.53, with the average value being 4.58; E_r^i value of Pb lies between 0.39 and 17.02, with the average value being 6.17; E_r^i value of Zn lies between 0.75 and 5.82, with the average value being 1.76; E_r^i value of Cd lies between 26.09 and 1636.96, with the average value being 205.18; E_r^i value of Cr lies between 0.02 and 2.18, with the average value being 0.87; E_r^i value of Hg lies between 5.00 and 55.00, with the average value being 18.75; E_r^i value of As lies

Table 5. Geo-accumulation indices of heavy metals in surface sediments of Jiulong River estuary.

Station	Cu	Pollution level	Pb	Pollution level	Zn	Pollution level
1	0.43	Slight	0.48	Slight	1.02	Moderate low
2	−3.81	None	−1.30	None	−0.96	None
3	−0.79	None	−0.22	None	−0.47	None
4	0.07	Slight	0.83	Slight	1.16	Moderate low
5	−1.23	None	−0.44	None	0.25	Slight
6	0.49	Slight	1.18	Moderate low	1.96	Moderate low
7	−0.24	None	−0.14	None	0.52	Slight
8	−1.73	None	−0.72	None	−0.67	None
9	−0.76	None	−0.55	None	0.10	Slight
10	−0.87	None	−0.30	None	0.32	Slight
11	−0.40	None	−0.06	None	0.27	Slight
12	−0.32	None	−0.64	None	0.31	Slight
13	−0.95	None	−4.28	None	−0.46	None
14	−2.87	None	−0.20	None	−1.01	None
15	−1.48	None	−1.04	None	−0.65	None
16	−2.43	None	−1.24	None	−0.92	None
17	−0.55	None	−0.90	None	−0.37	None
18	−1.36	None	−1.37	None	−0.95	None
Average	−0.71	None	−0.28	None	0.23	Slight

Station	Cd	Pollution level	Cr	Pollution level	Hg	Pollution level	As	Pollution level
1	2.57	Moderate	−1.66	None	−0.13	None	−0.46	None
2	−0.13	None	−2.84	None	−3.58	None	−3.05	None
3	0.16	Slight	−2.52	None	−3.30	None	−2.12	None
4	3.94	Moderate high	−0.62	None	−1.90	None	−0.98	None
5	0.81	Slight	−1.57	None	−2.26	None	−1.98	None
6	5.18	Severe	−1.51	None	−2.14	None	−0.85	None
7	1.77	Moderate low	−0.46	None	−1.97	None	−1.16	None
8	0.83	Slight	−2.59	None	−2.58	None	−1.98	None
9	0.88	Slight	−2.49	None	−2.14	None	−1.42	None
10	1.79	Moderate low	−2.21	None	−2.14	None	−1.62	None
11	2.01	Moderate	−2.59	None	−2.49	None	−1.62	None
12	0.03	Slight	−1.27	None	−1.61	None	−1.33	None
13	−0.20	None	−6.99	None	−1.44	None	−0.56	None
14	0.21	Slight	−3.45	None	−1.42	None	−1.09	None
15	−0.20	None	−2.04	None	−1.16	None	−1.42	None
16	−0.46	None	−1.89	None	−0.78	None	−1.85	None
17	−0.20	None	−1.05	None	−1.18	None	−0.91	None
18	−0.79	None	−1.80	None	−2.14	None	−0.82	None
Average	2.19	Moderate	−1.79	None	−1.68	None	−1.28	None

between 1.81 and 10.87, with the average value being 6.18. The potential ecological risk of the 7 heavy metals goes in the following order: Cd > Hg > As > Pb > Cu > Zn > Cr. The Ecological risk of Cd has reached the "High" level while that of other heavy metals is the "Low" level. regarding the potential ecological Risk Index (RI), the value lies between 52.80 and 1693.29, while the average value being 243.48. The potential ecological risk of the heavy metals in surface sediments of the whole Jiulong River estuary sea area is at the "Moderate" level.

Table 6. Potential ecological risk indices of heavy metals in surface sediments of Jiulong River estuary and risk level.

Station	E_r^i Cu	Pb	Zn	Cd	Cr	Hg	As	RI	Potential ecological risk level
1	10.08	10.45	3.03	266.74	0.95	55.00	10.87	357.12	High
2	0.54	3.05	0.77	41.22	0.42	5.00	1.81	52.80	Low
3	4.33	6.42	1.08	50.35	0.52	6.07	3.44	72.22	Low
4	7.86	13.30	3.36	691.30	1.96	16.07	7.61	741.46	Very high
5	3.19	5.53	1.79	78.91	1.01	12.50	3.80	106.74	Low
6	10.53	17.02	5.82	1636.96	1.05	13.57	8.33	1693.29	Very high
7	6.33	6.82	2.15	153.91	2.18	15.36	6.70	193.45	Moderate
8	2.26	4.55	0.94	80.22	0.50	10.00	3.80	102.27	Low
9	4.42	5.12	1.60	82.83	0.53	13.57	5.62	113.69	Low
10	4.11	6.09	1.87	155.87	0.65	13.57	4.89	187.06	Moderate
11	5.67	7.21	1.80	180.65	0.50	10.71	4.89	211.43	Moderate
12	6.03	4.80	1.86	45.91	1.25	19.64	5.98	85.47	Low
13	3.89	0.39	1.09	39.20	0.02	22.14	10.14	76.87	Low
14	1.03	6.53	0.75	52.17	0.27	22.50	7.07	90.32	Low
15	2.69	3.66	0.95	39.13	0.73	26.79	5.62	79.57	Low
16	1.39	3.16	0.80	32.61	0.81	35.00	4.17	77.93	Low
17	5.11	4.03	1.16	39.13	1.45	26.43	7.97	85.28	Low
18	2.92	2.90	0.77	26.09	0.86	13.57	8.51	55.62	Low
Average	4.58	6.17	1.76	205.18	0.87	18.75	6.18	243.48	Moderate

Table 7. Comparison of assessment results from potential ecological risk index and geo-accumulation index.

Station	RI	RI grade	Average Igeo	Total grade of Igeo
1	357.12	High	0.32	7
2	52.80	Low	−2.24	0
3	72.22	Low	−1.32	1
4	741.46	Very high	0.36	8
5	106.74	Low	−0.92	2
6	1693.29	Very high	0.62	11
7	193.45	Moderate	−0.24	3
8	102.27	Low	−1.35	1
9	113.69	Low	−0.91	2
10	187.06	Moderate	−0.72	3
11	211.43	Moderate	−0.70	4
12	85.47	Low	−0.69	2
13	76.87	Low	−2.12	0
14	90.32	Low	−1.40	1
15	79.57	Low	−1.14	0
16	77.93	Low	−1.37	0
17	85.28	Low	−0.74	0
18	55.62	Low	−1.32	0

6 CONCLUSION

Geo-accumulation index and potential ecological risk index are two important methods to assess the heavy metal environment quality of the sediments. Table 7 gives a comparison of the results of the two methods, from which it can be seen that the assessment results of both methods are consistent in reflecting the overall trend of the heavy metal pollution in the sediments. I_{geo} grade and RI level correspond to each other at most of the stations.

Geo-accumulation index is an assessment method based on total heavy metal concentration and heavy metal background value, which takes into consideration the influence of human activities on the environment and the factor that the natural diagenesis might bring about background value change. But it can not indicate the chemical toxicity and biological activity of the heavy metals in the sediments. Potential ecological risk index considers the concentration, distribution and biological toxicity of different heavy metals in the sediments. And in combination with spatial heterogeneity of the background value, it is capable of reflecting the hazard the heavy metals pose for the environmental biology. It takes into account the toxicity and migration & transformation rule of the pollutants. The assessment result of geoaccumulation index shows that the 7 heavy metals have the following order of pollution level: Cd > Zn> Pb > Cu > As > Hg > Cr. The assessment result of potential ecological risk index (RI) shows the 7 heavy metals have the following order of potential ecological hazard: Cd > Hg > As > Pb > Cu > Zn > Cr. With different lines of thinking and emphases, the results of the two methods are somewhat different. However, the overall assessments by both methods are consistent in that Cd pollution is the severest. By using two different methods, a more objective assessment is obtained to assess the pollution situation of local sea areas.

From the angle of potential ecological risk index (RI), the potential ecological risk of the heavy metals in surface sediments of the whole Jiulong River estuary sea area is at the "Moderate" level, which is consistent with the research findings by Li Guihai (2007) and Lin Cai (2011). The assessment results of both methods suggest Cd pollution is the severest in surface sediments of this water area. Cd mainly comes from industrial waster water discharge. It can be concluded that the heavy metals in the sediments of Jiulong River estuary are mainly from drainage basin. Therefore, comprehensive improvement shall be greatly promoted in Jiulong River drainage basin, so as to reduce the pollution brought in by the drainage basin.

REFERENCES

Adamo, P., Arienzo, M. & Imperato, M., et al. 2005. Distribution and partition of heavy metals in surface and sub-surface sediments of Naples city port. *Chemosphere* (61): 800–809.
Attrill, M.J. & Thomesm, R.M. 1995. Heavy metal concentrations in sediment from the Thames estuary, UK. *Marine Pollution Bulletin* 30(11): 742–744.
Caplat, C., Texier, H. & Barillier, D., et al. 2005. Heavy metals mobility in harbour contaminated sediments: The case of Port-en-Bessin. *Marine Pollution Bulletin* 50(5): 504–511.
Fukue. M., Yanai, M. & Sato, Y., et al. 2006. Background values for evaluation of heavy metal contamination in sediments. *Journal of Hazardous Materials* (136): 111–119.
Hakanson, L. 1980. An ecological risk index for aquatic pollution control: A sediment ological approach. *Water Research* 14(8): 975–1001.
Huang, K. & Lin, S. 2003. Consequences and implication of heavy metal spatial variations in sediments of the Keelung River drainage basin, Taiwan. *Chemosphere* (53): 1113–1121.
Li, G., Lan, D. & Cao, Z., et al. 2007. Specificity and Potential Ecological Risks of Heavy Metals in the Sediments of the Xiamen Sea Area. *Marine Science Bulletin* 26(1): 67–72.
Lin, C., Lin, H. & Chen, J., et al. 2011. Pollution assessment of heavy metals in the sediment of Jiulong River Estuary. *Marine Sciences* 35(8): 11–17.
Lin, J. & Chen,Y. 1998. The relationship between adsorption of heavy metal and organic matter in river sediments. *Environment International* 24(4): 345–352.
Liu, C., Wang, Z. & He, Y., et al. 2002. Evaluation on the Potential Ecological Risk for the River Mouths around Bohai Bay. *Research of Environmental Sciences* 15(5): 33–37.
Liu, W., Luan, Z. & Tang, H. 1999. Environmental assessment on heavy metal pollution in the sediments of le an river with potential ecological risk index. *Acta Ecologica Sinica* 19(2): 206–211.
Ma, L., Li, J. & Lu, Z. 2013. Pollution characteristics and potential ecological toxicity assessment of heavy metals in surface sediments of Xinglin Bay, Xiamen. *Ecological Science* 32(2): 212–217.
Mackey, A.P. & Hodgkinson, M.C. 1995. Concentrations and spatial distribution of trace metals in mangrove sediments from the Brisbane River, Australia. *Environmental Pollution* 90(2): 181–186.
Martinol, M., Turner, A. & Nimmo, M., et al. 2002. Resuspension, reactivity and recycling of trace metals in the Mersey Estuary, UK. *Marine Chemistry* 77: 171–186.
Rubio, B., Nombela, M.A. & Vilas, F. 2000. Geochemistry of Major and Trace Elements in Sediments of the Ria de Vigo (NW Spain): an Assessment of Metal Pollution. *Marine Pollution Bulletin* 40(11): 968–980.
Turner, A. 2000. Trace metal contamination in sediments from UK estuaries: an empirical evaluation of the role of hydrous iron and manganese oxides. *Estuarine, Coastal and Shelf Science* 50: 355–371.

Hydraulic Engineering III – Xie (Ed)
© 2015 Taylor & Francis Group, London, ISBN 978-1-138-02743-5

Study and development of early warning system for groundwater hazard in the coal mine

P.P. Chen

Department of Coal Mining and Design, Tiandi Science and Technology Co. Ltd., Beijing, China

ABSTRACT: Early warning mechanism for groundwater inrush from roof was analyzed, based on research findings of the mechanism and fatalness evaluation for groundwater inrush from roof. Multifactorial early warning model for groundwater inrush from roof was put forward, which involved stratums, tectonic structure, hydro-geological condition and coal mining method. Taking Huaibei Zhuxianzhuang Coal Mine for example, hydrogeology and real data of water inrush from roof were analyzed. The main influencing factors include: the distance between working-face and fault, thickness of overburden rocks, water levels of aquifer IIII and V. With the inflow of groundwater as the goal, the early warning model was established based on the Artificial Neural Network techniques, and was checked up with the observational data of Zhuxianzhuang coal mine. According to the drainage capability and resistibility of groundwater hazard in Zhuxianzhuang coal mine, 3 early warning thresholds were determined.

1 INTRODUCTION

Groundwater hazard is one of the five major hazards in the coal product. Early warning system for groundwater hazard can monitor change of aquifers, and forecast the groundwater inrush from stratus, and give an alarm of groundwater hazard. It's very importance for safety of the coal mine production.

Since 1980, Chinese researchers have been studying the forecasting and early warning for groundwater hazard, in the coal mine. In the middle of 1990's, China Coal Research Institute studied the structure of early warning system for groundwater hazard. Researchers measured the displacement of seam floor before water inrush, hydraulic pressure, and elastic wave of bores; discovered the changes of physical quantity before water inrush; and developed the monitoring instrument of groundwater hazard premonitory, and built the discriminant model for groundwater hazard. Sui Haibo, Chen Jiulong studied logistic structure and macro-construction of early warning system for water inrush from seam floor in workface of the coal mine. According to the hydro-geological characteristics of Longkou coal mine under sea, the early warning system for sea water inrush was built, based on the monitoring of roof aquifer water level and water quality. Main index parameters were the water levers of aquifers, and threshold values for early warning were set based the change of water levers in early warning systems for water hazard in China. Early warning mechanism was studied incomplete, some factor wasn't be study, such as the destroy of overly seams and influence of aquifers with coal mining.

Early warning mechanism for groundwater inrush was analyzed in the paper, based on research findings of the mechanism theory and fatalness evaluation for groundwater inrush from roof, in the coal mine in China. Taking Huaibei Zhuxianzhuang Coal Mine as an example, in view of factors of the water inrush from coal seam roof, from the water source monitoring, master of overlying strata and tectonic conditions, the early warning model of roof water inrush based on artificial neural network is established, and combined with the mine waterproof performance, the warning thresholds are divided.

2 EARLY WARNING MECHANISM RESEARCH OF WATER INRUSH

2.1 Real-time monitor of water source

Existing early warning systems for groundwater hazard are mainly based on monitor of water source. Monitor parameters included water level, hydraulic pressure, and water temperature and water quality. Monitor parameters of water quality mainly include degree of mineralization, and diagnostic ions, such as Na^+ and Cl^-.

Early warning discriminance and threshold value are mainly according to perennial changes of water levels, distinctness of water temperature and quality between different aquifers, and experience of researchers. Along with application of real-time monitors, change rule of hydrologic information fore-and-aft water inrush from seam roof will be found step by step, and early warning discriminance will be more creditable based on the real-time monitors.

2.2 Exploration and discriminance of overburden seams and geological structure

Water inrush often happens at areas where overburden seams attenuation or overgrown failure for the little faults which were considered un-infection of mining. Then exploration & discriminance of overburden seams and geological structure are very important to prevention and cure of water hazards.

Now it is unrealizablereal that monitor technologie of failure and geological structure, but thinness of overburden seams and location of geological structure can be calculated from the mining engineering map. So in early warn system, the infection of overburden seams and geological structure should be considered.

2.3 Control of mining methods and cure measure of water hazards

Failure of overburden seams is different with different mining methods. In some area where were threaded by water, the change of mining method (such as filling mining) or cure measure (such as draining of aquifers) could provide safety of mining. If early warning systems give an alarm according to water levers only, mistakes might happen and arose confusion. So early warning systems should consider influence of mining method and cure measure, to increase accuracy and instruct production.

2.4 General intelligent analyzing based on multi-factors

Water hazard in the coal mine is a complicated mutational event which is affected by multi-factors, and controlled by mining method and cure measure, have many character such as complexity, illegibility and uncertainty. Modern non-linear technologies should be using to study of early warning of water hazard, such as intelligent science, expert system, and fuzzy mathematics. Early warning system of water hazard should have a general intelligent early warning model based on multi-factors, which would add veracity of early warning.

3 HYDROGEOLOGY AND MINING CONDITION OF ZHUXIANZHUANG THE COAL MINE

3.1 Hydrogeology of the coal mine

Mining method is full mechanized top caving, roof seam is long-wall caving in Zhuxianzhuang coal mine. Bedrock is overlaid by Tertiary and Quaternary loose bed. Thickness of loose bed is 246~260 m. There are multilayer aquifers and aquifuges that alternating aggraded in loose bed, four aquifers and three aquifuges are set off from top to bottom in loose bed, and not hydraulic connections between each aquifers. Aquifer-IV directly lays over coal measure strata, it contains abundant water, is one of the main source of water inrush in the coal mine. In northeast of the coal mine, there is Jurassic conglomerate aquifer (Aquifer-V) which has abundant karsts.

3.2 *Water yield in the coal mine*

Zhuxianzhuang coal mine put into production in 25-4-1983. Before production, water yield in the coal mine was 95 m³/h approximately. Aquifer-IV was connected by crannies with roof caving, when workface-741 mined. Water yield in the coal mine increased to 183.5 m³/h in short time. Water yield in the coal mine increased to 300.78 m³/h, with more workfaces mined. Then Aquifer-IV's water yield gradually reduced, with mined to deep area. Water yield in the coal mine mainly involved by fissured water of coal measure sandstone and goaf water, about 200~240 m³/h.

Since construction of Zhuxianzhuang coal mine, there have been 24 times statistical water inrushes which water yield more than 30 m³/h. There have been 6 workface water inrushes in 12 mined workface. Max water yield was 162 m³/h. It can be found from statistics, that water yield was more when Aquifer-IV was connected by faults.

4 EARLY WARNING MODEL FOR GROUNDWATER HAZARD BASED ON THE ARTIFICIAL NEURAL NETWORK

4.1 *Frame of model*

BP is the typical multilayer frame, it composed with the input layer, hidden layer and output layer. Layers often fully connect each other. The cells in the same layer don't connect each other.

The relationship between the input layer and the hidden layer is:

$$Hidden[j] = f\left(\sum_{i=1}^{n} W_{ij} a_i\right) \tag{1}$$

The relationship between the hidden layer and the output layer is:

$$out = f\left(\sum_{j=1}^{q} V_j b_j\right) \tag{2}$$

In the formulas, a_i is the ith input cell, b_j ($Hidden[j]$) is the jth hidden cell, W_{ij} are the weight values from the input layer to the hidden layer, v_j are the weight values from the hidden layer to the output layer; f is the transfer function.

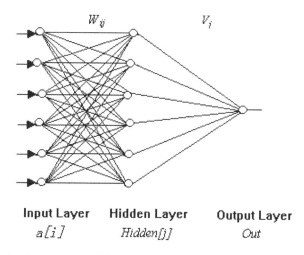

Input Layer   Hidden Layer   Output Layer
$a[i]$ *Hidden[j]*   *Out*

Figure 1. Frame of early warning model.

4.2 Arithmetic of BP

At first, the weights are random initialized, and the sample set is inputted. Then according to the Equation 1 and the Equation 2, the output is worked out. At the same time, the errors between calculated outputs with the outputs in the sample set are calculated.

If the errors beyond the error bound, the errors are back prorated, and the weights of each layer are adjusted until the errors are in the error bound. In order to assure the adjusting direction being accordant with the reducing errors, the error function (E_k) is formatted.

$$E_k = (out - c)^2 / 2 \tag{3}$$

In Equation 3, E_k is the mean square error, C is the target output in the sample set, then the adjustment value(ΔV) from the output layer to the hidden layer is:

$$\Delta V = \beta(-gradvE_k) = \beta d_i^k b_j \ (\beta \text{ is learning rate}) \tag{4}$$

The adjustment values (ΔW) from the hidden layer to the output layer is:

$$\Delta W_{ij} = \beta[\sum_{i=1}^{q} V_{ji} d_i^k] f'(s_j') a_i^k \tag{5}$$

4.3 Analyze and quantization of water inrush factors

By long-term observation, It was concluded that fault is the magistral factor of water inrush, lithology & configuration of overburden rocks primarily affected water yield in Zhuxianzhuang coal mine. So there were 3 factors in the model:

1. Fault: It was calculated with distant between fault and forecast area in workface as affect of fault. Spatial query of GIS (Geography Information System) was using to count distant between fault and forecast area in mining engineering plan.
2. Thickness of base rook: Thickness of base rook can be explored by drilling and 3D seismic, and be inputted into GIS. The data of forecast area could be queried and quantizing in model.
3. Source of water inrush: The water levels of Aquifer-IV and Aquifer-V can reflect the dynamic hydrological condition of the coal mine. In this early warning model, real time data of water levels were input and quantized in model.

There have been more than 900 group observing data of different aquifer' water levels since well construction in Zhuxianzhuang coal mine. If there was water inrush in a mining area, the water yield could be the observing data in water yield records. If there wasn't water inrush in a mining area, the water yield could be confirmed as 10 m³/h in working face. Removing repeated data, sample set were composed by 50 group data, in which 45 group data were samples for study, 5 group data were samples for testing of model.

4.4 Training and testing of model

Standardization sample set was designed by the observed data in Zhuxianzhuang coal mine. At the beginning of this training, the learning rate was assigned as 0.9. After training 4000 times, the learning rate was reduced to 0.7. After training 2532 times, the mean square error became 4.52E-04, so the training finished.

5 groups validate data were input into model, result was displayed in Table 1. The forecasted result was coincident to actual data when real water yield was more than 10 m³/h, but the error was bigger when real water yield was 10 m³/h. It was main reason that water yield was artificially confirmed as 10 m³/h when there wasn't water inrush in a mining area. It wouldn't affect the forecast of water inrush.

Table 1. Result of early warning model with validate data.

Water level of Aquifer-IV (m)	Water level of Aquifer-V (m)	Thickness of base rook (m)	Distant of fault (m)	Water yield (m³/h)	Result (m³/h)	Error
−29.29	−31.32	40	150	30	30.70456	0.023485
−21.85	−17.18	39	100	60	62.01747	0.033625
−12.5	−14.39	80	20	10	4.156694	0.58433
−10.7	−13.36	150	200	10	16.45142	0.645142
−42.65	−54.5	70	150	10	3.475953	0.6524

5 SETTING OF THRESHOLD VALUES FOR EARLY WARNING

In a early warning model, it's very important that the setting of threshold values. Because the condition of each coal mine is very different, setting of threshold values must analyses the particular condition of each coal mine. Flow capacity of Zhuxiangzhuang coal mine is 1200 m³/h, flowing yield is about 200~240 m³/h now. The fracture aquifer in coal measure strata is a weak aquifer, there are a little water flowing with mining always. When flowing yield in working-face reach 30 m³/h, the normal coal mining will be effected. When flowing yield in working-face reach 90 m³/h, the coal mining of this working-face may be stopped. When flowing yield in working-face reach 160 m³/h, this working-face may be flooded. According the actual conditions of Zhuxiangzhuang coal mine, four kinds of early warning grades below.

- Safety as Green, when flowing yield less than 30 m³/h.
- Security as Yellow, when flowing yield is 30~90 m³/h.
- Emergency alert as Orange, when flowing yield is 90~160 m³/h.
- Danger as Red, when flowing yield greater than 160 m³/h.

6 CONCLUSIONS

1. Early warning of groundwater hazard from seam roof should be studied with mechanism research and forecasting of water inrush from seam roof. Mechanism research of early warning should include real-time monitor of water source, exploration & discriminance of overburden seams and geological structure, control of mining methods and cure measure of water hazards.
2. Main influencing factors of water inrush were confirmed as the distance between working-face with fault, thickness of overburden rocks, water levels of aquifer IIII and V. Aimed at the inflow of water in workface, the early warning model was established based on the Artificial Neural Network, and was checked up with the observational data of water inrush in Zhuxianzhuang coal mine.

REFERENCES

Chen. P.P. & Liu, X.E. 2010. Research and Application of Water Inrush Early Warning System to Mine Roof. *Coal Science And Technology* 38(12): 93–96.
Sui, H.B. & Cheng, J.L. 2009. Building of early warning system for groundwater hazard from floor-seam in coal mining workface. *Mining Safety & Environmental Protection* 36(1): 58–60.
Wang, J.M. & Dong, S.N., etc. 2005. The principal of early warning for groundwater hazards in coal mine and its application. *Coal Geology & Exploration* 33(Supp): 1–4.
Wang, L.J. & Han, R.Q. 2006. Application of flooding forecasting system to coal mining under sea. *Coal Geology & Exploration* 34(6): 54–56.
Zhang, D.S. & Zhen, S.S, etc. 1994. *GIS technology and its applications in coal mine flood forecasting.* Xuzhou: China University of Mining and Technology press.

Hydraulic Engineering III – Xie (Ed)
© 2015 Taylor & Francis Group, London, ISBN 978-1-138-02743-5

Stratified cluster sampling and its application on Warner model for sensitive question survey

Sheng Wang
School of Public Health, Soochow University, Suzhou, Jiangsu, China
School of Medicine, Hangzhou Normal University, Hangzhou, Zhejiang, China

Ge Gao
School of Public Health, Soochow University, Suzhou, Jiangsu, China

ABSTRACT: To explore scientific sampling methods and corresponding formulas for sensitive question survey on stratified cluster sampling. Cochran W.G.'s sampling theories, Warner model, total probability formulas and properties of variance were used in this paper. Formulas for the estimate of the population proportion and its variance on Warner model in cluster sampling and stratified cluster sampling were deduced. With high reliability, our survey methods on Warner model may have an extensive application in sensitive question survey.

1 INTRODUCTION

Sensitive questions means of high private secret or most people think it's not convenient to state in public place, such as drugs, gambling, prostitution, drinking driving, personal income, tax evasion, premarital sex, venereal disease, AIDS and homosexual tendency etc. Sensitive questions have been classified as two kinds by the general characteristics, sensitive questions of quality characteristic and sensitive questions of quantity characteristic which is further classified as sensitive questions of binary choice and multiple choice.

If respondents were asked sensitive questions directly, they could refuse to answer the questions or lie intentionally for protecting herself or other purpose, which lead to bias of the result. To avoid this, RRT (Randomized Response Technique) has been considered as useful method to protect respondent's private and increase real response efficiency rate. Presently, the RRT sampling survey study of sensitive questions has only limited in simple random sampling, and the application of RRT is also confined to simple random sampling of Small-scale special populations, or make a mistake use of the related formula of RRT simple random sampling survey to analysis the RRT complex sampling survey data of sensitive question, moreover, there are rarely evaluation on the validity and reliability of the RRT sampling survey of the sensitive question.

In this article, the complex cluster sampling, stratified sampling survey of RRT Warner model of binary choice sensitive question have been designed, and the estimate population proportion and estimated variance formula of RRT Warner model by cluster sampling and stratified sampling have been deduced; meanwhile, combined with the survey example of premarital sex of college student in Soochow University, With high reliability, our survey methods on Warner model may have an extensive application in sensitive question survey.

2 SURVEY METHODS

2.1 *The RRT of Warner model*

RRT is put forward by Warner originally who is sociologist of U.S. The design principle of Warner model is to design two opposite questions based on the sensitive character, and need to design a random device (Wang Jianhua, 2003). For example, a certain number of red ball and white ball which have the same size, weight and tactility are put in the pocket. The one who selected draw the ball out from the pocket repeatedly when there is nobody nearby, if the ball which was drew out is red, answer the question: do you the one who has the characteristic A? Otherwise, answer the question: didn't you the one who has the characteristic A? The respondents just need to answer "Yes" or "No".

2.2 The cluster sampling of Warner model

The virtue of the cluster sampling is that the sampling frame is simple; research unit is centralized correspondingly, it is convenient to the carry on and organize the research, the cost of research of every unit is reduced which made the same cost can do more research of basic unit; in general, the defect is that the research unit is not well-distributed in population, sampling error is large correspondingly. It was widely applied in medical research as a sampling method for its economical and easy to practical virtue. The cluster sampling of Warner model is divided into three steps: the first step is divide overall into cluster (first-level unit), which including sub-unit; The second step is taking the cluster as a sampling unit, to random sample some clusters from the population; the third step is carrying out the binary choices sensitive question survey with RRT of Warner model in all the sampling sub-unit.

2.3 Stratified cluster sampling of Warner model

The virtue of stratified sampling is reducing the sampling bias. The stratified cluster sampling of Warner model can be divided into 4 steps, firstly divided the population into several layers based on some certain characters; secondly to divide layers into clusters (first-level unit) composed with sub-unit; thirdly regard the cluster as the sampling unit and random sampling some cluster from every layer separately; finally carry out the binary choice sensitive question survey to all the sub-unit of sampling clusters by RRT of Warner model.

3 FORMULA DEDUCTION

3.1 Cluster sampling of Warner model

3.1.1 Estimate population proportion and estimated variance
Given divide into N clusters of the population, and the i cluster include M sub-units, sample n clusters randomly. Given the rate of I group (i = 1,2, ..., n) have characteristic A is π_i, a_i sub-unit have characteristic A in I group, the rate of characteristic A in population is π_i.

3.1.1.1 If the size of the group is equal
If the size of all the sub-unit of group is M, the estimate population proportion of cluster sampling given by M, Cochran W.G. is:

$$\hat{\pi} = \frac{1}{nM}\sum_{i=1}^{n}a_i = \frac{1}{n}\sum_{i=1}^{n}\frac{a_i}{M} = \frac{1}{n}\sum_{i=1}^{n}\pi_i \qquad (1)$$

The Unbiased estimator of statistic $\hat{\pi}$ is:

$$v(\hat{\pi}) = \frac{1-f}{n(n-1)}\sum_{i=1}^{n}(\pi_i - \hat{\pi})^2 \qquad (2)$$

In which, $f = nM/(NM) = n/N$ is sampling ratio.

3.1.1.2 If the size of cluster is not equal

If the size of sub-unit of each cluster is not equal, the estimate population proportion π of cluster sampling given by Jianfeng Wang & Ge Gao is:

$$\hat{\pi} = \frac{\sum\limits_{i=1}^{n} a_i}{\sum\limits_{i=1}^{n} M_i} \tag{3}$$

Estimated variance of statistic $\hat{\pi}$ is (Cochran W.G. 1987):

$$v(\hat{\pi}) = \frac{1-f}{n\bar{M}^2} \frac{\sum\limits_{i=1}^{n} a_i^2 - 2\hat{\pi}\sum\limits_{i=1}^{n} a_i M_i + \hat{\pi}^2 \sum\limits_{i=1}^{n} M_i^2}{n-1} \tag{4}$$

In which $\bar{M} = \sum\limits_{i}^{n} M_i / n$ is the number of average sub-unit of every sample cluster, and $f = \sum\limits_{i=1}^{n} M_i / \sum\limits_{i=1} M_i$ is the sampling ratio.

A little bit more simple Mean Square Error (MSE) can be used in actual application:

$$MSE(\hat{\pi}) = \frac{1-f}{n\bar{M}^2} \frac{\sum\limits_{i=1}^{n} M_i^2 (\pi_i - \hat{\pi})^2}{n-1} \tag{5}$$

3.1.2 Deduction of π_i, a_i

Given the rate of red is P ($p \neq 0.5$) in advance, the rate of I cluster responsed "yes" in selected sample is expressed as λ_i, by the formula of total probability, we can get:

$$\lambda_i = \pi_i \cdot P + (1 - \pi_i)(1 - P)$$

$$So: \pi_i = \frac{\lambda_i - (1-P)}{2P-1}, \quad i = 1, 2, \ldots, n \tag{6}$$

$$And \ a_i = M_i \pi_i, \quad i = 1, 2, \ldots, n$$

3.2 The stratified cluster sampling of Warner model

Given that the population is divided into L layer, the h layer including N clusters, the I cluster in h layer contains M sub-units, and there is N sub-unit in all, sample n clusters from h layer randomly. Given the incidence rate of sensitive questions in sub-unit of the I cluster of h layer is π_{ih}, the number of A characteristic in second-level unit is a_{ih}.

3.2.1 If the size of each cluster in h layer is equal

3.2.1.1 Estimate population proportion and estimated variance of h layer

If the size of each cluster in h layer is M_h (cluster size of each layer could not equal), from formula (1), the estimate population proportion π_h of h layer by cluster sampling is

$$\hat{\pi}_h = \frac{1}{n_h} \sum\limits_{i=1}^{n_h} \pi_{ih}, h = 1, 2, \ldots, L \tag{7}$$

From formula (2), the unbiased estimate of variance $\hat{\pi}_h$ of h layer is

$$v(\hat{\pi}_h) = \frac{1-f_h}{n_h(n_h-1)} \sum\limits_{i=1}^{n_h} (\pi_{ih} - \hat{\pi}_h)^2, h = 1, 2, \ldots, L \tag{8}$$

In which f_h is the sampling rate of h layer.

3.2.1.2 Estimate population proportion and estimated variance

Estimate population proportion is:

$$\hat{\pi} = \frac{\sum\limits_{h=1}^{L} N_h M_h \hat{\pi}_h}{N} = \sum\limits_{h=1}^{L} W_h \hat{\pi}_h \tag{9}$$

In which $W_h = N_h M_h / N$ is the relative size accounted by sub-unit.

For the sample of each group is independent, to formula (9), according to the general characteristic of variance (Wang Yan, 2006), there is: $V(\hat{\pi}) = \sum\limits_{h=1}^{L} W_h^2 V(\hat{\pi}_h)$, in this case, the $V(\hat{\pi}_h)$ of unbiased variance can be get from formula (8), so the estimated variance is:

$$v(\hat{\pi}) = \sum\limits_{h=1}^{L} W_h^2 v(\hat{\pi}_h) = \sum\limits_{h=1}^{L} W_h^2 \frac{1-f_h}{n_h(n_h-1)} \sum\limits_{i=1}^{n_h} (\pi_{ih} - \hat{\pi}_h)^2 \tag{10}$$

3.2.1.3 The formula of π_{ih}

Given the rate of red is P (p ≠ 0.5) in advance, the rate of I cluster respondent's answer is "yes" in h layer is expressed as λ_{ih}, by the formula of total probability (SU Liangjun, 2007), we can get: $\lambda_{ih} = \pi_{ih} \cdot P + (1-\pi_{ih})(1-P)$

$$\text{So: } \pi_{ih} = \frac{\lambda_{ih} - (1-P)}{2P-1}, h=1,2, ..., L; i=1,2, ..., n_k \tag{11}$$

3.2.2 If cluster size not equal

3.2.2.1 The estimate of population proportion and estimated variance of the h layer

If cluster size not equal, the estimate of population proportion of h layer is:

$$\hat{\pi}_h = \frac{\sum\limits_{i=1}^{n_h} a_{ih}}{\sum\limits_{i=1}^{n_h} M_{ih}}, h=1,2, ..., L \tag{12}$$

From formula (4) we can get the $\hat{\pi}_h$ estimated variance of h layer:

$$v(\hat{\pi}_h) = \frac{1-f_h}{n_h \bar{M}_h^2} \cdot \frac{\sum\limits_{i=1}^{n_h} a_{ih}^2 - 2\hat{\pi}_h \sum\limits_{i=1}^{n_h} a_{ih} M_{ih} + \hat{\pi}_h^2 \sum\limits_{i=1}^{n_h} M_{ih}^2}{n_h - 1}, h=1,2, ..., L \tag{13}$$

From formula (4) we can get the $\hat{\pi}_h$ Mean Square Error (MSE):

$$MSE(\hat{\pi}_h) = \frac{1-f_h}{n_h \bar{M}_h^2} \cdot \frac{\sum\limits_{i=1}^{n_h} M_{ih}^2 (\pi_{ih} - \hat{\pi}_h)^2}{n_h - 1}, h=1,2, ..., L \tag{14}$$

3.2.2.2 The estimate of population proportion and estimated variance

According to formula (9), we can get the the estimate of population proportion is:

$$\hat{\pi} = \frac{\sum\limits_{h=1}^{L} \sum\limits_{i=1}^{N_h} M_{ih} \hat{\pi}_h}{N} = \sum\limits_{h=1}^{L} W_h^2 \hat{\pi}_h \tag{15}$$

In the formula (15), $W_h = \sum_{i=1}^{N_h} M_{ih}/N$, which is a relative size of h layer calculated as a sub unit.

As each layer is independently, according to the basic character of variance, in the formula (15), $V(\hat{\pi}) = \sum W_h^2 V(\hat{\pi}_h)$, at the same time, the estimate population proportion of $V(\hat{\pi}_h)$ can get from formula (13), so the estimated variance is:

$$v(\hat{\pi}) = \sum_{h=1}^{L} W_h^2 \frac{1-f_h}{n_h \bar{M}_h^2} \cdot \frac{\sum_{i=1}^{n_h} a_{ih}^2 - 2\hat{\pi}_h \sum_{i=1}^{n_h} a_{ih} M_{ih} + \hat{\pi}_h^2 \sum_{i=1}^{n_h} M_{ih}^2}{n_h - 1} \qquad (16)$$

If we induced a simple MSE form of $\hat{\pi}_h$, we can get:

$$v(\hat{\pi}) \doteq \sum_{h=1}^{L} W_h^2 \frac{1-f_h}{n_h \bar{M}_h^2} \cdot \frac{\sum_{i=1}^{n_h} M_{ih}^2 (\pi_{ih} - \hat{\pi}_h)^2}{n_h - 1} \qquad (17)$$

3.2.2.3 Deduction of π_{ih}, a_{ih}

π_{ih} can calculated according to formula (11), and the h layer i cluster sub unit size of specific A is: $a_{ih} = M_{ih}\pi_{ih}$.

4 APPLICATION

To illustrate the application of Warner methods Stratified cluster sampling, a survey of the premarital sex activity among college students was carried out. In our survey, the population is the college students in the new campus of Soochow University. We regard the total colleges students in new campus of Soochow University in 2007 as the survey population, which divided into two layer, firstly is undergraduate students (3456 students), and the other is graduated students (1890 students). We asked the sensitive question of "if you have a premarital sex", the indicator is a Binary Choice answer, Yes or No. We regard the class as a cluster, and use the method of large cluster divided into smaller ones and small clusters are combined to a larger one, so that the size of each cluster can be nearly the same. Used Warner model stratified cluster sampling, we select 20 undergraduate classes (1080 students) and 18 graduated classes (818 students), totally 38 classes and 1898 students.

In the survey, a random device is designed to generate random numbers, in this device, there are 6 red balls and 4 white balls with the same size, weight and tactility. Each student select a ball from the device when nobody nearby, if he selected the red ball, he should answer "do you have the premarital sex"? the white ball should answer "didn't you have a premarital sex"? each students only need answer "Yes" or "No", and select 2 round. The totally survey population is 3796 counts, the questionnaire recovery and acceptability is 100%. Database was established with Excel 2003, after logistic checking by handwork and computer, we analyze data by SAS9.13.

4.1 Result of premarital sex incidence in each class

We use Warner model stratified cluster sampling for 2 round to survey the premarital sex of 38 classes students of new campus Soochow University, from formula (11), we get the premarital sex incidence of 20 classes undergraduate students π_{i1} (i = 1, 2, ..., 20), π'_{i1} (i = 1,2, ..., 20) and which of 18 graduate students π_{i2} (i = 1, 2, ..., 18), π'_{i2} (i = 1, 2, ..., 18) in first round and second round. The result can be find in Table 1.

Table 1. Survey result on 38 classes students premarital sex of Soochow university by 2 round Warne model stratified cluster sampling.

Undergraduate classes serial number	π_{i1}	π'_{i1}	Graduated classes serial number	π_{i2}	π'_{i2}
1	0.2074	0.2606	1	0.1591	0.1859
2	0.1543	0.1543	2	0.1944	0.1944
3	0.2283	0.1739	3	0.2384	0.2665
4	0.2024	0.2024	4	0.1944	0.2300
5	0.1944	0.2100	5	0.2194	0.2504
6	0.1814	0.1524	6	0.2000	0.1700
7	0.1802	0.2084	7	0.2606	0.2374
8	0.1591	0.1959	8	0.2826	0.2583
9	0.0238	0.0633	9	0.1944	0.2200
10	0.1173	0.1173	10	0.2396	0.1975
11	0.0000	0.0455	11	0.2074	0.2306
12	0.0833	0.1189	12	0.2396	0.2075
13	0.0833	0.1289	13	0.2606	0.2474
14	0.1591	0.1959	14	0.2619	0.2324
15	0.1739	0.1739	15	0.4167	0.4167
16	0.1023	0.1023	16	0.2406	0.2134
17	0.2024	0.1729	17	0.2727	0.2559
18	0.1739	0.2083	18	0.2727	0.2445
19	0.1429	0.1624			
20	0.1944	0.1589			

4.2 The incidence estimated variance of premarital sex in each layer

According to the first round survey data, from formula (7), we get the undergraduate students premarital sex incidence estimated variance is:

$$\hat{\pi}_1 = \frac{1}{n_1}\sum_{i=1}^{n_1}\pi_{i1} = \frac{1}{20}\times(0.2074+0.1543+\cdots+0.1944)=0.1482$$

From formula (8), the estimated variance of $\hat{\pi}_1$ is:

$$v(\hat{\pi}_1) = \frac{1-f_1}{n_1(n_1-1)}\sum_{i=1}^{n_1}(\pi_{i1}-\hat{\pi}_1)^2$$

$$= \frac{1-1080/3456}{20(20-1)}\left[(0.2074-0.1481)^2 +(0.1543-0.1482)^2 +\cdots+(0.1944-0.1482)^2\right]$$

$$= 0.0001$$

According to the first round survey data, from formula (7), we get the graduate students premarital sex incidence estimated variance is:

$$\hat{\pi}_2 = \frac{1}{n_2}\sum_{i=1}^{n_2}\pi_{i2} = \frac{1}{18}\times(0.1591+0.1944 +\cdots+0.2727)=0.2420$$

From formula (8), the estimated variance of $\hat{\pi}_2$ is:

$$v(\hat{\pi}_2) = \frac{1-f_2}{n_2(n_2-1)}\sum_{i=1}^{n_2}(\pi_{i2}-\hat{\pi}_2)^2$$

$$= \frac{1-818/1890}{18(18-1)}\left[(0.1591-0.2420)^2 +(0.1944-0.2420)^2 +\cdots+(0.2727-0.2420)^2\right]$$

$$= 0.0001$$

4.3 The incidence estimated variance of premarital sex of colleges students in new campus of Soochow university

From formula(9), the incidence estimated variance of premarital sex of Soochow university new school students is: $\hat{\pi} = \sum W_h \cdot \hat{\pi}_h = W_1\hat{\pi}_1 + W_2\hat{\pi}_2 = 0.65 \times 0.1482 + 0.35 \times 0.2420 = 0.1810$.

From formula (10), we get the estimated variance of $\hat{\pi}$:

$$v(\hat{\pi}) = \sum_{h=1}^{L} W_h^2 \cdot v(\hat{\pi}_h) = W_1^2 \cdot v(\hat{\pi}_1) + W_2^2 \cdot v(\hat{\pi}_2)$$
$$= 0.65^2 \times 0.0001 + 0.35^2 \times 0.0001 = 0.0001$$

Therefore, we get the 95% CI of population proportion:

$$\hat{\pi} \pm 1.96 \times \sqrt{v(\hat{\pi})} = 0.1954 \pm 1.96 \times \sqrt{0.0001} = 0.1758 \sim 0.2150$$

4.4 Reliability evaluation of the survey

We make a relative analysis after the 2 round 20 classes undergraduate students data by square root inverse sine transfer with SAS9.13, the Pearson product moment correlation of coefficient r = 0.89394, P < 0.0001, which show that the 2 round survey result is high correlation in the first layers. With the same transfer of the 2 round 18 classes graduate students survey data, we get the relative analysis Pearson product moment correlation of coefficient r = 0.91645, P < 0.0001, which show that the survey result is high correlation of all the 38 classes, and our survey method have a high reliability.

5 DISCUSSIONS

5.1 Practicability of the research

It is very important and prevalent used of the sensitive question survey in the health work and medical research, especially in the AIDS prevention work. In China, there are four stages of the prevalence of HIV/AIDS, the introduction stage (1985–1988), diffuse stage (1989–1994), increasing stage (1995–2001) and faster increasing stage (2002–2007), and now we are faced the serious situation of rapid increasing or break up. The national prevention and plan need the such exact data as the HIV infection rate, the population of the homosexual and lesbian, the number of prostitute, the number of prostitute customers and the counts of wenching, the average sexual partner number, the number of drugger etc. And the exact data needs the scientific survey method and statistic formula. The result of this research could provide a scientific and reliable data for the national and provincial health and related management department during the plan and policy making, as well of the practical application value in the disease prevention especially in the HIV/AIDS and venereal disease.

5.2 Creativity of the research

Mata analysis on the 38 related literature from 1965–2000 collected by the foreign scholars, it is showed that compared with traditional survey method, there are significance advantages in the result accuracy and reliability of RRT method in sensitive question survey (Gerty J.L.M., 2004). National and international statistic scholars have been studied and proposed a lot of sampling investigation method about the sensitive question sample survey design, however, the sampling survey design research of these scholars are limited to simple random sampling, furthermore there are few research on the validity and reliability evaluation of the sensitive question sample survey.

In this article, it is the first deduction of the estimate population proportion and estimated variance of the sensitive questions in the complex cluster sampling and stratified sampling with the binary sensitive question Warner model, which have great creative meaning and can fill lacuna of sensitive question sampling survey design in health statistic, biostatistics, demography, economic statistic, technology statistic, social statistic, environment and zoology statistic etc.

5.3 *Reliability of the research*

Combined with the students premarital sex survey example of Soochow University, this article get a success application effect in the Warner model the cluster sampling and stratified sampling of binary sensitive questions. Meanwhile, we also evaluate the reliability of repeat survey, which is high correlation of the 2 round surveys. This can be a evidence the survey method and statistic formula in this article is higher reliability.

REFERENCES

Cochran W.G. Sampling technical. Translated by Zhang Yaoting, Wu Hui. Beijing, China statistic publishing company. 1987, 93–95.
Gerty J.L.M., Lensvelt-Mulders J.J.H. & Peter G.M. Meta-Analysis of Randomized Response Research: Thirty-Five Years of Validation. Sociological Methods & Research, 2004, 33: 319–348.
Su Liangjun. Advanced Mathematical Statistics. Beijing. Pecking University publishing company. 2007, 3.
Wang Jianhua. Practical medical research method. Beijing. people health publishing company, 2003, 444–447.
Wang Yan, Sui Silian & Wang Aiqing. Mathematical statistics and MATLAB engineering data analysis. Beijing, Tsinghua University publishing company, 2006, 10–11.

Hydraulic Engineering III – Xie (Ed)
© 2015 Taylor & Francis Group, London, ISBN 978-1-138-02743-5

Quantitative risk assessment of domino effect in LPG tank area

Q.C. Ma
China University of Petroleum, Beijing, China

W.J. Qu
China Huanqiu Contracting Engineering Corporation, Beijing, China

ABSTRACT: Oil storage tank is flammable, explosive or toxic substances accumulation area. For the fire explosion accident, it's likely to cause secondary or more accidents, which are known as the "domino effect". This paper summarizes the definition of the "domino effect". LPG storage tank is selected as the research object. The study is focused on the domino effect mechanism and the improvement of criteria of escalation consequence severity. Damage threshold and probability model are established. The calculation method of chain effect of tank area is introduced into the safety evaluation. Through a case study, the calculation steps and result of the method is described. The accident severity and the probability of domino effect were obtained based on fire and explosion model. The quantitative evaluation of tank explosion damage model is realized according to different escalation vectors. By comparing the result between primary and domino accident, domino effect needs more consideration.

1 INTRODUCTION

Tank area is a flammable toxic concentration enclave, belonging to major hazard area. Storage tank area is usually compact layout. Once a fire, explosion accident happened in one device, the tank farm most likely causing secondary or even several accidents, which are called "domino effect", resulting in multi-tank fire explosion, causing heavy casualties and property losses. It is of great significance to study on fire and explosion accidents in storage tank area.

Inspired by the domino theory, the accident chain effects (also called domino effect) are getting people's attention; many scholars in their various literatures give a different definition.

The domino effect is defined by American Institute of Chemical Engineers Center for Chemical Process Safety (AIChE-CCPS) as: from the initial accident caused by thermal radiation, shock and explosion debris, etc. acting on the close device, resulting in some of the series of serious accidents and serious consequences. Delvosalle (1996) understanding a domino effect as a simple Superposition of accident consequences overlap, he pointed out that the source of the accident consequences due to secondary the happening of the accident severity, in space or time led to the greater impact. Khan & Abbasi (2001) understands a domino effect as a unit operation accident caused the failure of one or more units, led to a secondary or higher level of accident. Cozzani (2006) the domino effect of understanding for the spread of destructive power, the initial event destructive energy spread to neighboring units, lead to one or more secondary events, causing more serious accident than the initial event. Darbra (2010) pointed that a relatively minor accident can initiate a sequence of events that cause damage over a much larger area and lead to far more severe consequences.

Domino effect may be initiated by one or more of these causative events:

a. Fire: pool fire, flash fire, fireball, and jet fire.
b. Explosion: Confined Vapor Cloud Explosions (CVCE), Boiling Liquid Expanding Vapor Explosion (BLEVE), vented explosion, vapor cloud explosion, and dust explosion.

c. Toxic release: instantaneous or continuous release of toxic light-as-air-gases; lighter-than-air gases, and heavier-than-air gases; release of toxic liquids.

At present, it is generally understood a domino effect as: accident happened in an initial unit (equipment), destructive physical effect produced by the incident spread to adjacent equipment or devices. It will trigger a neighboring one or more devices have secondary or more accidents, accident consequences severity increases, the influence area is expanded.

2 MECHANISM OF A DOMINO EFFECT ACCIDENT OF LPG TANK AREA

2.1 Types of domino effect accidents in LPG tank area

The explosion limit of Liquid Petroleum Gas (LPG) is wide, and the ignition temperature is low. It is flammable and explosive dangerous material. LPG has a narcotic effect, easily lead to nausea, vomiting, loss of consciousness and even stopped breathing. Accident statistics data of port and tank area show that fire accidents accounted for 59.5%, 6% for the explosion, toxic spill occupy 34.5% (Khan, 2001). Petroleum product transportation accidents, 65% for the fires, 24% for the explosion, and the other 11% belong to the toxic substance leakage (Lees,1996). Therefore, there are three types of accidents of oil tank farm: fire, explosion, leakage of poisons. But a fire, explosion are very easily leads to a subsequent accident, the physical effects of fire and explosion accidents cannot only trigger domino sequence, but also a sharp increase in accident severity (Junjie, 2008).

Therefore, there are three main types accidents of tank area can trigger a domino effect: fire, explosion, and fire—explosion occurred in both cases simultaneously cross. The first accident has devastating effects of thermal radiation, and other physical effects like overpressure effects may work on equipment close to the first unit, resulting in the close tank rupture, fire, explosion, that is the secondary accident. Under certain conditions, the secondary accident may lead to higher levels of three or more accident, causing extremely serious consequences of the accident. In simple terms is that two or more times of accidents caused by the initial Tank accident (mainly fire and explosion), and became a serious consequences phenomenon is called Tank domino effect. Figure 1 shows the mechanism of Domino effect in LPG tank area accidents.

2.2 Expansion factor of LPG tank area accidents

In essence, tank fire and explosion caused a domino effect caused primarily by three factors: thermal radiation, blast overpressure and projectile fragments. Fire's damage effect on

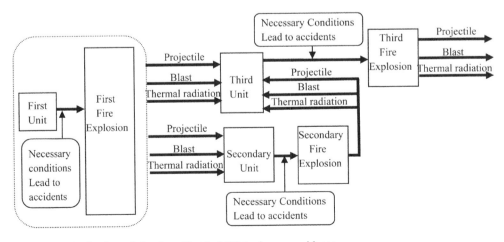

Figure 1. Mechanism of domino effect in LPG tank area accidents.

personnel and equipment in the affected area by thermal radiation; explosion caused a chain of accidents forms mainly through overpressure, blast fragmentation and thermal radiation.

Due to the heat radiation generated by the Flash fire is not sufficient to cause target device failure. The literature (Junjie, 2011) pointed out that Flash will not lead to a domino effect. The duration of the fireball is much smaller than 60 s, generally between 1 s–20 s, so the fireball will not cause Domino accidents (Planas-Cuchi, 1997). More than 100 surveys has been did for Domino effect accidents (Ronza, 2003). According to the findings of the expansion factor of accidents, we found that the explosion energy is mainly manifested in three forms: the shock wave energy, fragments energy and container residual deformation energy in pressure vessel explosion. Energy consumed by the latter two is only a very small portion of total energy blast. The explosion shock wave energy plays a main job in the form of performance. Therefore, two physical factors are mainly considering in this paper during the fire, explosion triggered a domino effect process as: the shock wave, thermal radiation.

3 QUANTITATIVE SAFETY ASSESSMENT OF TANK DISTRICT DOMINO EFFECT

The domino effect is introduced into the safety assessment of tank area; we do qualitative and quantitative evaluation of interaction between the evaluation units. Determine the likelihood of secondary accidents and accident losses. Safety evaluation of Domino effect is more focused on analyses the interactions between units, as well as the additive effects of the consequences of the accident. It is a more objective method of safety evaluation.

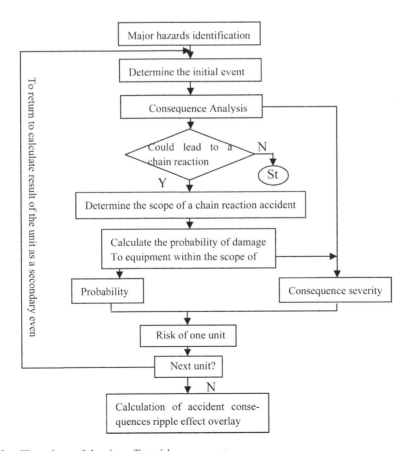

Figure 2.   Flow chart of domino effect risk assessment.

271

4 CASE STUDY

4.1 Object introduction

The petrochemical company has 19 sets of spherical tanks and two lie tanks. The layout of tanks is shown as follows. Tanks from G611 to G613 are propylene tanks with the volume of 2000 cubic and the rest are liquefied petroleum gas tanks. Tanks named G614–G619 have the volume of 2000 cubic, G605, G606 of 1000 cubic, G601–G604 of 400 cubic and G607–610 of 200 cubic.

Basic parameters of these tanks are shown in Table 1.

4.2 Analysis of tank accident's chain effect

4.2.1 Hazards analysis

LPG has strong volatility, lower flash point and wide explosion limits. Its minimum ignition energy is 0.2~0.3 mJ, and it is easy to get fire and even lead to secondary disasters once encounter fire. Subjected to external heat pressure vessels' wall will destructed suddenly and liquid is overheated and then released and lighted therefore the fireball emerged. Consequence analysis of first accidents.

There are two effects may be happening after leakage accident of LPG tanks. If the leakage liquid was ignited immediately, there will be a pool fire; if the ignition is delayed, flammable vapor cloud will form and then vapor cloud explosion accident may happen. By thermal radiation effects of pool fires, the nearby tanks' temperature rise and intensity drops even lead to the breakdown of tank, which causing secondary accident of vapor cloud explosion. The explosion shock wave can also impact around mechanical equipment, leading the failure of adjacent equipment, which we call this phenomenon as primary domino accidents. By analogy, the secondary domino accidents may also lead to new domino accidents.

4.2.2 Domino probability

According to equipment damage probability, the expansion probability of the chain effect introduced by tank G613 is shown in Table 2.

Only if extended probability is large enough and there are other target devices within influence area it is likely to cause a chain effect of accident.

4.2.3 Accident consequences severity

Choosing propylene tank of 2000 cubic as the research object, the leakage is 0.5% of the total, environment temperature is 283.15 K, fire dike covers 2800 m². Population density in the spherical tank area is 0.001/m². Casualties due to domino effect are as shown in Table 3.

From the Table 3, we learned that initial accident caused chain accidents by means of extension factors and the result are greater than the single accident. The result shows that domino effect needs more consideration in the risk assessment of tank area.

Table 1. Basic parameters of tanks.

No	Name	Volume	Material	Service condition Temperature °C	Pressure (MPa)	Texture
G601-G604	LPG tank	400 m³	LPG	50	1.6	SPV36
G605-G606	LPG tank	1000 m³	LPG	50	1.6	16MnR
G607-G610	LPG tank	200 m³	LPG	50	1.6	16MnR
G611-G613	Propylene spherical tank	2000 m³	Propylene	50	2.2	07MnCrMoVR
G614-G619	LPG spherical tank	2000 m³	LPG	50	1.6	15MnNbR
G605-G606	LPG lie tanks	1000 m³	LPG	Normal temperature	1.2	15MnNbR

Table 2. Extension probability of domino effects.

First object	Accident	Radius	Second accident	Distance (m)	Thermal radiation/over pressure	Extension probability
G613	Vapor cloud explosion	137.1561	G612/G616/G614	30	24.35476	1
			G611/G615	42.42	15.21008	0.7959
			G617/G619	58.30	10.08697	0.1789
			G618	50	12.27378	0.4057
			G601	111.4	4.535723	0.007969
			G602	129.3	3.784333	0.001434
			G603	128	3.831108	0.001621
			G604	110	4.606016	0.009141
			G606	120	4.143718	0.003465
	Pool fire	97.36507	G612/G614/G616	30	157.9993	0.9986
			G611/G615	42.42	79.02351	0.1674
			G617/G619	58.30	41.837	0.005921
			G618	50	56.87975	0.03617

Table 3. Accident consequences severity of domino effect.

Consequences severity	First accident	Second accident		
		G612	G614	G616
Death radius (m)	87.86347	344.45	341.097	341.097
Death toll (person)	24.24	372.54	365.33	365.33
Serious radius (m)	108.0064	437.814	433.744	433.744
Seriously injured (person)	12.38	229.33	225.41	225.41
Minor radius (m)	162.9326	493.789	490.209	490.209
Minor injured (person)	46.72	163.74	163.81	163.81

5 CONCLUSION

A method of escalation sequences was carried out. Based on the study of domino accident mechanism, specific thresholds for domino effect in LPG tank area were obtained according to different escalation vectors (heat radiation, overpressure and shock wave). Damage threshold and probability model are established. At the same time, the calculation method of chain effect of tank area is introduced into quantitative risk assessment.

ACKNOWLEDGEMENTS

The paper is supported by the Natural Science Foundation of China (Grant No. 51104167), Science Foundation of China University of Petroleum, Beijing (No. 2462012 KYJJ0422).

REFERENCES

Center for Chemical Process Safety. 2000. *Guidelines for Chemical Process Quantitative Risk Analysis: 2nd Edition.* New York.
Cozzani V, Antonioni G, Spadoni G, et al. 2005. The assessment of risk caused by domino effect in quantitative area risk analysis. *Journal of Hazardous Materials,* 127(1–3), 14–30.
Darbra, R.M, Adriana Palacios, Joaquim Casal. 2010. Domino effect in chemical accidents: Main features and accident sequences. *Journal of Hazardous Materials,* 183(1), 565–573.
Delvosalle, C. 1996. *Proceedings of the European Seminar on Domino Effects,* Leuven, p. 11–12.

Faisal, I.K. & Abbasi, S.A. 2001. An assessment of the likelihood of occurrence and the damage potential of domino effect (chain of accidents) in a typical cluster of industries. *Journal of Loss Prevention in the Process Industries,* 14, 283–306.

Khan F.L, Abbasi S.A. 2001. Estimation of Probabilities and likely consequences of a chain of accidents (domino effect) in Manali Industrial Complex. *Journal of cleaner production,* (9), 493–508.

Lees, F.P. 1996. *Loss Prevention in the Process Industries.* Butterworth-Heinemann, Oxford, UK.

Li Junjie, Li Qi, Jiang Weiwei. 2008. CFD-base fire risk assessment of large crude oil tanks. *Safety, Health and Environment,* 8(9), 30–32.

Ma Kewei, Zhu Jian xin, Bao Shiyi. 2011. Research on quantitative risk assessment in chemical park based on domino effect model [J]. *Journal of Zhejiang University of Technology,* 39(1), 29–33.

Planas-Cuchi E, Montiel H, Casal J, et al. 1997. Survey of the origin, type and consequences of fire accidents in process plants and in the transportation of hazardous materials. *Process Safety and Environ-mental Protection,* 75 (B1), 3–8.

Ronza, A.S. Felez, Darbra, R.M. et al. 2003. Predicting the frequency of accidents in ports areas by developing event trees from historical analysis. *Journal of Loss Prevention in the Process Industries,* 16, 551–560.

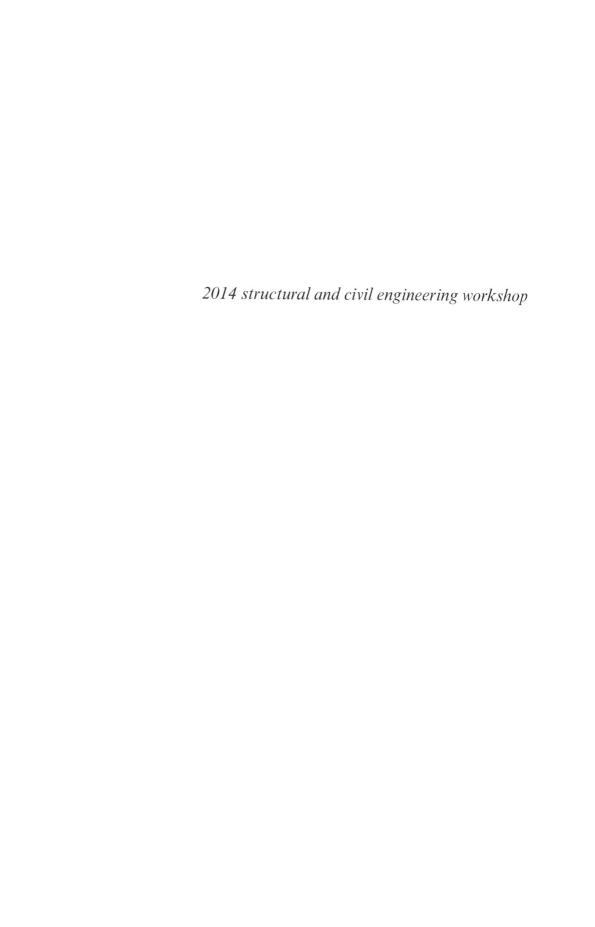

2014 structural and civil engineering workshop

Hydraulic Engineering III – Xie (Ed)

Numerical simulation of heat storage of solar-ground concrete pile

Lisha Liu
College of Construction Engineering, Jilin University, China
Changchun Institute of Technology, China

Dajun Zhao
College of Construction Engineering, Jilin University, China

Qiuling Lang
Changchun Institute of Technology, China

ABSTRACT: In this paper, established the mathematical model of the process of thermal storage in the solar-ground concrete pile. It adopted finite element methods to numerically simulate the unsteady-state temperature field of concrete pile that around the underground vertical tube, given the calculated format of different boundary condition, analyzed the temperature variation rule of the concrete around heat exchanger. This paper provided reference basis of ascertain method about the bury depth of the vertical U-tube and the mix proportion of the concrete pile.

1   INTRODUCTION

At present, solar energy is the best one of all source energy storage, it is stability and high heat. The relative reasonable method is to store energy in the ground. Now, most of the underground storage methods and principles are the development of new thermal storage materials and the use of phase-change latent heat storage and chemical storage, but these methods are often complex and high technical requirement and costly.

In order to explore the feasibility of construction a concrete pile in the ground then storage solar thermal high efficiently. So research a new technique of underground storage and extraction thermal—heat storage of solar-ground concrete pile. In this paper, on the basis of the research of underground soil heat storage, given the numerical simulation during the heat storage of underground concrete pile and analysis of the result of heat storage in order to explore the feasibility of this method.

2   ESTABLISHED MATHEMATICAL MODELS

2.1   *About of heat storage of solar-ground concrete pile*

It heats the recycled mediator through the solar collectors, then injected into the underground exchange well through the circulating pump. Exchanged heat through the heat exchanger, then stored the energy in the heat storage barn that has been insulated.

2.2   *Established mathematical models of storage process*

It is very complex about the heat exchangers and sounding concrete, relate to the type and thickness of the concrete, the form and install way of buried pipe, the tightness of the pipe and concrete, and so on. However, the heat transfer performance mainly depends on the heat transfer coefficient about the pipe and concrete, on the thermal conductivity of concrete and the change of the temperature field inside, and so on. In order to theoretical analysis make assumptions as following:

1. Assumed the established model use single U-tube heat exchanger, because of the small pitch in two pipes, will interfere each other, could cause heat change different of the outside pipe and concrete. To simplify the calculation, instead the two pipe of U-tube of an equivalence pipe. Its equivalent diameter $D_{eq} = \sqrt{2}D$ (Bose JE,1993), D is the diameter of one of the U-tube. Consider that the heat change of equivalence pipe wall and concrete is uniform;
2. The buried pipe and concrete contact well, ignored the contact resistance;
3. The hot properties of concrete and fluid are constant;
4. The flow rate of the heating fluid is constant.

3 NUMERICAL SIMULATION OF HEAT STORAGE

3.1 Differential equations of heat conductivity

After regarded U-vertical pipe as an equivalent pipe, soil temperature field which around it is a cylinder temperature field, in the circular direction there is no temperature gradient, the temperature distribution of the concrete around can be regarded as an axisymmetric problems. The styles of differential equations of heat conductivity are axisymmetric and unsteady state:

$$\rho C_p \frac{\partial T}{\partial t} = \lambda \left(\frac{\partial^2 T}{\partial x^2} + \frac{\partial^2 T}{\partial y^2} \right) \tag{1}$$

3.2 Initial conditions

The temperature distribution of concrete around the pipe in the initial moments may determined by the value of concrete temperature test before heat storage. Temperature value of node that not installed point for measuring temperature may determined by linear interpolation methods.

3.3 Determine the boundary conditions

Figure 1 $ABCD$ indicate the research area, side AD located in the pipe wall of the equivalent tube ($x = 0.028$ m), side CD is the horizontal plane ($y = 0$ m), side AB is the end of equivalent tube ($y = 12$ m), that is 12 m under ground, side BC is the concrete edge where 1 m from the center of the tube ($x = 1$ m). The boundary conditions of calculation area $ABCD$ as follows:

① Boundary AD regarded as second boundary condition

$$-\lambda \frac{\partial T}{\partial x} \Big|_{x=0.028} = q \tag{2}$$

Approximate consider that q is a definite value.

Because of the pipe wall of vertical U-tube is very thin (4 mm), could neglect the influence of pipe side, consider that the heat of concrete transferred form the pipe equal to the heat that circulating solution inside the pipe spread to the bury tube. So the change heat of per unit area underground tube and concrete, also as heat flow density q is:

$$q = \frac{Q}{F} \tag{3}$$

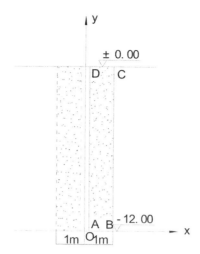

Figure 1. Calculation and boundary.

Table 1. The physical properties parameters of C25 concrete piles.

Length/m	12
Diameter/m	2
Thermal conductivity/W/(m · °C)	1.28
Specific heat/KJ/(kg · °C)	1.13
Density/kg/m³	2400

② Boundary AB regarded as adiabatic boundary condition

$$-\lambda \frac{\partial T}{\partial y}\Big|_{y=12} = 0 \tag{4}$$

③ Boundary BC regarded as adiabatic boundary condition

$$-\lambda \frac{\partial T}{\partial x}\Big|_{x=1} = 0 \tag{5}$$

④ Boundary CD regarded as third boundary condition

Could neglect the radiation heat transfer of pile surface and surrounding environment, hence the energy exchange heat balance equation of concrete surface and air is:

$$-\lambda \frac{\partial T}{\partial y}\Big|_{y=0} = \alpha_w (t - t_w) \tag{6}$$

3.4 Physical parameters of concrete

In order to solve these mathematical models, it is also necessary to determine the physical properties parameters of concrete that according to the proportion of concrete sample measured by the experiment (Dian, 1990). The physical properties parameters of C25 concrete piles in Table 1.

3.5 Finite element method analysis of numerical simulation

Finite element method of numerical calculation divided calculation area into several triangular elements to research. The shape and density of units could change arbitrary, thus would use less points to make the area achieve better approximation, with a great deal of flexibility and adaptability (Maoyu, 1995). Application of finite element method to solute the unsteady-state temperature field of sounding the vertical U-tube, where the temperature changes dramatically (like around the pipe) was divided into smaller units, but far away from the pipe the unit may be appropriate to enlarge, this may under the condition that not increase the number of nodes improve accuracy and reduce the amount of calculation. According to the basic principles of finite element method (Xianqian, 1986) and mathematical models of unsteady-state temperature field to numerical calculation.

For easy to solution supposed as follows:

1. To consider the temperature of following into the entrance of heat exchanger as a constant, that is, use the average of the entire storage period.
2. Because of heat storage process is a periodical process, so night is the process for the temperature field of concrete pile restoration. When consider the boundary condition of pipe side the heat transfer coefficient of night to regard as 0.
3. To see the concrete pile as a uniform material.

For the area ABCD can be divided into any of the triangular unit with finite element method. Each node has a corresponding order number 1, 2, ... ; each unit also has its own code ①, ②, ... ; for each unit every apex are numbered i, j, m by counter clockwise. Coordinates of three apexes determined by the unit division, so the three sides s_i, s_j, s_m of the three apexes i, j, m and the triangle area are also known. In the finite element method the temperature T of any point in the triangle (x,y) was discrete into the three nodes of unit, that is used T_i, T_j and T_m value to indicate the temperature field T of the unit, that is:

$$T = f\left(T_i, T_j, T_m\right) \tag{7}$$

Use of matrix inversion method could have a solution:

$$\begin{bmatrix} a_1 \\ a_2 \\ a_3 \end{bmatrix} = \frac{1}{2\Delta} \begin{bmatrix} a_i & a_j & a_m \\ b_i & b_j & b_m \\ c_i & c_j & c_m \end{bmatrix} \begin{bmatrix} T_i \\ T_j \\ T_m \end{bmatrix} \tag{8}$$

For the boundary jm of unit, the temperatures of two nodes are T_j and T_m, could construct the interpolating function on the boundary:

$$T = (1-g)T_j + gT_m \tag{9}$$

In formula $0 \le g \le 1$, as a parameter, $g = 0$ corresponds to the node j, $g = 1$ corresponds to the node m, so:

$$S_i = \sqrt{b_i^2 + c_i^2} \tag{10}$$

① For the internal unit and adiabatic unit, the format of finite element method for equation (1) is:

$$\begin{Bmatrix} \dfrac{\partial J^e}{\partial T_i} \\ \dfrac{\partial J^e}{\partial T_j} \\ \dfrac{\partial J^e}{\partial T_m} \end{Bmatrix} = \begin{bmatrix} K_{ii} & K_{ij} & K_{im} \\ K_{ji} & K_{jj} & K_{jm} \\ K_{mi} & K_{mj} & K_{mm} \end{bmatrix} \begin{Bmatrix} T_i \\ T_j \\ T_m \end{Bmatrix} + \begin{bmatrix} N_{ii} & N_{ij} & N_{im} \\ N_{ji} & N_{jj} & N_{jm} \\ N_{mi} & N_{mj} & N_{mm} \end{bmatrix} \begin{Bmatrix} \dfrac{\partial T_i}{\partial t} \\ \dfrac{\partial T_j}{\partial t} \\ \dfrac{\partial T_m}{\partial t} \end{Bmatrix} - \begin{Bmatrix} P_i \\ P_j \\ P_m \end{Bmatrix} = [K]^e \{T\}^e + [N]^e \begin{Bmatrix} \dfrac{\partial T}{\partial t} \end{Bmatrix}^e - \{P\}^e \tag{11}$$

② For the second boundary unit, the calculation format is: q_2 is the inflow into the boundary (W/m^2).

$$
\begin{cases}
K_{ii} = \varphi\left(b_i^2 + c_i^2\right) \quad (i, j, m \text{ circulant}) \\
K_{ij} = K_{ji} = \varphi\left(b_i b_j + c_i c_j\right) \quad (i, j, m \text{ circulant}) \\
N_{ii} = \dfrac{\Delta}{6}\rho c_p \quad (i, j, m \text{ circulant}) \\
N_{ij} = N_{ji} = \dfrac{\Delta}{12}\rho c_p \quad (i, j, m \text{ circulant}) \\
P_i = 0 = \varphi = \lambda/(4\Delta) \\
P_j = P_m = -\dfrac{q_2 s_i}{2}
\end{cases}
$$

③ For the third boundary unit, In it the calculation format is:

$$
\begin{cases}
K_{ii} = \varphi\left(b_i^2 + c_i^2\right) \\
K_{jj} = \varphi\left(b_j^2 + c_j^2\right) + \dfrac{\alpha s_i}{3} \\
K_{mm} = \varphi\left(b_m^2 + c_m^2\right) + \dfrac{\alpha s_i}{3} \\
K_{ij} = K_{ji} = \varphi\left(b_i b_j + c_i c_j\right) \\
K_{im} = K_{mi} = \varphi\left(b_i b_m + c_i c_m\right) \\
K_{jm} = K_{mj} = \varphi\left(b_j b_m + c_j c_m\right) + \dfrac{\alpha s_i}{6} \\
N_{ii} = \dfrac{\Delta}{6}\rho c_p \quad (i, j, m \text{ circulant}) \\
N_{ij} = N_{ji} = \dfrac{\Delta}{12}\rho c_p \quad (i, j, m \text{ circulant}) \\
P_i = 0, P_j = P_m = \dfrac{\alpha}{2} s_i T_f \\
\varphi = \lambda/(4\Delta)
\end{cases}
$$

Used synthetic equation $\dfrac{\partial J^D}{\partial T_l} = \sum\limits_{e=1}^{E} \dfrac{\partial J^e}{\partial T_l} = 0 \quad (l = 1, 2, \ldots, n)$, could obtain the synthesis calculation formula:

$$
[K]\{T\}_t + [N]\left\{\frac{\partial T}{\partial t}\right\}_t = \{P\}_t \tag{12}
$$

So could obtain the temperatures of nodes $T_1, T_2, \ldots T_n$.

4 THE RESULTS OF NUMERICAL SIMULATION

The concrete pile is an axsymmetric form, thus taking a half of the pile as investigated subject, the division of finite element as shown in Figure 2.

The water temperature of the heat exchangerentrance selected 80°C; The initial temperature of the points in the concrete pile selected 15°C. Figure 3 indicated the numerical results of the temperature distribution that inside the pile after heat storage has been 10 hours.

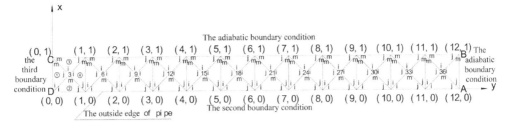

Figure 2.   The division of finite element.

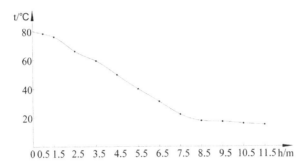

Figure 3.   The temperature distribution inside the pile after heat storage has been 10 hours.

5   CONCLUSION

Through the theoretical analysis and numerical simulation for the temperature field of concrete pile could obtain the following conclusions:

1. Use the equivalent tube method established the physical and mathematical models for the temperature field surrounding of the vertical U-tube, and used the finite element method for solving the mathematical model.
2. the temperature distribution inside the pile is uneven and unsteady state caused by the heat exchange between the concrete and water through the stage of storage.
3. For the use of calculation of computer programming provided a theoretical basis. These parameters can be changed; though the research of the predictive effect of heat storage may give the best concrete mix through adjust it.

REFERENCES

Bose JE, Parker JD. Ground coupled heat pump research [J]. Ashrae Transactions, 1993, 175–180.
Dian Luo. Application and analysis of heat transfer [M]. Beijing: Tsinghua University Press, 1990.67–68.
Maoyu Zheng. Energy-saving and heating of building [M]. Harbin: Heilongjiang Science and Technology Press, 1995.43–45.
Xianqian Kong, Edited. The application of finite element method in heat transfer [M]. Beijing: Science Press, 1986.

Hydraulic Engineering III – Xie (Ed)
© *2015 Taylor & Francis Group, London, ISBN 978-1-138-02743-5*

Study on the characteristics of temperature for concrete box girder

Lai-Jun Liu, Wei-Gang Sun, Fu-Tao Ni & Yong-Rui Chen
School of Highway, Chang'an University, Xi'an, China

Qun-Hu Wu
The No. 6 Engineering Co. Ltd., China Railway 20th Bureau Group, Xi'an, China

ABSTRACT: Analysis of temperature variations for concrete box-girder is the prerequisite to control its strains and cambers, and the law of temperature variations of concrete box girder is different in various regions. To study on the rules of temperature variation of concrete box girder in the areas with large temperature difference, the concrete box girder of Liutiao River Bridge of Yongdeng-Gulang Highway in Gansu province was considered as an engineering example, the on-site environmental parameters and temperature variations of box girder for a year were observed. Contrastive analysis has been made between the box-girder and environmental temperature variations. It suggested that the law of concrete box girder temperature variations is parallel to that of environmental temperature as a sinusoidal way, and box-girder temperature's change is hysteretic compared to the ambient temperature's change. Finally, the distribution with maximum temperature difference of per-month was cluster analyzed, and it can be divided into two categories initially.

1 INTRODUCTION

Temperature effect is one of the main causes of cracks in concrete box-girder bridge, the temperature gradient mode and value of concrete box girder did not gain consensus in different countries or different regions. Based on heat transport theory and test data, many specialists at home or abroad performed lots of research on the temperature distribution, influence factor and analytical method. Some temperature observations for concrete box-girder bridge at home were performed, and temperature distribution form of tested bridges were presented (Lei Xiao, Ye Jianshu, Wang Yi, et al. (2010), Zhang Liangliang, Yang Lei, Yang Zhuanyun, et al. (2011) and Fang Zhi, Wang Jian (2007).). Based on 2.5 a measured data for temperature field, Roberts-Wollman (2002) presented a prediction method for vertical temperature gradient of box-girder, and compared the measured value to the gradient value of the code. According to more than one years tested observation for temperature variations of a concrete beam, and analysis on the vertical and horizontal temperature gradient in various environments, Lee J, h and Kalkan (2012) showed estimation formula to estimate the maximum temperature gradient, and calculated deflection of the beam effected by the vertical and horizontal temperature gradient using 1-D beam theory. Temperature variations of concrete box girders is various in different areas with the changing of sunshine, wind speed, and others environment condition etc. Study on the laws of temperature in the specific area, there is some value to calculate the temperature effect on bridge structure in the similar environmental area.

2 PROJECT PROFILE

Gulang, that is a county whose geographical coordinates: Northern latitude 37°09′~37°54′, East longitude: 102°38′~103°54′, average annual temperature: 4.9 °C, average annual rainfall: about 300 mm, average annual evaporation capacity: above 2500 mm, average annual sunshine hours: 2852.3 hours, frost-free season: about 142 days, and who belongs to the Qilian mountains in Hexi cold temperature arid areas. The test bridge is located in Gufeng village, Gulang County, and its upper structure is constructed with 30 m prefabricated prestressed concrete box girder. According to the Research Report of Natural Zoning for Highways in Gansu Province and Environmental Parameters, the bridge is located in cold plateau climate zone, which is climate type I. There is large temperature difference between day and night in spring and fall seasons, dry and hot in summer, but cold in winter.

3   TESTING PROGRAM

3.1   Layout of measured points

The test section that is 1 meter from middle of 9th span was selected, that is shown in Figure 1.

3.2   Testing instrument

1. Ambient temperatures were monitored by using the HTC-1 temperature and humidity meter with temperature measuring range: −50 °C~70 °C, and accuracy: ±1 °C.
2. Surface temperatures of concrete box girder (A-) were monitored by using the UT302B non-contact infrared thermometer with temperature measuring range: −32 °C~550 °C and display precision: ±1.8 °C and ±1.8%.
3. Cross section test points (B-) were measured by using WRN-type K thermocouple with temperature measuring range: −50~400 °C, and temperature measurement accuracy: 0.75%±2.5 °C. TES-1310 digital temperature meter was used as a monitor with resolution 0.1 °C, temperature measuring range: −50 °C~199.9 °C, and temperature measurement accuracy: ± (0.3%+1 °C).
4. To measure temperature of points (C-), JMZX-215 AT strain gauges with measuring range: ±1500 με were used.

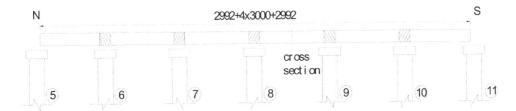

Figure 1.   Test cross section.

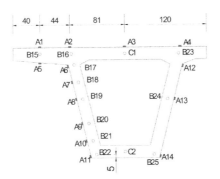

Figure 2.   Test points layout (cm).

4 TEMPERATURE-VARIATIONS ANALYSIS

The temperature distribution is changing due to change of external environment, and by comparing the relationship between ambient temperature variations and box-girder temperature variations, the effects of environment on box-girder temperature variations was analyzed.

Temperature variations in March 29, 2011 is showed in Figure 3 as an example.

As it can be seen from Figure 3, the temperature variations on the roof surface is small, temperature-time curves are uniform. Comparing the temperature on the roof surface and of environment, the consistent change trends were found and all of which appear rising and falling stages. However, the environmental temp-time curve looks flat, and all the temperature comes to the practically uniform trends for the time from 18:00 to 20:00. That indicates the gap between the roof-surface temperature and environmental temperature become small during that time. The average daily temperature difference was 32 °C.

Figure 4 shows that the temperature variation on the outside surface of right web is small, and the maximum temperature (2.2 °C) appears at 13:00, and the ambient temperature is

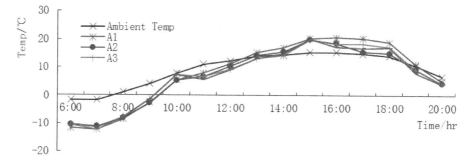

Figure 3.   Temperature history on the roof surface.

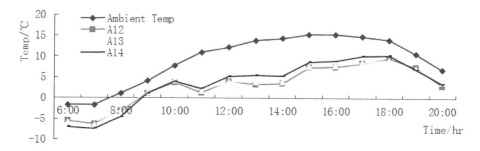

Figure 4.   Temperature history on right outer surface of webs.

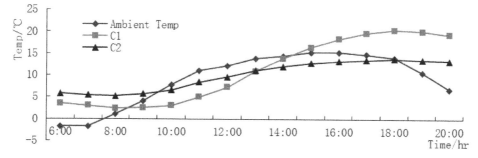

Figure 5.   Temperature history in the proof and bottom plate.

285

higher than the temperature on the outside surface of right web. The ambient temperature comes to the maximum at 15:00–16:0, while the test point temperature comes to the maximum at 18:00, The average diurnal range of temperature on the outside surface of right web was 16.4 °C.

From Figure 5 it can be seen that the temperature trends inside the roof and bottom plate are different. The temperature inside the roof plate was lower than the ambient temperature firstly, but which start to be higher than ambient temperature at about 13:00. At the end, the temperature inside the roof plate goes up feebly, while the ambient temperature decline, and the average diurnal range of temperature inside the roof plate was 8.6 °C.

5 CLUSTER ANALYSIS

Cluster analysis is an exploratory analysis, but in the process of classification, it is not necessary to give a classification standard, nonetheless, cluster analysis can proceed from the sample data, and classify automatically. Different cluster analysis methods often come to a different conclusion.

Assume there is little change for environmental conditions in one month, and then the distribution with maximum temperature difference can be considered as one group, so all the test data can been divided into 7 groups in advance. It is showed in Table 1.

The cluster method: Between-groups linkage, interval: Euclidean distance, Transform Values-standardize: Range 0–1.

The dendrogram was shown as Figure 6 and Figure 7.

From Figure 6, It can be seen that 7 groups of positive temperature difference can be divided into 2 categories, winter sample data (From December 2010 to March 2011), and summer sample data (From April 2011 to August). Among them, the Euclidean distance in May and June is the smallest, and there is high degree of aggregation.

It can be seen from Figure 7 that 7 groups of negative temperature difference also can be divided into 2 categories, winter sample data (From December 2010 to March 2011), and summer sample data (From April 2011 to August). Among them, the Euclidean distance from May to June is the smallest, and there is high degree of aggregation.

Table 1. Groups for distribution with maximum temperature difference.

Group	1	2	3	4	5	6	7
Time	2010–12	2011–01~03	2011–04	2011–05	2011–06	2011–07	2011–08

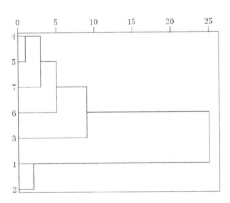

Figure 6. Dendrogram for the distribution with maximum positive temperature.

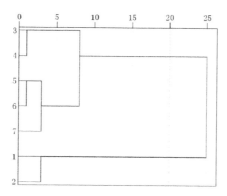

Figure 7.   Dendrogram for the distribution with maximum negative temperature.

6   CONCLUSION

1. The law of concrete box girder temperature variations is parallel to that of environmental temperature as a sinusoidal way, and their change cycle is basically the same. Box-girder temperature's change is hysteretic compared to the ambient temperature's change.
2. By using cluster analysis, the distribution with maximum temperature in one month can be divided into two categories.

REFERENCES

Fang Zhi, Wang Jian. Sun light thermal difference effect on long-span PC continuous box girder bridge [J]. China Journal of Highway and Transport, 2007, 20(1):62–67.

Lei Xiao, Y.E. Jianshu, Wang Yi, et al. Analysis of concrete box-girder temperature and strain based on long term observation [J]. Journal of Jiangsu University: Natural Science Edition, 2010, 31(2): 230–234, 239.

Lee J.H., Kalkan L. Analysis of Thermal Environmental Effects on Precast, Prestressed Concrete Bridge Girders: Temperature Differentials and Thermal Deformations [J]. Advances in Structural Engineering, 2012, 15(3):447–460.

Roberis-Wollman C.L., Breen J.E., Cawrse J. Measurements of thermal gradients and their effects on segmental concrete bridge [J]. Journal of Bridge Engineering, 2002, 7(3):166–174.

Zhang Liangliang, Yang Lei, Yang Zhuanyun, et al. Temperature Field Analysis of Long-span Concrete Box Girders [J]. Journal of Civil, Architectural and Environmental Engineering, 2011, 33(1):36–42.

Hydraulic Engineering III – Xie (Ed)
© 2015 Taylor & Francis Group, London, ISBN 978-1-138-02743-5

Study of the earthquake response of self-anchored suspension bridges

Chang-Huan Kou
Department of Architecture and Urban Planning, Chung Hua University, Hsinchu, Taiwan

Jin-Ja Fan & Meng-Wei Lai
Department of Civil Engineering, Chung Hua University, Hsinchu, Taiwan

ABSTRACT: The main cables of self-anchored suspension bridge are directly anchored at the two ends of the main girder with the axial pressure transferred from the main cable to the main girder used as a source of pre-stressed force that indirectly enhances the bending resistance capability of the reinforced concrete of the main girder, which then enables the self-anchored suspension bridge to be an economically realistic long span structure. This paper will analyze and explore when a self-anchored suspension bridge's main cables, main girders, towers, hangers, and other structural elements' material elastic modulus is individually altered, what the effect on the bridge's earthquake responses in the direction along the main girder would be in order to clearly understand the self-anchored suspension bridge's dynamic characteristics. The results of this research indicate that the effects of an alteration in the main girder elastic modulus on the self-anchored suspension bridge's maximum internal forces and displacements due to an earthquake along the direction of the main girder is the largest, followed by the effect of an alteration in the elastic modulus of the towers, main cables, and hangers.

1 INTRODUCTION

Because the main cables of self-anchored suspension bridges are already directly anchored at the two ends of the main girder, this structure saves the cost of installing other anchoring devices; furthermore, the axial pressure is transferred from the main cable to the main girder and used as a source of pre-stressed force that indirectly enhances the bending resistance capability of the main girder, which then renders the self-anchored suspension bridge worthy of further research and widespread use as a water-spanning structure (Wollmann, G. 2001, Wollmann, G. et al. 2001). Therefore, this paper, upon completing a stochastic static analysis of the self-anchored suspension bridge (Kao, C.S. et al. 2006), further explores a self-anchored suspension bridge's earthquake responses in order to clearly understand the dynamic characteristics of the self-anchored suspension bridge in addition to the impact that different material characteristics may have on the earthquake responses of this very structure.

2 THEORETICAL ANALYSIS OF THE DYNAMIC CHARACTERISTICS OF SELF-ANCHORED SUSPENSION BRIDGES

In considering the self-anchored suspension bridge's large displacement generalized potential energy function, this paper will be based on the following assumptions (Zhang, H.J. 2004):

− All the materials must comply with Hook's law.
− At the complete stage, the statical load is evenly distributed along the span, and the cable forms a parabolic curve.

- At its points of connection with the towers, the main girder receives vertical support.
- Given that the hangers are distributed densely, they are treated as even membranes with only vertical resistance and malleability is not taken into account.
- Without considering the bending resistance rigidity and axial compression deformation in the direction along the main girder of the tower, the paper will assume that the rigidity in the transverse direction is infinitely large.
- The paper considers only the warping deformation of the twist in the main girder but does not consider the cross-section's distorted deformation and will assume the diaphragm is distributed densely and its shearing rigidity is infinitely large.

Assume the displacements along the main girder are u, vertical displacements are v, displacements in the transverse direction are w, twisted angles are θ, displacements along the direction of the main girder, in the vertical direction, and transverse direction of the left side cables of the main girder are u_l, v_l, w_l, displacements along the direction of the main girder, in the vertical direction, and transverse direction of the right side cables of the main girder are u_r, v_r, w_r, $H_l(t)$, $H_r(t)$ represent the left and right side increment horizontal forces instigated by the vibration inertial force, E_c, A_{cl} are the cross-sectional areas and the elastic modulus of the main cables, q_s is the statical load density of the main girder, q_c is the statical load density of the main cables, H_q is the horizontal force of the two main cables under statical load conditions, L_c is the length of each span, λ_l, λ_r are the Lagrange Multipliers, f_l, f_r are the forces constraining the extension of the left and right hangers, g is gravity acceleration, $h' = $ the cable length along the x-axis, A_s, I_y, I_z are the cross-sectional area, lateral, and vertical flexural inertial moments of the main girder, respectively, E is the elastic modulus of the main girder, μ is the shearing influence coefficient, $\beta E_l J_w$ is the main girder's warping rigidity, $\beta = J_p/(J_p - J_t)$, $E_l = E/(1 - v^2)$, J_p is the pole moment inertia of the cross section, J_t is St. Venant's torsional constant.

The large displacement generalized potential energy function (Zhang, H.J. 2004, Fan, J.J. 2007):

$$\Pi^* = \int_{t_1}^{t_2} \left\{ \frac{1}{2} \sum_{i=1}^{3} \int_0^{L_i} \frac{q_s}{gA_s} \left[I_y \left(\frac{\partial^2 w}{\partial x \partial t} \right)^2 + I_z \left(\frac{\partial^2 v}{\partial x \partial t} \right)^2 + \left(I_y + I_z \right) \left(\frac{\partial \theta}{\partial t} \right)^2 + A_s \left(\frac{\partial u}{\partial t} \right)^2 + A_s \left(\frac{\partial v}{\partial t} \right)^2 \right. \right.$$

$$+ A_s \left(\frac{\partial w}{\partial t} \right)^2 \right] dx_i + \frac{1}{2} \sum_{i=1}^{3} \int_0^{L_i} \frac{q_c}{2g} \sqrt{1 + h'^2} \left[\left(\frac{\partial u_l}{\partial t} \right)^2 + \left(\frac{\partial v_l}{\partial t} \right)^2 + \left(\frac{\partial w_l}{\partial t} \right)^2 + \left(\frac{\partial u_r}{\partial t} \right)^2 + \left(\frac{\partial v_r}{\partial t} \right)^2 \right.$$

$$+ \left(\frac{\partial w_r}{\partial t} \right)^2 \right] dx_i - \sum_{i=1}^{3} \frac{L_c}{E_c A_{cl}} \left[\frac{1}{2} H_q (H_l + H_r) + \frac{1}{2} (H_l^2 + H_r^2) \right] - \frac{1}{2} \sum_{i=1}^{3} \left[\int_0^{L_i} \beta E_l J_w \left(\frac{\partial^2 \theta}{\partial x^2} \right)^2 dx_i \right.$$

$$\int_0^{L_i} GJ_t \left(\frac{\partial \theta}{\partial x} \right)^2 dx_i + \int_0^{L_i} EI_y \left(\frac{\partial^2 w}{\partial x^2} \right)^2 dx_i + \int_0^{L_i} EI_z \left(\frac{\partial^2 v}{\partial x^2} \right)^2 dx_i + \int_0^{L_i} EA_s \left(\frac{\partial^2 u}{\partial x^2} \right)^2 dx_i$$

$$+ \int_0^{L_i} N_x \left(\frac{\partial v}{\partial x} \right)^2 dx_i + \int_0^{L_i} \mu G A_s \left(\frac{\partial u}{\partial y} + \frac{\partial v}{\partial x} \right)^2 dx_i + \int_0^{L_i} \mu G A_s \left(\frac{\partial u}{\partial z} + \frac{\partial v}{\partial x} \right)^2 dx_i \right]$$

$$- \sum_{i=1}^{3} \left[\int_0^{L_i} q_s v dx_i + \int_0^{L_i} \frac{1}{2} q_c (v_l + v_r) dx_i \right] - \sum_{i=1}^{3} \int_0^{L_i} \lambda_l f_l dx_i - \sum_{i=1}^{3} \int_0^{L_i} \lambda_r f_r dx_i \right\} dt \tag{1}$$

where H_q is the sum of horizontal forces triggered by the statical load of the two main cables.

The vibration equations of the self-anchored suspension bridges can be obtained from following operation:

$$\delta \Pi^* = \frac{\partial \Pi^*}{\partial \phi_i} \delta \phi_i = 0 \tag{2}$$

3 BASIC DATA ANALYSIS

In this paper's analysis, the choice of bridge corresponds to that in reference (Zhang, H.J. 2004), namely a double tower, suspension bridge with a main span of 160 meters, two side spans of 76 meters, an entire length of 312 meters, tower height of 42.4 meters, rise-span ratio of 1/6, hanger to hanger distance of 5 meters, a reinforced concrete box girder as the main girder with a standard cross-section of 5 single box cells, girder width of 41 meters, height of girder's geometric center of 2.5 meters, and main girder height-span ratio of 1:64. Furthermore, the construction site that this paper considers is located in a strong earthquake zone where the site's underlying geological plates are of category 3. The earthquake response spectrum is displayed in Figure 1.

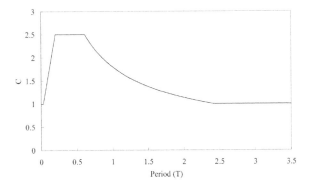

Figure 1.   Category 3 plates earthquake response spectrum.

Table 1.   The effects of a change in the material elastic modulus on the structure's maximum internal forces during an earthquake in the direction along the main girder.

Item	Main cable's maximum hanger force (N)	Hanger's maximum hanger force (N)	Tower base's maximum shear force (N)	Tower base's maximum bending moment (N-m)	Main girder's maximum bending moment (N-m)
Main cable E value					
Reduce to 0.7 times the original value	+0.13%	+0.11%	−0.12%	−0.07%	−0.18%
Increase to 1.3 times the original value	−0.17%	−0.14%	+0.09%	+0.04%	+0.10%
Main girder E value					
Reduce to 0.7 times the original value	−0.24%	−0.23%	−0.30%	−0.29%	+2.37%
Increase to 1.3 times the original value	+0.19%	+0.018%	+0.23%	+0.22%	−2.01%
Tower E value					
Reduce to 0.7 times the original value	+0.11%	+0.09%	+0.28%	+0.21%	−1.70%
Increase to 1.3 times the original value	−0.17%	−0.14%	−0.32%	−0.27%	+1.78%
Hanger E value					
Reduce to 0.7 times the original value	+0.02%	+0.04%	−0.02%	−0.01%	−0.02%
Increase to 1.3 times the original value	−0.01%	−0.02%	+0.01%	+0.01%	+0.01%

Table 2. The effects of a change in the material elastic modulus on the structure's displacement during an earthquake in the direction along the main girder.

Item	Main girder midpoint horizontal displacement (m)	Main girder vertical displacement			Tower roof horizontal displacement (m)
		Left side span midpoint (m)	Main span midpoint (m)	Right side span midpoint (m)	
Main cable E value					
Reduce to 0.7 times the original value	+1.92%	−0.21%	−0.53%	−0.19%	−0.05%
Increase to 1.3 times the original value	−1.76%	+0.11%	+0.29%	+0.10%	+0.03%
Main girder E value					
Reduce to 0.7 times the original value	+1.38%	+26.32%	+27.65%	+26.32%	−0.27%
Increase to 1.3 times the original value	−0.84%	−25.04%	−26.23%	−25.04%	+0.12%
Tower E value					
Reduce to 0.7 times the original value	+22.54%	−1.78%	−3.75%	−1.78%	+25.89%
Increase to 1.3 times the original value	−21.15%	+1.87%	+4.17%	+1.86%	−24.09%
Hanger E value					
Reduce to 0.7 times the original value	+0.19%	−0.02%	−0.03%	−0.02%	−0.01%
Increase to 1.3 times the original value	−0.11%	+0.01%	+0.01%	+0.01%	+0.01%

4 THE EFFECT OF A CHANGE IN THE MATERIAL ELASTIC MODULUS ON THE EARTHQUAKE RESPONSE IN THE DIRECTION ALONG THE MAIN GIRDER

Tables 1 and 2 both indicate that a change in the main girder's elastic modulus has the greatest effect on the structure's internal forces and displacements generated by an earthquake in the direction along the main girder followed by a change the elastic modulus in the towers, main cables, and hangers.

5 CONCLUSION

A double tower self-anchored suspension bridge with bridge length 312 m, main span 160 m, two side spans of 76 m, tower height 42.4 m, and hanger to hanger distance of 5 m, this paper's analysis yields the following conclusions:

– During an earthquake in the direction along the main girder, when the main cables' elastic modulus is reduced, this will cause the maximum hanger force of the main cables' and hangers to increase; when the main girder's elastic modulus is reduced, this will cause the main girder's maximum bending moment to increase. Also, when the towers' elastic modulus is reduced, this will cause the maximum hanger force of the main cables and hangers and the tower base's maximum internal force to increase; finally, a reduction in the hangers' elastic modulus will cause the maximum hanger force of the main cables and hangers to increase.

– During an earthquake in the direction along the main girder, when the main cables' elastic modulus is reduced, this will cause the main girder's midpoint horizontal displacement to increase; when the main girders' elastic modulus is reduced, this will cause the main girder's midpoint horizontal displacement and its vertical displacement to increase. Also, when the towers' elastic modulus is reduced, this will cause the main girder midpoint horizontal displacement and tower roof's horizontal displacement to increase; finally, a reduction in the hangers' elastic modulus will cause the main girder's midpoint horizontal displacement to increase.

– A change in the main girder's elastic modulus has the largest effect on the structure's internal forces and displacements generated by an earthquake in the direction along the main girder followed by a change the elastic modulus in the towers, main cables, and hangers.

REFERENCES

Fan, J.J. 2007. *Study of the Dynamical Response of Self-Anchored Suspension Bridges*. Hsinchu, Taiwan: Master Thesis, Department of Civil Engineering, Chung Hua University.

Kao, C.S., Kou, C.H., Tsai, J.L. & Shao, Y.M. 2006. Nonlinear Stochastic Static Analysis of Self-Anchored Suspension Bridge. *Asia Pacific Review of Engineering Science and Technology* 4(1): 653–668.

Wollmann, G. 2001. Preliminary Analysis of Suspension Bridges. *Journal of Bridge Engineering, ASCE* 6(4): 227–233.

Wollmann, G., Ochsendorf, J. & Billington, D. 2001. Self-Anchored Suspension Bridges. *Journal of Bridge Engineering, ASCE* 6(2): 156–158.

Zhang, H.J. 2004. *Mechanical Performance Analysis and Experimental Research on Concrete Self-Anchored Suspension Bridge*. Dalian: Master Thesis, Department of Bridge Engineering, School of Civil & Hydraulic Engineering, Dalian University of Technology.

Hydraulic Engineering III – Xie (Ed)
© 2015 Taylor & Francis Group, London, ISBN 978-1-138-02743-5

The analysis and ponder of instability and failure cause of composite subgrade in a railway project under construction

Wentian Gu, Guonan Liu & Yueying Liang
China Academy of Railway Sciences, Beijing, China

ABSTRACT: In Pearl River Delta, the jet grouting composite subgrade was used to improve some range of embankment in a new railway project, which lied in a range of 8~12 m marine soft clay; but this range embankment whose designed elevation was 6.4 m occurred damage under constructing filling to 4.6 m. According this actual case, the calculation method which was supplied by code (China Academy of building research, 2012) and manual was used to check the problem; but it was great difference between the analysis result of using the method of combined strength and actual situation. Through the stress ratio of pile to soil of back analysis, the calculation result of using the method of load distribution was close to actual situation. The analysis result shown that the calculation result of using the method of combined strength is bigger than actual situation on condition of more different strength between the strengthening soil and natural soil, so it was unsafe relatively; but the result of load distribution method was reasonable relatively. To sum up the stress ratio of pile to soil was an essential parameter in design of composite foundation.

1 INTRODUCTION

In the area of Pearl River delta, 8~12 m soft clay in foundation soils was very widespread, so it was necessary of improving this type soil before construction building. Composite subgrade was a construction method which has many advantages such as: shorter period, higher subsoil bearing capacity after improving, especially there has many success case of using composite subgrade of soil-cement columns in Pearl River delta.

The jet grouting composite subgrade was used to improve some range of embankment in a new railway project, which lied in a range of 8~12 m marine soft clay; but this embankment whose designed elevation was 6.4 m occurred damage under constructing filling to 4.6 m. In this actual case, the calculation method which was supplied by code and manual was used to check the problem; and the method of load distribution was used to calculate the stress ratio of pile to soil on composite subgrade. Lastly, the result of analysis and discussion was given.

This case showed the fill which was under the foundation of composite subgrade have property of flexibleness; and the deformation which was occurred in the process of distributing load from subgrade made the relation of stress between piles and soil become more complex (Guonan Liu, 2013). In the process of composite subgrade design, designer needed pay more attention to use piles-soil stress ratio and monolithic stability method rationally when code was vided.

2 CASE INTRODUCTION

2.1 *Topography of site*

This project locate in the area of marine sedimentation. The original topography was inter-tidal belt; and after human movement, this area became a fishing pool and marine land.

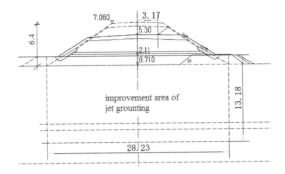

Figure 1. Standard section of design embankment (unit: m).

In this site, there were many ditches, which were filled by water. The ground elevation was about 0.0~1.0 m; and depth of water was 1.0~3.0 m; the elevation of small dams which distribution in many fishing pools was 2.0~2.5 m.

2.2 Geology of site

In this site, the stratigraphic distribution from up to down was summarized as follows.

1. Surface soil: was 0.5 to 1.0 m of thickness, locate in bottom of ditches, and was rich in organic matter. Moisture content was greater than 100%.
2. Soft clay: was 8.5 to 12.0 m of thickness, locate below surface soil, and was widespread distribution; moisture content was greater than 100%; and has other property such as: liquid-plastic state, under consolidated, low strength, high compensation, and thixotropy.
3. Salty clay: was flood alluvial soil, was about 6.0 m thickness, and was soft plastic soil.
4. Medium coarse sand: was nonuniform thickness, and max thickness was 4.0 m; then had the property of slightly dense to dense.
5. Severely-weathered rock: was moorstone.

2.3 Embankment and improvement plan

Design section of embankment, as shown in Figure 1, has 6.4 m height, and top of embankment elevation was about 7.08 m. In this Section, the jet grouting composite subgrade was used to improve embankment. The design diameter of jet grouting pile was 0.5 m; and triangle distribution distance was 1.4 m to 1.6 m. The strength of jet grouting pile was 1.3 Mpa, value of the bearing capacity of a single jet grouting pile was 130 kN. Bearing capacity of soft clay was 60 kPa.

3 PROCESS OF CONSTRUCTION AND STATE OF GEOLOGIC HAZARD

The process of construction can be summarized as follows. The first step: draining, constructing 0.5 m to 1.0 m cushion, and Cushioning surface clay lying the range of embankment. The second step: constructing jet grouting piles, and checking quality of jet grouting piles. The third step: filling and rolling embankment. In the process of construction, core drilling method and indoor compression strength test were used to check the quality of jet grouting piles, but the third step would not start unless passing the check. In the process of third step, part surface of embankment cracked when embankment height was up to 4.0 m, and occurred damage under constructing filling to 4.6 m, as illustrated in Figure 2 to 3. When damage occurred, embankment height below design stilled be at about 1.5 m.

Figure 2. Photo (1) from site of instability embankment.

Figure 3. Photo (2) from site of instability embankment.

4 APPLICABLE CODES AND STANDARDS CHECK AND CALCULATION

4.1 *Composite subgrade bearing capacity check*

According applicable Chinese codes-Technical code for ground treatment of buildings (JGJ79-2012, item: 7.1.5-2), the composite subgrade bearing capacity using jet grouting piles should be show as follow.

$$f_{spk} = \lambda \frac{R_a}{A_p} + \beta(1-m)f_{sk} \tag{1}$$

$$m = (d/d_e)^2 \tag{2}$$

where: β = soil among pile bearing capacity discount coefficient; λ = a single pile bearing capacity effective coefficient; f_{spk} = characteristic value of composition subgrade bearing capacity; f_{sk}—characteristic value of foundation bearing capacity; R_a = characteristic value of a single pile bearing capacity; A_p = area of a single pile section; m = displacement ratio; d_e = equivalent diameter; d_e = 1.05 s (triangle distribution) or d_e = 1.13 s (square distribution); s = distance of distribution.

Let the parameters of the formula (1) to (2) have the value according to Figure 4 and code, $\beta = 0.95$, $\lambda = 0.8$, $R_a = 130$ kN, $A_p = 0.196$ m², s = 1.5 m, d = 0.5 m, $f_{sk} = 60$ kPa.

So, characteristic value of subgrade bearing capacity f_{spk} = 111.4 kPa. Moreover, displacement ratio m = 0.088. While, mix load value from embankment was 110.0 kPa. From above-mentioned, it could be get that: f_{spk} = 111.4 (kPa) ≥ 111.0 kPa (passed according to code).

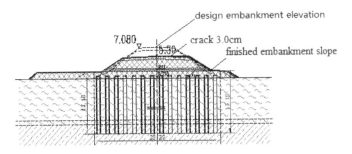

Figure 4.   Section of calculation.

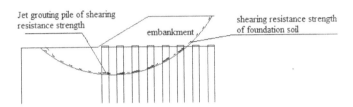

Figure 5.   Calculation model of combined strength.

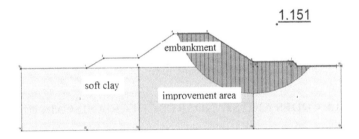

Figure 6.   Calculation and analysis sketch of combined strength.

However, the height of filled embankment was 4.5 m when it occurred problem. The loads from embankment was 82.8 kPa, which was less than subgrade bearing capacity according to code. This analysis result showed stable stability of embankment was not only rely on subgrade bearing capacity.

4.2   Safety factor analysis by composite strength method

Composite strength slices method was recommended by applicable codes to solve the embankment stability of composite subgrade. Composite strength slices method assumed that strengthening and soft clay were rigid-plastic material, which could bring to skid resistance simultaneously, as illustrated in Figure 5.

In this case, the shearing strength of composite foundation was

$$\tau_{sp} = m\tau_p + (1-m)\tau_s \tag{3}$$

where: τ_{sp} = combined strength of composite foundation. τ_p = Jet grouting pile of shearing resistance strength. τ_s shearing resistance strength of foundation soil.

The quick shear test value of soft clay was that c was 5.5 kPa and frictional angle was 5.00; the average shearing resistance strength of soft clay layer was 12 kPa; the shearing resistance strength of jet grouting pile was 0.18 MPa. When all above parameter got into the formula (2), value of shearing resistance strength of Foundation soil was 38.4 kPa. Then though Sweden slices method, the safe factor of design embankment monolithic stability was 1.15, when embankment was filled to design height.

But this range of road embankment whose designed elevation was 6.4 m occurred damage under constructing filling to 4.6 m. This actual situation showed that the method of combined strength was dissatisfy for this type composite subgrade.

5   STABILITY ANALYSIS BY DISTRIBUTE LOAD METHOD

5.1   Introduction

In overseas codes (British standard BS8006, 1995; Nordic handbook, 2002; The Design and Construction Handbook of Mixing Piled foundation, 2001), the distribution load method was recommend to analyze composite subgrade such as semi-rigid pile of cement-soil type. This method assumed that part of uploads born by strengthening, and others born by soil. So the monolithic stability of embankment was controlled by upload of forcing soil. The distribution load method required to know a definite stress ratio of pile to soil firstly. Then according to the principle of the distribution load, upload of forcing soil was calculated to evaluate monolithic stability of embankment.

5.2   Analysis of the stress ratio of pile to soil

In this case, the detail path of using the distribution load method was as followed by: Selected an accident embankment section, assumed that safe factor of monolithic stability was 1.0, and analyzed stress ratio of pile to soil.

The distribution load of composite foundation was as followed by Figure 7.

The relation of total stress P and stress of forcing soil was as followed:

$$P = \left[1 + m\left(n - 1\right)\right]\sigma_s \qquad (4)$$

$$H = \frac{\sigma_s}{\gamma_s} \qquad (5)$$

where: σ_s = the soil stress from upload; n the stress ratio of pile to soil; m displacement ratio H equivalent height of embankment; γ_s = volume weight of embankment fill.

Assumed diverse stress ratio of pile to soil, and calculated homologous equivalent height of embankment according to the formula (4) and (5). Then analyzed monolithic stability of embankment using slice method, which was based on equivalent height. When the safe factor of monolithic stability equaled 1.0, the actual stress ratio of pile to soil could be obtained.

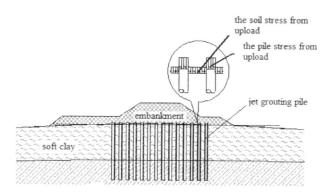

Figure 7.   Sketch of pile-soil load distribution.

5.3 Calculation result of the stress ratio of pile to soil

Selected two section of accident embankment to analyze the stress ratio of pile to soil.

1. Section one
   As illustrated in Figure 8, the elevation of top of embankment was 7.08 m, and the elevation of surface of clay was 0.7 m. When embankment was filled to elevation 5.3 m, the top of embankment occurred crack, and collapsed subsequently. Therefore, a conclusion could be obtained by actual situation: the safe factor of monolithic stability was smaller or equal to 1.0, when embankment collapsed.

   In critical state, the total upload of composite subgrade was equivalent earth column height 4.8 m-earth stress 91.2 kPa. Assumed n = 2, 2.5, 3.0, 3.5, 4.0, 4.5, and calculated homologous stress of foundation σ_s, as illustrated in Table 1.

   Though profession software, six calculation model were built, which based on slice method. After calculating, six relevant safe factor could be obtained, as illustrate in Table 2.

   When the safe factor was 1.0, the value of stress ratio of pile to soil was 3.3 according the Figure 9.

2. Section two
   As illustrated in Figure 10, the elevation of top of embankment was 7.32 m, and the elevation of surface of clay was 0.25 m. When embankment was filled to elevation 6.2 m, the top of embankment occurred crack, and collapsed subsequently. Therefore, a conclusion could be obtained by actual situation: the safe factor of monolithic stability was smaller or equal to 1.0, when embankment collapsed.

   In critical state, the total upload of composite subgrade was equivalent earth column height 6.0 m-earth stress 113.6 kPa. Assumed n = 2, 2.5, 3.0, 3.5, 4.0, 4.5, and calculated homologous stress of foundation σ_s, as illustrated in Table 2.

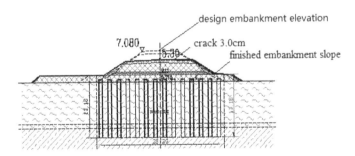

Figure 8.   Section 1 of calculation model.

Table 1.   Relationships between piles-soil stress ratio and stress of foundation for section 1.

n	2	2.5	3.0	3.5	4.0	4.5
σ_s/kPa	82.9	79.3	76.0	72.9	70.1	67.5
H/m	4.4	4.2	4.0	3.8	3.7	3.6

Note: H was equivalent earth column height.

Table 2.   Relationships between piles-soil stress ratio and safe factor for section 1.

n	2	2.5	3.0	3.5	4.0	4.5
Fs	0.968	0.970	0.979	1.025	1.045	1.046

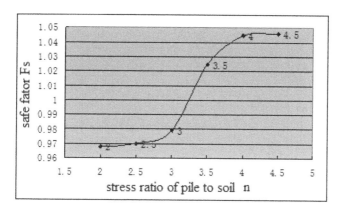

Figure 9.   Relationships between piles-soil stress ratio and safe factor for section 1.

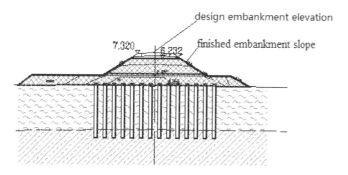

Figure 10.   Section 2 of calculation model.

Table 3.   Relationships between piles-soil stress ratio and stress of foundation for section 2.

n	2	2.5	3.0	3.5	4.0	4.5
σ_s/kPa	103.3	98.8	94.7	90.9	87.4	84.1
H*/m	5.4	5.2	5.0	4.8	4.6	4.4

*H = equivalent earth column height.

Table 4.   Relationships between piles-soil stress ratio and safe factor for section 2.

n	2	2.5	3.0	3.5	4.0	4.5
Fs	0.948	0.958	0.964	0.981	0.996	1.007

Though profession software, six calculation model were built, which based on slice method. After calculation, six relevant safe factor could be obtained, as illustrate in Table 2.

When the safe factor was 1.0, the value of stress ratio of pile to soil was 3.3 according the Figure 11. The stress ratio of pile to soil was about 3.3 to 4.2, which closed to conclusion of literature (Jifu Liu, 2003).

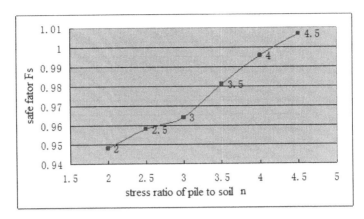

Figure 11.   Relationships between piles-soil stress ratio and safe factor for section 2.

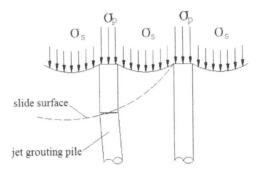

Figure 12.   Sketch of load distribution at the piles and soil.

6   ANALYSIS AND DISCUSS OF ENGINEERING ACCIDENT

After embankment collapse, the first thing: checked the quality of construction, and used the coring and indoor compression test to check the problem. But the result of test showed that the strength and length of piles matched the design generally.

The calculation result of using applicable codes showed uploads from filled embankment were less than the bearing capacity of composite foundation. But it was be found that the piles-soil stress ratio of composite subgrade between the practical situation and design had a great difference. The bearing capacity of single jet grouting pile was 130 kN. After calculation, it was easy got that the characteristic value of bearing capacity of pile was 662 kPa, and the characteristic value of bearing capacity was 60 kPa. It was supposed when the bearing capacity of piles and soil got limit state in the same time, the stress ratio of pile to soil could close to 11. This data was more than the data from back analysis result–3.3 to 4.2. It showed that the upload which distributed the soil was more than designer's expectation.

According the result of calculation and analysis, the process of the embankment collapse is that:

1. The upload which distributed to the foundation soil was more than the bearing capacity of soft clay;
2. The soft clay occurred lateral movement when its' stress was up beyond the limit bearing capacity;
3. The jet grouting piles were broken when the soft clay occur lateral movement, as illustrated in Figure 12;
4. The fill embankment occurred collapse.

302

When embankment occurred collapse, the upload which distributed soil of composite foundation was more than design. The stress ratio of pile to soil from back analysis was 3.3 to 4.2, but the stress ratio of pile to soil from calculation of using applicable codes was 11.0 to 13.0. Therefore, the major embankment collapse reason was that between design and actual situation have a great difference.

So improving the bearing capacity of cement-soil piles could not limitless strengthen the bearing of composite foundation. According actual situation and experience of Guangdong province, it is not suitable when the stress ratio of pile to soil was great 4.0 in soft clay area.

7 CONCLUSION

Though this case, some conclusion and experience could be got, as follow by:

1. On the condition of soft clay foundation, the calculation result was lean to unsafe, when combined strength method was used to analyze monolithic stability of cement-soil pile composite foundation.
2. On the condition of known the stress ratio of pile to soil, the distribution load method was used to analyze monolithic stability of composite foundation.
3. According to back analysis result of actual situation, the stress ratio of pile to soil was about 3.3 to 4.2, which equaled other literatures in general. Therefore, the stress ratio of pile to soil was not suitably greater than 4.0.
4. Because the deformation of embankment fill changed the distribution of stress of composite foundation, it was pay more attention to select reasonable discount coefficient of piles and effective coefficient of soil.
5. It was suggested to pay more attention of the stress of pile to soil sit test and research (Guonan Liu, 2010).

REFERENCES

British Standard Institution. 1995. British standard BS8006, Code of Practice for Strengthened/ Reinforced Soils and Other Fills.
China Academy of building research. 2012. (JGJ 79-2012) Chinese Technical code for ground treatment of buildings. Beijing: China architecture and building press.
Guonan Liu et al. 2010. *The mechanism and design method of pile-Net Foundation*, Ground Treatment (in China) 21(4):4–18.
Guonan Liu et al. 2013. *Some problems of embankment composite Foundation*, Ground Treatment (in China) 2(3):3–17.
Jifu Liu et al. 2003. *The analysis of stress ratio of pile to soil under composite subgrade*, Chinese Journal of Rock Mechanics and Engineering 22 (4):4–18.
Nordic Geosynthetic Group. 2002. Nordic handbook, Reinforced soils and fills.
The Design and Construction Handbook of Mixing Piled foundation. Tokyo; Railway Technology Research Institute, 2001, in Japanese.

Hydraulic Engineering III – Xie (Ed)
© 2015 Taylor & Francis Group, London, ISBN 978-1-138-02743-5

Parametric study of blind bolted end-plate connection on structural square hollow section

S. Ivy
Faculty of Civil Engineering, Universiti Teknologi Malaysia, Johor, Malaysia

P.N. Shek
UTM Construction Research Centre, Universiti Teknologi Malaysia, Johor, Malaysia

ABSTRACT: Blind bolts are defined as bolts that can be access for installation from one side of the connection only, where the application can referring to the case of connecting the end-plate of a beam to a hollow column. Blind bolts offer many advantages including strengths and sizes comparable to ordinary bolts and therefore it has potential use in tension applications as well as in moment resisting connections. This paper presents a parametric study of blind-bolted end-plate connections between tubular columns and open beams. Firstly, the experimental study used for validation purposes is briefly described with focus on the connection details. Secondly, two theoretical models, based on Gomes method and component method are then proposed and a detailed description of the model assumptions and model validation is presented. The proposed theoretical model with higher accuracy is used to perform a parametric investigation into the key factors influencing the behavior of blind-bolted end-plate connections. It is shown that the blind-bolt type and size, end-plate thickness, column face thickness and gauge distance have a significant influence on the connection behavior.

1 INTRODUCTION

Structural hollow section is a type of metal profile with a hollow tubular cross section. Square and circular hollow sections are suitable to be used as column member mainly due to its closed geometry that contributes to good structural efficiency, reliability and high aesthetical values. Until recently, welding has been the popular method to connect open beams to structural hollow sections. The high costs and labor-intensive procedures involved with welding have limited the use of hollow section as structural members. Further, welding is difficult to inspect, needs high quality control and must be performed in dry weather. Recent development of new fastening systems offer several connection alternatives, which allows access for installation is from one side of the connection only, as in the case of connecting the end-plate of a beam to a tubular column. Blind fasteners offer many advantages including strengths and sizes comparable to ordinary bolts and therefore it has potential use in tension applications as well as in moment resisting connections. There are many types of blind fasteners including Lindapter Hollo-bolt, flowdrilled bolts, blind-bolt and Ultra-twist blind fasteners. Each type of fastener differs in the bolt components, resistance mechanism and method of installation. Many studies have been carried out on blind-bolted moment connections. France (1997) compared the behavior of blind-bolts and flowdrilled bolts, and suggested that flowdrill systems can provide relatively higher stiffness and capacity. Barnett et al. (2000, 2001) performed a review of different blind-bolting options and carried out an experimental study on blind-bolted T-stubs and connections using Hollow-bolts. Wang et al. (2009) carried out four tests on end-plate connections between open beams blind-bolted to concrete filled tubular columns by means of Hollow-bolts, and good rotation capacity for the connections was reported. Elghazouli et al. (2009) performed an experimental investigation into the monotonic and cyclic behavior of blind-bolted top and seat and top, seat and web angle connections, and then proposed a component based mechanical models for blind-bolted angle connections (Malaga-Chuquitaype, 2010). In contrast, analytical research on the response prediction of blind-bolted connections to tubular columns is scarce. Ghobarah et al. (1996) suggested and analytical model for the evaluation of the initial stiffness and plastic moment capacity of blind-bolted end-plate connections between open beams and tubular columns. Silva et al. (2003) presented an analytical model for the estimation of the column face stiffness in end-plate connections with concrete filled tubular column components; representative expressions were proposed based on finite element models as well as experimental results. However, available models on end-plate connections cannot be directly applied to different blind bolting system due to the significant influence of contact phenomena as well as the complex interactions between the blind-bolts, column face and end-plate. Accordingly, there is a need for a dedicated model that provides a faithful characterization of the response of blind-bolted end-plate connections.

This paper presents a parametric study of blind-bolted end-plate connections between tubular columns and open beams using theoretical models. The proposed models are based on the component method to predict the initial stiffness and moment resistance of blind-bolted end-plate connections. After validation against experimental results, the model is employed in a parametric investigation into several key factors influencing the behavior of blind-bolted end-plate connections, namely: the blind-bolt type and size, end-plate thickness, column face thickness and gauge distance (defined here as the horizontal distance between two bolts). Based on these findings, optimization can be carried out for the configuration of blind-bolted end-plate connections by considering the influence of each key factor.

2   EXPERIMENTAL PROGRAMME

The experimental study comprised eight tests performed on blind-bolted end-plate connections between open beams and tubular columns under monotonic loading conditions. Column sections are square hollow section of size $200 \times 200 \times 10$ and beam sections are universal beams with a depth varies from 400 mm to 600 mm. The connection configurations utilised are flush end-plate and extended end-plate adopted from the standardized connection tables referred to in the SCI publication (SCI, 1996). M20 and M24 blind-bolts were employed for connecting the end-plate to the tubular column. The grade of steel used for the beam and column was S275. Typical setup of the experimental study is illustrated in Figure 1. Overall, the experimental study concluded that blind-bolted connections possess significant ductility and hardening characteristics under monotonic loading.

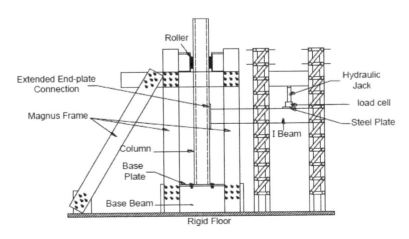

Figure 1.   Arrangement of test specimen.

3 THEORETICAL MODEL

The design moment resistance, $M_{j,Rd}$ and initial rotational stiffness, $S_{j,ini}$ of the connection are determined from equations adopted from Eurocode 3, Part 1–8. With the presence of both design moment resistance and initial stiffness, the moment-rotation (M-R) curve can be plotted by using equation 1.

$$M = M_{j,Rd} \left(1 - e^{\left(\frac{-S_{j,ini}}{M_{j,Rd}} \right) \phi} \right)$$

(1)

4 COMPARISON WITH TEST RESULTS

Table 1 shows the comparison between theoretical predictions to experimental results. From the comparison, it is shown that theoretical initial stiffness is higher than the experimental results. The ratio of theoretical values to experimental results is in the range of 3.93 to 6.02. On the other hand, theoretical moment resistance is lower than the experimental results. The ratio of theoretical values to experimental results is in the range of 0.54 to 0.96.

5 PARAMETRIC ASSESSMENTS

Parametric study is carried out by manipulating blind bolt size ranging from 8 mm to 24 mm, column thickness ranging from 8 mm to 40 mm, end plate thickness ranging from 6 mm to 15 mm and gauge length ranging from 60 mm to 140 mm. Figure 2 to Figure 5 showed the

Table 1. Comparison to experimental results.

	$S_{j,ini\ Theory}$	Ratio to $S_{j,ini\ Exp}$	$M_{j,Rd\ Theory}$	Ratio to $M_{j,Rd\ Exp}$
N1 FEP1	20.82	4.85	34.45	0.54
N2 FEP2	39.92	6.02	61.21	0.96
N3 FEP3	29.8	4.17	62.56	0.56
N4 FEP4	58.66	5.15	106.87	0.57
N5 EEP1	35.66	5.33	70.12	0.52
N6 EEP2	62.1	3.93	113.75	0.58
N7 EEP3	51.16	4.85	107.83	0.65
N8 EEP4	86.35	3.94	189.69	0.7

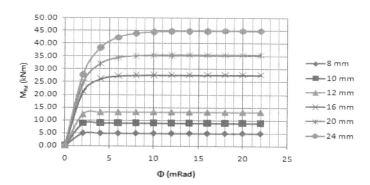

Figure 2. Comparison between different blind bolt sizes.

307

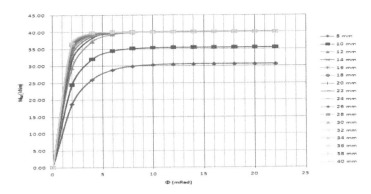

Figure 3.   Comparison between different column face thicknesses.

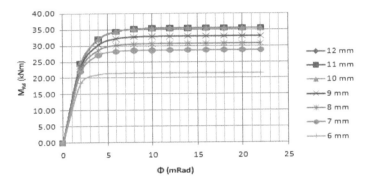

Figure 4.   Comparison between different end-plate thicknesses.

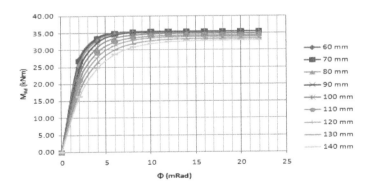

Figure 5.   Comparison between different gauge lengths.

comparison results from the parametric study. From the comparison, it is obvious that blind bolt sizes are the key factor that influences the moment resistance of the connection. Comparison between different column thicknesses shows that no change in moment resistance when the column thickness is equal to the end-plate thickness as the failure criterion is due to the mechanism of end-plate yielding. Comparison between end-plate thicknesses and gauge lengths showed that the influences of these two factors are relatively low as compared to bolt sizes and column thicknesses.

6 CONCLUSIONS

Several conclusions can be drawn from this study:

1. The comparison to the experimental results shows that component approach has over predicted the initial stiffness of the connection and under estimated the moment resistance of
the connection as those examined in this study.
2. The parametric study shows that column thickness and bolt sizes are the main key factors
that influence the moment resistance and initial stiffness of the connections.

ACKNOWLEDGEMENT

The authors would like to thank Universiti Teknologi Malaysia (Q.J130000.2522.03H77) for
funding this research, Faculty of Civil Engineering and UTM Construction Research Centre
for providing assistance and laboratory facilities.

REFERENCES

Barnett, T, Tizani, W, Nethercot, D. 2000. Blind bolted moment resisting connections to structural
hollow sections. Connections in Steel Structures IV, Virginia.
Barnett, T, Tizani, W, Nethercot, D. 2001. The practice of blind bolting connections to structural hollow
sections: a review. Steel and Composite Structures: 1; 1–16.
Elghazouli, A, Malaga-Chuquitaype, C, Castro, J, Orton, A. 2009. Experimental monotonic and cyclkic
behavior of blind- bolted angle connections. Engineering Structures. 31(11); 2540–2553.
France, J. 1997. Bolted connections between open section beams and box columns. Ph.D. thesis. UK:
Department of Civil and Structural Engineering, University of Sheffield.
Ghobarah, A, Mourand, S, Korol, R. 1996. Moment-rotation relationship of blind bolted connections
to tubular columns. Journal of Constructional Steel Research. 50; 1–14.
Grove, A.T. 1980. Geomorphic evolution of the Sahara and the Nile. In M.A.J. Williams & H. Faure
(eds), *The Sahara and the Nile*: 21–35. Rotterdam: Balkema.
Malaga-Chuquitaype, C, Elghazouli, A. 2010. Component-based mechanical models for blind-bolted
angle connections. Engineering Structures. 32; 3048–3067.
Silva, L, Neves, L, Gomes, F. Rotational stiffness of rectangular hollow sections composite joints.
Journal of Structural Engineering, American Society of Civil Engineering. 129; 487–494.
Steel Construction Institute and British Constructional Steelwork Association. (1996). Joints in Steel
Construction—Volume 1: Moment Connections.
Wang, J, Han, L, Uy, B. 2009. Behaviour of flush end-plate joints to concrete-filled steel tubular columns.
Journal of constructional steel research. 65(4); 925–939.
Weynand, K. and Jaspart, J. P. 2003. Application of the Component Method to Joints Between Hollow
and Open Sections. CIDECT Draft Final Report: 5BM.

Hydraulic Engineering III – Xie (Ed)
© 2015 Taylor & Francis Group, London, ISBN 978-1-138-02743-5

Elasto-plastic analysis on seismic response of PC continuous box-girder bridge with corrugated steel webs based on section fiber model

X. Li, L. Liang & X. Wang
Northeastern University, Shenyang, China

ABSTRACT: In order to study the seismic response of PC continuous box-girder bridge with corrugated steel webs, seismic responses of Shenzhen Nanshan Bridge are analyzed under rare earthquake. Time history method based on section fiber model is used. Seismic responses of bridge under Taft wave, El Centro wave and and artificial wave are obtained. Through researching the seismic responses of bridge under El Centro waves with different peak acceleration, the results show that the pier is designed on the safe side and there is room for further optimization. The results and conclusion can provide reference and experience for seismic design of similar bridges.

1 INTRODUCTION

Prestressed Concrete (PC) composite box-girder with corrugated steel webs is a new type of steel-concrete composite structure, which was studied from the beginning of this century in China. The structure has the advantages of light weight, simple structure, beautiful shape, etc (Wang et al. 2009, Xu 2010). In recent years, seismic performance of bridge structures is getting more and more attention. Therefore, it is necessary to study the seismic response of this new bridge with corrugated steel webs.

The theory of elasto-plastic time history analysis based on section fiber model has been rapid developed in recent years, especially in seismic technology developed country such as American and Japan (D'Ambrisi & Filippou 1999, Spacone et al. 1999). In this paper, Shenzhen Nanshan Bridge is taken as the engineering background. Time history method based on section fiber model is used to analyze the nonlinear seismic responses of the bridge under rare earthquake.

2 ENGINEERING BACKGROUND

Shenzhen Nanshan Bridge is a PC continuous box-girder bridge with corrugated steel webs. Its span arrangement is 80 m + 130 m + 80 m. The general arrangement and standard section of the bridge are shown in Figure 1 and Figure 2 (Chen et al. 2008). The construction drawing of pier is shown in Figure 3.

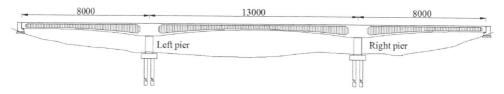

Figure 1. General arrangement of Nanshan Bridge (Units: cm).

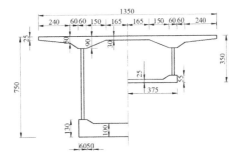

Figure 2.   Standard section of Nanshan Bridge (Units: cm).

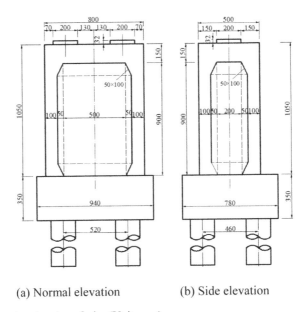

(a) Normal elevation   (b) Side elevation

Figure 3.   Construction drawing of pier (Units: cm).

3   FINITE ELEMENT MODEL AND MATERIAL CONSTITUTIVE MODEL

3.1   Finite element model of the bridge

The general finite element software Midas Civil is used to build the finite element model of the bridge structure. Spatial beam element is used to simulate girder and substructure of the bridge. Combine element is used to simulate bridge bearing. The linear equivalent soil spring is used to simulate pile-soil interaction. The calculation diagram and finite element model of pile foundation is shown in Figure 4. The finite element model of full bridge is shown in Figure 5.

3.2   Material parameter and constitutive model

3.2.1   Concrete

The concrete strength grade of main girder is C50, and its elastic modulus is 3.45×10^4 MPa. The concrete strength grade of bridge substructure is C30, and its elastic modulus is 3.00×10^4 MPa. Posson's ration of concrete is 0.2. Mander model (Zhang 2003) is used to define the constitutive model of pier concrete, which is the constitutive model of confined concrete.

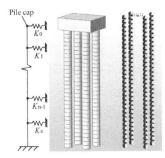

Figure 4.   Pile foundation model.

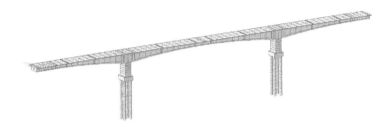

Figure 5.   Finite element model of full bridge.

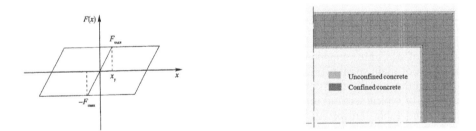

Figure 6.   Restoring force model of bearing.   Figure 7.   Fiber division of pier section.

3.2.2   Steels

The steel grade of corrugated steel web is Q345, and its elastic modulus is 2.1×10^5 MPa. Prestressed reinforcement uses $\Phi^s 15.24$ strand with high strength and low relaxation, and its characteristic value of tensile strength and elastic modulus is 1860 MPa and 1.95×10^5 MPa, respectively. The steel bar of pier uses HRB335, and its elastic modulus is 2.0×10^5 MPa. Bilinear model (Zhang 2003) is used to define the constitutive model of pier steel bar. Posson's ration of steel is 0.3.

3.3   Restoring force model of bridge bearing

Pot rubber bearings with nonlinear characteristic are used in the bridge. Nonlinear combine element is used to simulate movable bearing and the ideal elastoplastic model is used to describe the restoring force model of the bearing as shown in Figure 6.

3.4   Plastic hinge model of pier

Fiber hinge model is used to define the plastic hinge of pier. The fiber division of pier section is shown in Figure 7, considering result accuracy and computer ability.

313

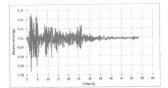 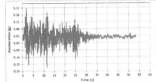 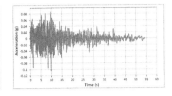

(a) East-west direction   (b) South-north direction   (c) Vertical direction

Figure 8.   El Centro seismic wave.

3.5   Selection and adjustment of seismic wave

According to the bridge site condition (class II), Taft wave and El Centro wave are selected and a group of artificial wave is generated. Moreover, each group of seismic wave contains two horizontal and one vertical component. El Centro seismic wave is shown in Figure 8. In time history analysis, the intensity of seismic wave needs to be adjusted. For the 7 degree earthquake intensity area, the peak acceleration of horizontal seismic wave under rare earthquake is 220 cm/s² (Qin & Zhang 2004, Qu 2011). In the elasto-plastic time history analysis, Rayleigh damping is used and damping ratio is taken as 4%. In the numerical integration, Newmark-β method is selected and time step is 0.02 s.

4   RESULTS AND ANALYSIS

4.1   Seismic responses of the bridge when peak acceleration is equal to 0.22g

4.1.1   Influence of vertical seismic wave
In the case of El Centro seismic wave, the vertical displacement time history curves of mid-span section of main span of bridge girder under the seismic wave with two horizontal components and three components are shown in Figure 9. The displacement results and internal force results of typical sections under two-direction and three-direction earthquake are shown in Table 1 and Table 2, respectively.

From the tables it can be seen that the errors of displacement and internal force results are within 10% in general, when vertical seismic component is not considered. For the displacement results, the effect of vertical seismic component on the vertical displacement of main girder is larger and the error of vertical displacement of mid-span section of side span is 6.51%. However, the effect of vertical seismic component on longitudinal and transverse displacement of the piers is smaller and the maximum error is 3.09%.

For the internal force results, the effect of vertical seismic component on shear force of main girder and piers is smaller and the maximum error is 1.49%. The effect of vertical seismic component on moment of main girder is larger and the maximum error is 8.50%. But the effect on the moment of piers is smaller.

In a word, the effect of vertical seismic component on moment and vertical displacement of main girder is larger. However, the effect on displacement and internal force of piers is smaller and can be ignored.

4.1.2   The maximum moment and curvature of pier
The maximum moment and curvature of the bottom sections of piers under El Centro seismic wave are shown in Table 3. By comparison, the longitudinal direction and transverse direction of the piers don't arrive to the yield condition. This shows that there is no plastic hinge in the piers and the bridge is still in elastic state under design rare earthquake of actual site. The same conclusion can be drawn under Taft seismic wave and artificial seismic wave.

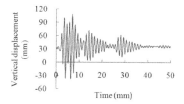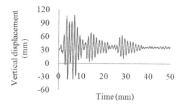

(a) Two horizontal components   (b) Three component

Figure 9.   Vertical displacement of mid-span section of main span under El Centro waves.

Table 1.   Displacement results of typical sections under two-direction and three-direction earthquake.

Section	Analysis case	DX* (mm)	DY* (mm)	DZ* (mm)
Mid-span	Two-direction	76.70	149.83	111.24
section of	Three-direction	75.24	148.64	105.44
main span	Difference	1.94%	0.80%	5.50%
Mid-span	Two-direction	52.81	45.23	56.23
section of	Three-direction	52.50	45.24	59.89
side span	Difference	0.59%	0.02%	6.51%
Top section	Two-direction	51.96	54.96	6.40
of left pier	Three-direction	50.40	54.47	6.91
	Difference	3.09%	0.90%	7.89%
Top section	Two-direction	13.71	53.88	6.67
of right	Three-direction	13.69	53.39	6.79
pier	Difference	0.13%	0.93%	1.76%

*DX, DY and DZ mean longitudinal displacement, transverse displacement and vertical displacement, respectively.

Table 2.   Internal force results of typical sections under two-direction and three-direction earthquake.

Section	Analysis case	Shear force (kN)	Moment (kN · m)
Mid-span	Two-direction	1551.65	110802.02
section of	Three-direction	1552.64	119592.64
main span	Difference	0.06%	7.93%
Mid-span	Two-direction	3991.01	63656.30
section of	Three-direction	4011.09	67267.30
side span	Difference	0.50%	5.67%
Middle	Two-direction	24466.62	143449.34
support	Three-direction	24832.11	155641.98
section	Difference	1.49%	8.50%
Bottom	Two-direction	14639.17	144684.75
section of	Three-direction	14522.81	143772.73
left pier	Difference	0.80%	0.63%
Bottom	Two-direction	4281.73	37005.89
section of	Three-direction	4281.93	37042.02
right pier	Difference	0.00%	0.10%

4.1.3   *Moment-curvature relationship curves*

The longitudinal moment-curvature relationship curves of the bottom section of left pier under each seismic wave are shown in Figure 10. From the figures, each curve passes through the origin and there is no obvious hysteresis loop. This shows that the pier is still in elastic state.

Table 3. The maximum moment and curvature of piers under El Centro wave.

Section	Direction	The max. moment (kN · m)	Yield moment (kN · m)	The max. curvature (1/m)	Yield curvature (1/m)
BSLP	Longitudinal	1.42×10^5	2.09×10^5	2.54×10^{-4}	4.67×10^{-4}
	Transverse	2.51×10^5	3.26×10^5	1.95×10^{-4}	2.99×10^{-4}
BSRP	Longitudinal	3.69×10^4	2.09×10^5	3.48×10^{-5}	4.67×10^{-4}
	Transverse	2.46×10^5	3.26×10^5	1.81×10^{-4}	2.99×10^{-4}

*BSLP means the bottom section of left pier; and BSRP means the bottom section of right pier.

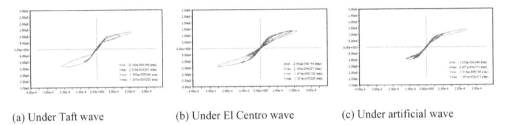

(a) Under Taft wave   (b) Under El Centro wave   (c) Under artificial wave

Figure 10. The longitudinal moment-curvature relationship curves of the bottom section of left pier under different seismic waves.

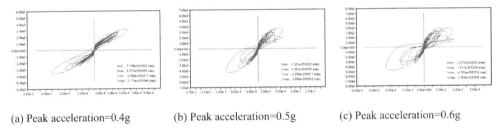

(a) Peak acceleration=0.4g   (b) Peak acceleration=0.5g   (c) Peak acceleration=0.6g

Figure 11. The transverse moment-curvature relationship curves of the bottom section of left pier under different peak accelerations.

4.2 Time history analysis under seismic waves with different peak accelerations

4.2.1 Moment-curvature relationship curves

In the case of El Centro seismic wave, the transverse moment-curvature relationship curves of the bottom section of left pier under different peak accelerations are shown in Figure 11. From the figures, when peak acceleration is equal to 0.4 g~0.6 g, hysteresis loop area of moment-curvature relationship curve and the plastic deformation of the pier gradually increase.

4.2.2 The maximum plastic rotation angle of the pier

The ratio of the maximum plastic rotation angle (θ_p) to ultimate plastic rotation angle (θ_u) of plastic hinge region of the piers under different peak accelerations is shown in Figure 12. From the figure, when peak acceleration is equal to 0.5 g, $\theta_p/\theta_u \approx 0.80$ and it means that the pier is close to failure. When peak acceleration is equal to 0.6 g, $\theta_p/\theta_u > 1.00$ and it means that the pier damages. Therefore, the maximum peak acceleration that the bridge can bear is 0.5 g.

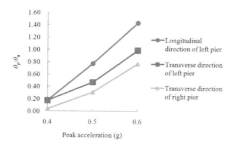

Figure 12.   Ratio of the maximum plastic rotation angle (θ_p) to ultimate plastic rotation angle (θ_u) of the piers under different peak accelerations.

5   CONCLUSIONS

When research seismic responses of PC continuous box-girder bridge with corrugated steel webs, the effect of vertical seismic component on moment and vertical displacement of main girder is larger. However, the effect on displacement and internal force of piers is smaller and can be ignored.

The pier of the bridge studied in this paper is still in elastic state and doesn't produce plastic hinge under design rare earthquake of actual site.

Through researching the seismic responses of the bridge under El Centro waves with different peak acceleration, the maximum peak acceleration that the bridge can bear is 0.5 g. It means that the bridge is designed on the safe side and the section size and reinforcement ratio of pier can be reduced properly.

ACKNOWLEDGEMENTS

This work was financially supported by the Fundamental Research Funds for the Central Universities (N120301003).

REFERENCES

Chen, Y.Y. et al. 2008. Design of Shenzhen Nanshan prestressed concrete continuous girder bridge with corrugated steel webs. The proceedings of 2008 national bridge academic conference 160–165.

D'Ambrisi, A. & Filippou, F.C. 1999. Modeling of cyclic shear behavior in RC members. Journal of Structural Engineering 125(10): 1143–1149.

Qin, C.L. & Zhang, A.H. 2004. Nonlinear time history analysis based on section fiber model. Journal of Zhejiang University (Engineering Science) 39(7): 1003–1008.

Qu, H. 2011. Selections of ground motions in seismic history analysis for bridges with high piers. Dalian: Dalian Maritime University.

Spacone, E. et al. 1999. Fiber beam-column modeling for non-linear analysis of R/C frames. Journal of Earthquake Engineering and Structural Dynamics 25: 711–725.

Wang, S. et al. 2009. Application of prestressed concrete composite box-girder structure with corrugated steel webs in bridge engineering in China. Journal of Architecture and Civil Engineering 26(2): 15–20.

Xu, Z.N. & Yang, X.Y. 2010. Development status and characteristics of composite girder bridge with corrugated steel webs. Inner Mongolia Science Technology & Economy 1: 70–72.

Zhang, X.P. 2003. Nonlinear analysis of RC seismic structure. Beijing: Science Press.

Hydraulic Engineering III – Xie (Ed)
© 2015 Taylor & Francis Group, London, ISBN 978-1-138-02743-5

Key technologies of construction control on long-span variable cross-section aqueduct with reinforced concrete box arch using cantilever erection method

Zhongchu Tian, Wenping Peng & Xiaoli Zhuo
Changsha University of Science and Technology, Changsha, Hunan, China

ABSTRACT: Cable-stayed buckle hang rope hoisting construction technology is not widely used in the construction of long-span cross-section of aqueduct construction application. In this paper, the biggest span of arch aqueduct in China's water conservancy industry which is the Longchang aqueduct in Guizhou province as engineering background. Cast-in-site arch foot section reversely-pulled bailey truss, cable and anchorage system are summarized in the process of construction. Focusing on the study of control methods and parameters in the process of cable erection, the content of the control methods and parameters include arch ring alignment, abutment foundation displacement, arch ring stress and reinforcement stress, displacement of buckled tower, cable or anchor cable force and stability. It is of guiding function for engineering practice.

1 INTRODUCTION

Cable erection method because of its great spanning capacity, large lifting capacity, and the advantages of small terrain in the absence of bracket construction is widely used in the construction of arch bridge. The United States, for example, Mike Mr Callaghan-pat tillman memorial bridge successfully using the cable erection construction method; Built in 1990 at the end of the sichuan wangcang east river bridge; Built in 2004 in Guangxi nanning yonghe bridge; Built in 2007 Hunan xiangjiang fouth river bridge; In March 2012 to traffic aizhai bridge in Hunan province; Built in 2012 in Sichuan a yangtze river bridge of cable erection has been adopted and implemented smooth closure of the arch bridge. These bridges are mostly belongs to the concrete filled steel tubular arch bridge have been completed, but the segmental precast arch ring, using the cable erection with cable-stayed buckle construction of arch aqueduct are rare. The cable erection and cable-stayed buckle construction technology is complex, difficult to locate the elevation, structure system have more conversion times, cable tension control is difficult, higher risk. Carry on the reasonable and effective construction control is necessary to study (Xiang Zhongfu 2011). In this paper, the main arch ring of Longchang aqueduct construction control of the key technologies are introduced, providing reference for similar bridge construction.

2 ENGINEERING BACKGROUND

Longchang aqueduct is a single span arch aqueduct, the first phase of central Guizhou water conservancy key project in Guizhou, one of the biggest span arch aqueduct in China's water conservancy industry. Long chang aqueduct's main arch ring is a reinforced concrete box arch of single box with double chamber, with a clear span of 200 m. A net arrowheight of 40 m and a rise-span ratio of 1/5 (Fig. 1). The arch axis of the main arch ring is set up to a catenary with arch-axis coefficient m = 2.240. The section of arch box is a variable

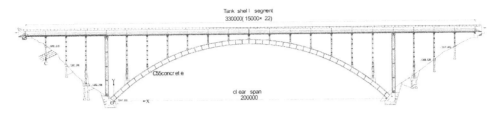

Figure 1. Elevations layout of Longchang aqueduct (unit: mm).

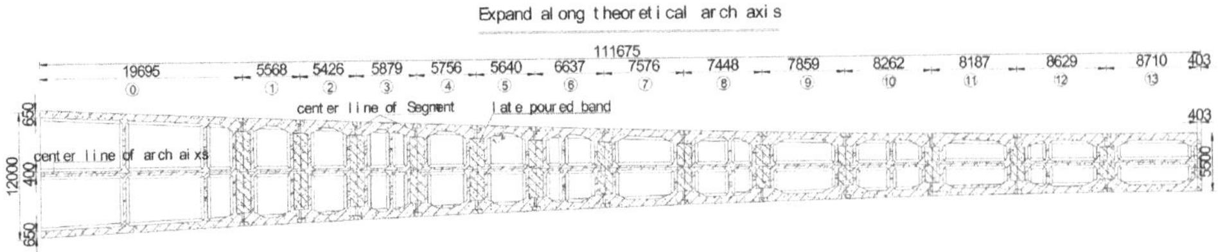

Figure 2. Floorplan of Longchang aqueduct (unit: mm).

cross-section, has a width from 12 m gradually change to 5.5 m (Fig. 2), a height of 3.5 m, and a thickness of roof and floor of 40 cm. Boundary web plate is 60 cm,and medium web plate is 40 cm. The main arch ring is made of C55 concrete, 0# cast-in-situ section is poured on reversely-pulled bailey truss. The rest is divided to 13 sections per half-span, constructed by cable erection and strayed knotting method. Mid-span closure segment has a length of 0.8 m is cast-in-place.

3   CONSTRUCTION TECHNOLOGY OF LONGCHANG AQUEDUCT

Longchang aqueduct's main arch ring is devided into 29 sections and constructed by the method of prefabrication and lifting. 0# cast-in-situ section is poured on reversely-pulled bailey truss. The rest is constructed by whole prefabrication and cable erection method. After one segment lifting complete, it would be cable-stayed buckled for a short time, and then, beginning the next cycle. The steel buckled towers are build on the two borderline piers with steel pipe truss structure, have a height of 27 m,its main compressed structure is made up of eight Φ630 × 20 mm steel tubes. The tension anchor beam is constituted by I50 joist steel. 0#~6# supporting cable layered hang on the borderline pier, 7#~13# supporting cable layered hang on steel buckled towers' each anchor point. Cable erection system and cable-stayed buckle system (Fig. 3).

3.1   *Reversely-pulled bailey truss on 0# cast-in-site segment*

Cast-in-site segment arc arch ring 0# 19.695 m, adopt bailey truss counter loop construction (Fig. 4). Bailey truss assembly along the axis of the aqueduct, The bailey truss articulated on arch foot, finish rolled threaded reinforcing bar as a rigid connection counter pull bailey truss on the borderline pier. The back anchor cable on borderline pier adopts steel strand anchor in mountain anchor cable. Counter loop rigid bar is divided into two parts. One part arrangement near the end of the outside bailey truss called 1# bar, composed of eight set of tie bar, another part arrangement in the central of the bailey truss called 2 # bar, is made up of six groups of tie bar. The borderline pier anchor cable made up of five groups of prestressed steel strand, corresponding tie rod boundary pier at the front two rows of to ensure pier deviation under controlled.

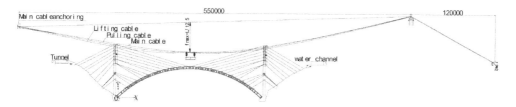

Figure 3. Arrangement drawing of Longchang aqueduct (unit: mm).

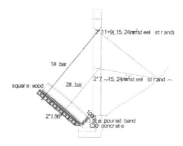

Figure 4. Cast-in-place segment arrangement drawing of Longchang aqueduct (unit: mm).

Figure 5. Fragmentary view of cable system.

3.2 *Cable erection system*

Longchang aqueduct using cable-stayed buckle hang rope hoisting construction method. The heaviest section of arch ring precast segmental 2413 kN in weight, the lightest segment weight of 1942 kN, cable erection system design of lifting 300 kN in weight. Hoisting rope across distributed from import to export 550 m + 120 m in turn, 550 m of hoisting cable is main cable, and 120 m is the export side balance cable (Fig. 5).

4 THE CONSTRUCTION CONTROL TECHNOLOGY OF LONGCHANG AQUEDUCT

4.1 *Construction control calculation analysis*

According to the Longchang aqueduct design drawings and site construction and implementation plan, do the simulation optimization calculation for the construction process of arch ring cable erection and cable-stayed buckle. The main arch ring, border pier, tower of steel tube columns, steel bracing and brace are simulated by beam element when modeling. Cable, anchor cable are simulated by only-tensile truss element. The connection of buckle point and main girder, anchor point and tower are simulated by rigid connection of elastic connection (Fig. 6).

321

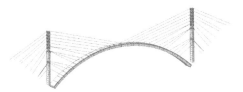

Figure 6.   Finite element model.

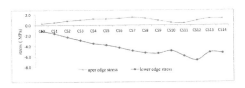

Figure 7.   The stress of both the top and bottom sides during installing main arch ring.

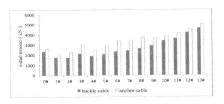

Figure 8.   Buckle and anchor cable initial tension.

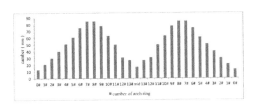

Figure 9.   Camber of each section of arch ring.

According to the Longchang aqueduct construction procedures of construction organiza-tion design, the arch ring section cable hoisting suspension hanging construction simulation analysis was made on the whole process of research. Analyses the stress of arch ring section, anchor cable tension control cable force. The arch ring segment construction camber control parameters, CS0 ~ CS14 is the current segment in place after the edge of arch ring of arch foot section on stress, as shown in Figure 7. The beginning anchor cable force of each section as shown in Figure 8, arch ring of all segmental construction camber is shown in Figure 9.

In Figure 8, in the process of the cable hoisting suspension hanging and the button in the main arch ring segments, anchor rope tension and arch ring segment, edge of arch ring section on tensile and compressive stress change is stable. In the process of each arch ring segment installation, wet joint top casting show the tensile stress increases. By cable, anchor cable ten-sion make the arch ring of tensile and compressive stress in the installation process meet the specification requirements. The biggest compressive stress in each arch ring edge of segmental arch foot section of the cross section after the arch ring #12 segment installation is complete, to 6.7 MPa. The margin of maximum tensile stress in the arch ring # 7 segment after the installation is complete, is 1.5 MPa. Construction control practice shows that the C55 arch ring cross section on the edge of compressive stress control in force within 10 MPa, C55 edge of arch ring section on tensile stress control within 1.5 MPa, arch ring structure is safe.

The cable force control is the key in the main arch ring cable hoisting suspension hanging construction process. Based on the theory of the button shown in Figure 8, anchor cable and cable beginning force, controlling the cable force in the process of arch ring of aqueduct segmental hoisting cable. According to the Figure 9 arch ring of the segmental construction camber, adjust the aqueduct arch ring segment elevation of the model.

4.2 The arch ring construction control method

Adjust the model elevation is the most direct means of the main arch ring alignment adjustment. In actual construction process, the state of structure and theoretical calculation has the difference. We need adjust the main arch elevation changes value caused by these parameters error through adjusting the model evolution in later stage. Different cable force in the process of the arch ring construction will directly affect the structure of the alignment and internal force distribution (Yan Donghuang et al. 2004). And it's critical for the structure constructed by the cable hoisting suspension hanging method to control the cable force (Tian Zhongchu et al. 2004). The achievement of elevation of the main arch ring in the process of hoisting position can be adjusted by lifting segmental arc line elevation.

4.2.1 The arch ring segments lifting alignment control

The goal of arch ring segments alignment control is as far as possible consistent with alignment design, satisfy the arching state design and specification requirements. On the one hand, the length of the precast arch ring segmental error need adjust elevation through lifting position. Analysis, on the other hand, on the theory of value and the measured elevation changes of the completed construction beam section caused by parameter error, guide and predict the next segment of the elevation of the model.

4.2.2 The abutment foundation displacement control

In long-span arch bridge construction, in terms of aqueduct, the more live load relative to highway bridges, the influence of foundation settlement more not allow to ignore (Lu Xueping et al. 2010). In actual engineering, mostly adopt the way of setting settlement observation point to abutment foundation displacement monitoring, but this method greatly influenced by temperature and other factors. The bridge adopts multi-point displacement meter as shown in Figure 10 and 11, can accurately considering the change of the vertical and horizontal displacement. By comparison with the measured values and the theoretical value analysis to ensure stability and safety of engineering projects.

Figure 10.   The displacement measuring point (unit: mm).

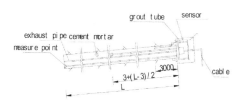

Figure 11.   Multi-point extensometer layout of arch abutment (unit: mm).

323

4.2.3 The arch ring, steel bar stress control

The main arch ring on the cable hoisting suspension construction stage is strongly influenced by the construction cable force. Through analyzing the data of main arch ring cross control section stress, control section steel bar stress in a timely manner, to ensure the arch ring construction stage actual bearing meet specification requirements. As shown in Figure 12, reinforcement stress test arrangement is the same as arch ring stress test arrangement, strain gauge test equipment has been widely used, reinforcement stress test device as shown in Figure 13.

4.2.4 The displacement of buckled tower and borderline pier

On the one hand, in the construction control, we need do the arch ring elevation correction according to the actual displacement of the buckled tower and borderline pier. Through the geometric relation can be derived $\Delta z = l\Delta\Phi\cos\theta$, Δz is the arch rib section height variation, l is the arch ring segments' chord length, $\Delta\Phi$ is the angle changes between arch ring section chord and horizontal, θ is the angle between arch ring section chord and the horizontal (Chen Shuhong. 2011). In construction process, on the other hand, as the main compression member of tower, borderline pier, along the bridge to the state of actual deformation is vital for the safety of the structure. In the process of the arch ring segmental hoisting buckle to hang, giving the current lifting section's theory influence value and the actual deviation in tower and borderline pier and for reasonable evaluation and correction is the core content of the bridge construction control, as shown in Figure 14.

4.2.5 Buckle cable, anchor cable forces control

Segmental arch in the lifting process with alignment positioning adjustment arch segments, improving buckle tower deviation, arch's internal force optimization are closely linked with

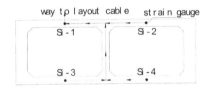

Figure 12. The strain monitoring of main arch ring.

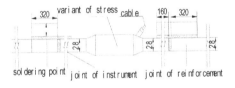

Figure 13. Steel bar meter layout (unit: mm).

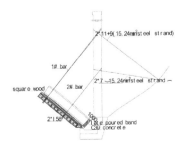

Figure 14. The deviation monitoring of tower and juncture pier.

324

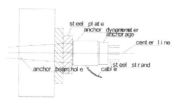

Figure 15. Cable tension sensor layout.

the cable tension control. On the basis of the theory of finite element calculation, reasonable evaluation of the actual state of the cable and anchor cable is a judgment that ensures the safety of arch ring segments. Button as shown in Figure 15 in arch ring point embedment cable force sensor and combining the method of "frequency", "oil", "elongation" lasso for cable, anchor cable tensioning cable force, compare to the previous cable force test method, the differential resistance type cable force sensor has good stability and higher precision. Located in arch ring segment hoisting, rope, anchor rope tension cable force control within the scope of the theoretical value of ±5% commonly, give priority to with elevation adjustment range, based on the actual state of structure is revised, to obtain accurate tension cable force.

4.2.6 *Stability control*

Structural stability generally has two forms: the first kind of stability and bifurcation point instability problem, the second category of stability, extreme value point instability problems. The first kind of stability problems in the actual construction is eigenvalue problems, the critical load is convenient to implement and approximation corresponds to the second kind of stability limit. Making arch ring maximum cantilever state as the most unfavorable conditions for the stability analysis, eigenvalue stability analysis according to the first kind of stability problem (Wu Hengli. 1979) show that the most unfavorable condition of the structure of the first-order stability coefficient is 38.46 specification requirements is greater than 4 ~ 5, arch ring during the installation process is guaranteed the stability of arch ring structure, add wind cable has little improvement on stability.

5 CONCLUSION

During the cable erection and cable-stayed buckle construction process of main arch ring, in order to approach to the design arch axis, make the main arch ring force more reasonable. Elevation adjustment is necessary to reduce the error caused by precast arch ring section length.

During the cable erection and cable-stayed buckle construction process of main arch ring, through the multi-point extensometer, steel bar stress test instrument, cable force sensors and other monitoring instrument, obtained more accurate monitoring data, provides the construction security guarantees.

During the cable erection and cable-stayed buckle construction process of main arch ring, the geometric correction on tower and pier deviation, reduced the main arch ring elevation error of the measured values.

During the cable erection and cable-stayed buckle construction process of main arch ring, The stability of the construction control calculation results show that, the maximum cantilever state stability of arch ring meet the requirements, do not need to set the wind cable, improved the economic benefit.

ACKNOWLEDGEMENT

This work is supported by Hunan Province Universities Innovation Platform of Open Fund Project of China (10 K008) and Hunan Provincial Natural Science Funds of China (14 JJ2075), which are gratefully acknowledged.

REFERENCES

Chen Shuhong. The research of cables of the reinforced concrete arch bridge construction technology [D]. Dissertation, Chongqing Jiaotong University, 2011.

Lu Xueping, Zhang Linhong, FanYong. The quadratic parabola hingeless internal force calculation under the action of arch feet displacements and arch structure [J]. Journal of Subgrade Engineering, 2010, 06 (3): 68–70.

Tian Zhongchu, Chen Deliang, Yan Donghuang, etc. In the process of the long-span arch bridge arch ring assembled buckle cable force and elevation in the process of the carrying amount to determine [J]. Journal of Railway, 2004, 26 (3): 81–87.

Wu Hengli. Arch system stability calculation [M]. People's Traffic Press, 1979.

Xiang Zhongfu. Bridge construction control [M]. Beijing: People's Traffic Press, 2011.

Yan Donghuang, Tang Dong, Tu Guangya, etc. Precast reinforced concrete arch bridge cantilever construction control [J]. Journal of Changsha University of Science and Technology (Natural Science Edition), 2004 (Z1): 18–22.

Hydraulic Engineering III – Xie (Ed)

Experiment research on flexural performance of reinforced concrete beam by high titanium blast furnace slag

Shuanghua Huang, Jie Wang, Shuangmei Xie, Kecai Chen & Bin Chen
Panzhihua University, Architecture and Civil Engineering College, Sichuan, China

ABSTRACT: Through the bending performance test of high titanium blast furnace slag reinforced concrete beams that with different reinforcement ratio, which verified the plane section assumption of high titanium blast furnace slag reinforced concrete beam. And the research also made test on deformation, cracks develop, cracking load, yield load, and ultimate bearing capacity. Compared with the current theoretical calculation, the results show that the structure of high titanium blast furnace slag reinforced concrete beams has a superior performance, can be used in structural engineering.

1 INTRODUCTION

The Pangang group in smelting vanadium titanium magnetite naturally cool or hot in air would form a dense slag, it contents TiO2 is as high as 22%, we named it high titanium blast furnace slag (Sun Jinku, 2006 and Huang Shuanghua, 2006).Up to now, the High titanium blast furnace slag of Pangang group have accumulated more than 6000 tons in east and west slag yard and Baguanhe slag yard, and it keeps growing at the speed of nearly 4 million tons per year. High titanium blast furnace slag acted as building material in engineering in since the1970s (Xiao Fei, 2004). In 2001, Panzhihua metallurgical slag ring industry development limited liability company was founded, the company is committed to provide the basis application for high titanium blast furnace slag in the field of architecture. Recent years, as the Panzhihua college of engineering structure mechanics experiment center in-depth development of its structural components, deformation properties and seismic performance researchigh. Since then, titanium blast furnace slag started widely used in industrial and civil building structures in Panxi region. through the contrast test, we carried research on Normal section flexural performance with different reinforcement ratio of High titanium blast furnace slag reinforced concrete beam.

2 TEST DESIGN

2.1 Experimental materials and properties

Coarse and fine aggregate with the slow cooling of The Pangang group's production of titanium slag blast furnace slag (namely High titanium blast furnace slag). The coarse aggregate satisfies the requirement of 5–31.5 mm grading. Fine aggregate, fineness modulus at 2.71, conforms to the requirement of medium sand.The applied rebar of hot rolled steel whose Strength grade is secondary which produced by The Pangang group, The applied cement which is P.O 32.5 R Portland cement produced by a cement factory in Panzhihua, Tap water from laboratory.

Through testing reserved $150 \times 150 \times 150$ mm^3 block of concrete and the tensile test of rebar, obtaining the mechanical performance parameters of concrete and steel material, listed as Table 1.

Table 1. Concrete and steel material mechanics performance.

Concrete compressive strength standard values fcu (MPa)	Steel type	Steel yield strength fu (MPa)	Ultimate strength steel fu (MPa)
46.28	Φ12	244.72	332.00
	Φ16	432.44	575.06
	Φ20	466.04	610.16

Table 2. Design of high titanium slag concrete beams mix.

Design grade	Water cement ratio	Sand coarse aggregate ratio %	Cement/kg	Water/kg	Coarse aggregate (kg)	Fine aggregate (kg)
C30	0.5	47	400	200	945	848

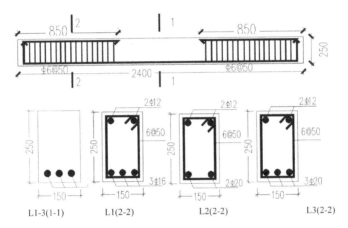

Figure 1. Reinforcement figure.

2.2 Mixture ratio design

According to the common concrete mixture ratio design regulation (JGJ 55-2011) and the institute of Panzhihua engineering structure experiment center of results for the research of concrete mixture ratio, there is a design for concrete beam mixture ratio of High titanium blast furnace slag, listed as Table 2.

2.3 Section and the reinforcement design

The test beams are named as L1, L2, L3. The cross section size b × h × l is 150 mm × 250 mm × 250 mm. In order to pure bending generated in the beam, handling reinforcement and stirrup would not arrange in the middle section. The reinforcement ratio of L1 is 1.86%, L2 is 2.00% and L3 is 2.93% (On account of the strength of high titanium blast furnace slag concrete is higher than ordinary concrete, the actual design of C30 concrete compressive strength standard values of 46.28 Mpa the original design of steel in L3 had exceeded, after correction, this problem no longer exists). The detail view of reinforcement shown in Figure 1.

2.4 Design load test

The experiment concentrated load on two symmetrical points of the test by assigning beam. Decorate a dial indicator in the bearing and across the beam. To check the test equipment

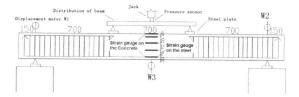

Figure 2.   Figure of test equipment 1.

parts are connected in good condition or not, do the preloading at 20% of the limit load (namely specification method), At the same time, ensure the preload is no more than 70% of the crack load (China building industry, 1992). Calculate crack load in the "specification method", "tongji law", "Liu Lixin method" respectively, take the minimum (Liu Lixin method) as the theoretical value. In the test, take observation for component deformation load for 10 minutes per level. Select the beam experiment device which was self designed and processed by Panzhihua college as loading equipments, manual hydraulic jack load. Meanwhile the computer which loaded the data acquisition system of YE6233 collect data in time. Loading device is shown in Figure 2.

3   THE EXPERIMENTAL PHENOMENON AND ANALYSIS

3.1   *The experiment phenomenon*

When the load on beam L1 applied to 21.7 KN, concrete cracks begin to appear at the tension zone of the test beam. And the cracks developed with the increase of load. But the cracks are in control. The steel yield until the load increase to 132.6 KN. But the concrete crack is still relatively thin which is less than 0.1 mm. At this point, the midspan deflection is 14.51 mm. Continue to increase the load after steel yielded, the deformation of the test beam speed, cracks width become obvious. When the load applied to 165.6 KN. The NO.4 crack vertical in the middle of the beam achieved rapid development. Its width reached 0.2 mm and its deflection reached 19.72 mm. When the load applied to 169.8 KN, the crack become clear, The width of NO.4 crack come to 0.5 mm, the width of the cracks beyond the limit. And the cracks in the style of eight extension show on both sidesof the beam. We concluded that the artifacts began to breakdown. When the load applied to 174.8 KN, the concretes in the compressive zone were completely splited. the test beam reached the ultimate load, its deflection come to 31.19 mm. We concluded that the artifacts were completely destroyed.

High titanium blast furnace slag reinforced concrete beam under the load, only small crack appeared before the reinforced yielded. and it mainly concentrated in the cross area. When the reinforced yielded, the crack's width significantly increased. Continue to increase load, the width of the cracks beyond the limit. The artifacts were completely destroyed finally. The failure mode of L1 shown in Figure 3.

The test beam's destroying process of L2, L3 are similar with L1, but with the increase of reinforcement ratio, the stiffness of the test beam increased. The ductility descend. The way that the test beam destroyed gradually develop in the direction of brittle failure. The damage of L3 in compressive zone of concrete happened suddenly. It's a kind of instantaneous burst, along with the noise. The main characteristics of the load shown in Table 3.

3.2   *The analysis of the test phenomenon*

3.2.1   *The flat section test*
We placed measuring strain along the height direction in the cross of the beam and collected information of concrete strain. It come to a conclusion that the deformation on the High titanium blast furnace slag concrete beam accord with flat section assumption. The concrete strain distribution along the height shown in Figure 4.

329

Figure 3. Damage condition of L1.

Table 3. Mechanical properties of concrete and steel reinforcement.

Beam made up	Theoretical calculate value (KN)				Measured value (KN)		
	Cracking load						
	Specification method	Tongji law	Liu Lixin method	Flexural capacity	Cracking load	Yield load	Ultimate load
L1	30.34	21.52	16.40	120.20	21.7	132.6	174.8
L2	30.66	21.67	16.09	130.74	20.6	145.5	186.6
L3	35.05	24.62	16.09	172.29	35.6	229.0	252.3

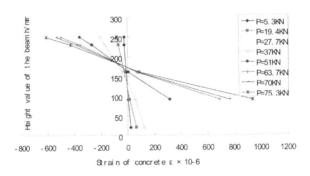

Figure 4. Figure of concrete strain distribution along the beam.

3.2.2 *The load—midspan deflection and the analysis of ductility*

Before the steel bar yield, deflection increases linearly with the increase of load. When the steel yielded, deflection increases sharply even the load constant. The midspan deflection of L1 is 14.51 mm when steel yielded. The midspan deflection of it is 64.58 mm when it come to ultimate load. Its ductility coefficient is 4.45. The midspan deflection of L2 is 21.64 mm when steel yielded. The midspan deflection of it is 60.42 mm when it come to ultimate load. Its ductility coefficient is 2.79. The midspan deflection of L3 is 26.84 mm when steel yielded. The midspan deflection of it is 49.37 mm when it come to ultimate load. Its ductility coefficient is 1.84. We conclude that with the increase of reinforcement ratio, the ultimate load, stiffness and deflection of the test beam increased. The midspan deflection on ultimate load and the ductility descend gradually. The relationship between Test load and deflection of the beam shown in Figure 5.

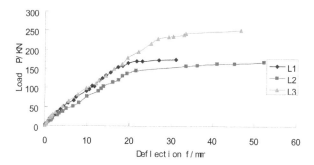

Figure 5. Figure of relationship between load and deflection.

3.2.3 *The analysis of crack propagation*

We measure the cracking load of the test beam. Compare with the theoretical value which Calculated separately in specifications for design of concrete structures GB50010-2011 calculation method (namely the specification method) (Blue building, 2011, China building industry, 2011 and Construction industry press, 2000), structure of tongji university, teaching and research section calculation method and Liu Lixin calculation method (Liu Lixin, 2010). We concluded that it is a bit conservative to calculate the cracking load of suitable reinforcement beam High titanium blast furnace slag in the specification method. The cracking load calculated in the specification method is 139.82% of the test value on L1. On L2 The proportion is as high as 148.84%. Structure of tongji university teaching and research section adopts the method of calculation value and experimental value basic consistent. The calculated value is test value 99.17% and 105.19% respectively the test value on L1 and L2. The calculated value gained in Liu Lixin method is lower than the experimental value. The calculated value of L1 and L2 is75.58% and 78.11% of the test value respectively. Relative to the super reinforced concrete beam, the theoretical value gained in specification method for the cracking load is 98.46% of the real terms. The theoretical value caluated in tongji university structure method of teaching and research section and Liu Lixin method is 69.16% and 45.20% of the real terms respectively. And there is a big gap between the test results and the actual value.

3.2.4 *The analysis of its ultimate bearing capacity*

The normal section flexural bearing capacity calculated in specification method of beams L1, L2, L3 is 120.2 KN, 130.74 KN, 172.29 KN separately. Meanwhile the practical flexural bearing capacity of High titanium blast furnace slag is 174.8 KN, 186.6 KN, 252.3 KN respectively. The results is 145.42%, 142.73%, 146.44% calculated value gained by specification for design of concrete structures (GB50010-2011). The normal section bearing capacity is strengthened with the increase of reinforcement ratio. And generally the normal section flexural bearing capacity of high titanium blast furnace slag reinforced concrete beam is generally higher than that of ordinary reinforced concrete beam. The measured value is about 1.45 times of standard values. There is a large surplus coefficient using the specification method.

4 CONCLUSION

The normal section flexural bearing capacity based on three different reinforcement ratio reinforced concrete blast furnace slag beams. We get the conclusion as follows.

1. High titanium blast furnace slag reinforced concrete beam accord with the assumption named plane hypothesis.

2. The yield deflection and the limit deflection of high titanium blast furnace slag reinforced concrete beam meet the requirements. The ductility of the beam decreased with the increase of test beam reinforcement ratio.
3. The structure of tongji university, teaching and research section calculation method is more suitable to predict the cracking load of the reinforced concrete high titanium blast furnace slag optimal balanced-reinforced beam. And the specification method is more suitable to forecast cracking load of the over-reinforced beam. As high titanium blast furnace slag is a kind of porous structure. When we concrete it in the structure, cement mortar will fill the pore. It come into being effect of bolt. So the strength of high titanium slag concrete is higher than ordinary concrete. And the wideh of crack developed slower and more narrow.
4. The limitation of reinforced concrete beam bending bearing capacity of blast furnace slag is about 1.45 times the value caluated in specification for design of concrete structures GB50010-2011 method. The beams made of high titanium blast furnace slag own more excellent performance and carrying capacity. There is a large surplus using the theory of the current specification for bearing capacity to predict. All the indicator of the high titanium blast furnace slag reinforced concrete can meet the specification requirements, but also has obvious advantages. We can drow a conclusion that the high titania bearing blast furnace slag can be applied in structural engineering.

REFERENCES

Blue building. Concrete structure (I) [M]. Beijing: China power press, 2011.7:188–189.

Huang Shuanghua, Chen wei, Sun Jinkun, etc. The application of high titanium slag in concrete materials [J]. Journal of new building materials, 2006, 11:71–73.

Industry standard editorial committee of the People's Republic of China. Test method standard GB 50152-92 concrete structure [S]. Beijing: China building industry press, 1992.

Liu Lixin Ye Yanhua. Principle of concrete structure [M]. Wuhan. Wuhan university of science and technology press, 2010.8:68.

Sun Jinkun. All high titanium heavy slag concrete applied basic research master degree theses of master of [in] [D]. Chongqing: chongqing university, 2006.

The national standard of the People's Republic of China. Concrete structure design code GB50010-2011 [S]. Beijing: China building industry press, 2011.

Tongji university teaching and research section of concrete structure. The basic principle of concrete structure [M]. Beijing. Construction industry press, 2000.8:75.

XiaoFei. Titanium slag concrete performance study of master degree theses of master of [in] [D]. Chongqing: chongqing university, 2004.

Hydraulic Engineering III – Xie (Ed)
© 2015 Taylor & Francis Group, London, ISBN 978-1-138-02743-5

Static analysis of pier base of hydroturbine with ring beam and columns

Fuqiang Chen & Xiangyu Liu
Guangdong Research Institute of Water Resources and Hydropower, Guangzhou, Guangdong, China
The Geotechnical Engineering Technology Center of Guangdong Province, Guangzhou, Guangdong, China
The Emergency Technology Research Center of Guangdong Province for Public Events Guangzhou, Guangdong, China

Zhixing Huang
Department of Civil Engineering, South China University of Technology, Guangzhou, Guangdong, China

Yucheng Zhang
Guangdong Research Institute of Water Resources and Hydropower, Guangzhou, Guangdong, China
The Geotechnical Engineering Technology Center of Guangdong Province, Guangzhou, Guangdong, China
The Emergency Technology Research Center of Guangdong Province for Public Events Guangzhou, Guangdong, China

ABSTRACT: Pier base is the supporting system of generator and hydroturbine, which bears very large steady and dynamic loads, therefore sufficient stiffness, strength and stability are required. In regular static analysis the structure of the ring beam is usually simplified to a planar multi-linear beam. In this paper, the simplification is verified by analyzing the simply supported beam by elastic theory. The verification shows that the simplification has inevitable defect; a finite element model is built for comparison, the result indicates that the moment of the beam calculated from the simplified analysis has a 65% deviation. It is recommended to consider the error and side effect due to the simplification in engineering design, so as to guarantee the safety of structure.

1 INTRODUCTION

Pier base supported by ring beam and column is widely used in small hydraulic plant due to its advantages such as: it facilitates installation and reparation of auxiliary equipment; the wheel of hydroturbine is able to be inspected and maintained in place by taking it out in between columns; the space for plant building can be compact and small; construction materials can be less consumed compared with the cylindrical style pier base[1]. The stress state of pier with ring beam and column is relatively complex, rigorously solving the internal stresses of pier base will be a very time-consuming work; also, it is unnecessary to solve it in hydraulic engineering problems[2]. Normally, firstly, the ring beam is considered as a multi-linear beam fixed on columns; and then the fixed end torque is calculated existing in the connecting area of multi-linear beam and columns; at last, the internal stress of the column can be derived by assuming this torque as the moment of the top end of the column. A crucial assumption of this method is that the torsional stiffness of the ring beam is considered negligible compared with the column, hence the ring beam is regarded as multi-linear beam with fixed end on columns[3], and then the reinforcement can be designed based on stress analysis of linear beam. However, one should notice that the size of the cross-section of ring beam is not far less than its span in practice, so debates arise if the ring beam was analyzed as a linear beam.

This paper discusses the simplification mentioned above, and a comparison is made against results from a finite element analysis.

2 PROJECT DESCRIPTION

A pier base with ring beam and column supporting system is analyzed. The pier based consists of air shield, ring beam and columns; the air shield and beam locate on top of the four columns, and the columns stand on the pump well. The width of the air shield is 0.325 m and the height is 1.75 m. On top of the air shield is the load transferred from the slab of generator floor, and the bottom of the air shield connects to the ring beam and they form an entirety. The width of the ring beam is 0.7~1.1 m, and the height is 0.7 m; the arc columns have lengths of 1.65~2.2 m, widths of 0.7 m and height of 1.2 m. The top view and cross-section of the pier base are shown in Figure 1 and Figure 2, respectively.

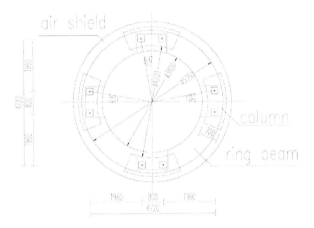

Figure 1. Top view of the pier base.

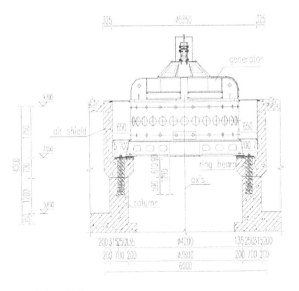

Figure 2. Cross-section of the pier base.

334

3 REGULAR ANALYSIS OF PIER BASE

In the regular analysis and reinforcement of the ring beam, the beam is simplified to a multi-linear beam for modeling. The required assumptions are:

1. Simplification of the structural system: the 3 dimensional structure in practice is simplified to planar multi-linear beam.
2. Simplification of the beam element. The ring beam is calculated based on its axis, the connecting area of beams are simplified to nodes, the length of the beam is assumed to be the distance between nodes, the point of action is transferred to the axis.
3. The connecting areas between ring beam and columns are simplified to fixed end.

The illustration of the loads, simplifications and results are shown in Figure 3 to Figure 5, respectively.

The resulting maximum negative moment of the ring beam is 107.6 KN · m.

Based on the stress analysis of simply supported beam under uniform surcharge in elastic theory (as shown in Fig. 6), one can derive that $\sigma_x = M/I \, y + q \, y/h \, (4 \, y^2/h^2 - 3/5)$, where the first term is the solution from material mechanics, however in elastic theory, a calibration term is needed, which is the second term. The second term cannot be ignored for beams having large height of cross-section. When the span of the beam is two times the height of the cross-section, the maximum stress σ_x should be modified by 1/15, and when the span of the beam is four times the height of the cross-section, the maximum stress σ_x should be modified by 1/60. It is obvious that only the beam with ratio of span over height larger than 4 is

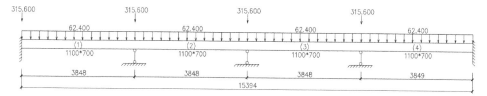

Figure 3.   Illustration of the loads (unit: mm).

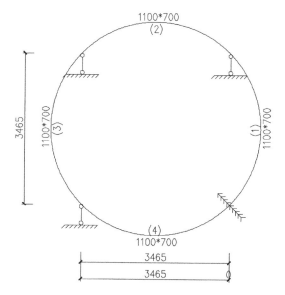

Figure 4.   Calculation diagram of the multi-linear beam (unit: mm).

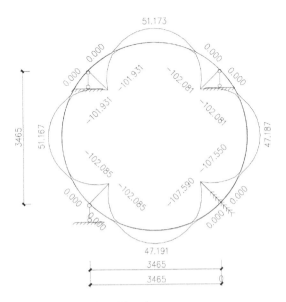

Figure 5. Moment envelop diagram (unit: KN · m).

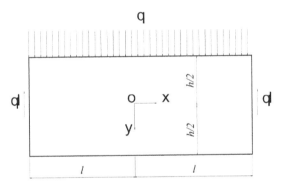

Figure 6. Uniform surcharge on simply supported beam.

able to be simplified to beam element with sufficient accuracy. However, in the case discussed here, the ratio is 2.75, which means that it is not suitable to simplify the ring beam to beam element, the resulted stresses will be smaller than the reality.

4 FINITE ELEMENT ANALYSIS

The 3 dimensional model consisting of the air shield, ring beam and columns is built by MIDAS/GTS based on the real size, as shown in Figure 7. The load from the slab of generator floor is, the Young's modulus of the reinforced concrete is 28000 MPa, the Possion's ratio is, $v = 0.2$, $\gamma = 25$ kN/m³, linear elastic model is applied. The bottom ends of the columns are fully fixed; The gravity is applied to the entire model.

The results are shown in Figure 8 and Figure 9. The moment of the cross-section of the base of air shield and ring beam is 178 KN · m, the torque is 9.2 KN · m, the shear force is 26.2 KN. It can be seen that the error due to the simplification of the ring beam is 65% (i.e. (178–107.6)/107.6 = 65%).

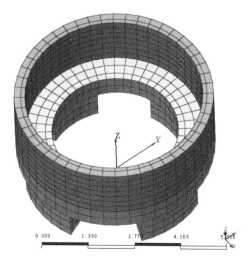

Figure 7.   3D model by MIDAS/GTS.

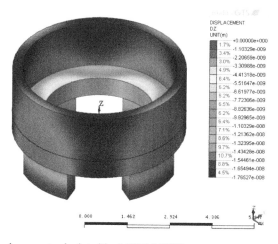

Figure 8.   Vertical displacement calculated by MIDAS/GTS.

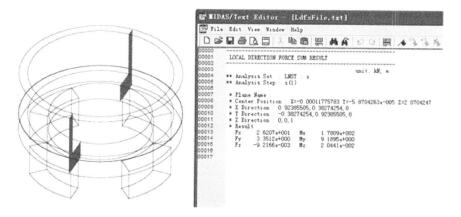

Figure 9.   Moment of the cross-section of ring beam and air shield.

5 CONCLUSION

In this paper, the author compares the analysis of pier base with ring beam and columns by regular method and finite element method. It is found that simplifying the ring beam to multi-linear beams is not accurate enough, which leads to an error of 65%. Hence, it is recommended to take the side effect of this simplification into account during engineering design, so as to guarantee the safety of the structure.

REFERENCES

Jiabo, Hu. Structural calculation of pier base with ring beam and columns [J]. Water Resources and Hydropower Engineering. 1981, (4):27–32. (in Chinese).

Liang, Zhou. The second development system of generator support with ring type beam in hydroelectric power station [D], Master thesis Nanchang university, 2005.6 (in Chinese).

Zhibin, Zhang. Discussion on the opening angle of pier base with ring beam and columns [J]. Journal of Chengdu university of science and technology 1983, (1):100–102. (in Chinese).

Hydraulic Engineering III – Xie (Ed)
© *2015 Taylor & Francis Group, London, ISBN 978-1-138-02743-5*

Numerical prediction for the deformation of adjacent building foundation in Beijing metro construction

Xiyuan Sun, Chaoyang Heng & Zhi Zhou
Institute of Foundation Engineering, China Academy of Building Research, Beijing, China

ABSTRACT: With the growing demand for public transportation in big cities, metro construction is developed rapidly. It is much important to keep the adjacent buildings safe in the course of metro construction. In this paper, a numerical simulation method for predicting the deformation of adjacent building foundation influenced by metro construction (Beijing Subway Line 10) in Mengjiacun, Beijing is suggested. The maximum simulation settlement of ground is 7.6 mm, which is less than the permissible value (10 mm). With simulation results and previous assessment, the safety of engineering design can be evaluated. Furthermore, field measurements are carried out to verify the numerical simulation method after construction and a satisfying agreement between them is obtained.

1   INTRODUCTION

As a rapid, convenient and high-capacity public transportation system, metro is playing a more and more significant role in unban communication now. Since metro lines are generally located in downtown districts or residential areas of cities, keeping the deformation of adjacent building foundation in an allowable range is a key point for design before construction (Chen, 2011). The numerical simulation technique is widely used in structural engineering practice, but not commonly in geotechnical engineering. For accumulating experiences in this field, a three-dimensional numerical simulation using the general-purpose finite element analysis package MADAS GTS is taken in an actual metro engineering (the tunnel of Beijing Subway Line 10 in Mengjiacun) and subsequent measurement verification is also carried out to evaluate the validity of this numerical simulation.

2   ENGINEERING SITUATION

2.1   *Project description*

The subsurface excavation of Metro tunnel is located in mengjiacun, which is a part of Beijing Subway Line 10. There are 4 existing residential buildings completed in 1990 over the tunnels (left and right line). The spatial relationships between them are shown in Figure 1 ~ 3.

2.2   *Engineering geology*

According to the report by *Beijing Aerospace Geotechnical Engineering Institute*, the engineering geology of relative area is described in Table 1.

2.3   *Construction design*

Keeping the foundation deformation in an allowable range is very important to the safety of both adjacent building and metro tunnel construction. Based on the previous assessment by *Institute of Foundation Engineering China Academy of Building Research*, the residual

Figure 1.   Locale photo of Mengjiacun.

Figure 2.   Satellite-picture of Mengjiacun.

Figure 3.   The vertical position relationship diagram (unit: meter).

Table 1.   Main engineering geology.

Serial number	Soil type	Unit weight	Cohesive strength	Friction angle	Modulus of compression	Poisson's ratio	Layer thickness
1	Fill	19.5 kN/m³	27.7 kPa	26.7	5.61 MPa	0.2	5.5 m
2	Round gravel	19.8 kN/m³	0	35	35 MPa	0.22	5.5 m
3	Grait	19.8 kN/m³	0	40	44 MPa	0.2	14 m
4	Erratic boulder	19.8 kN/m³	0	45	55 MPa	0.15	20 m

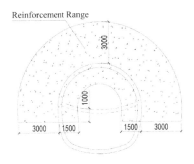

Figure 4.   The cross sectional drawing.

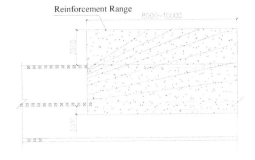

Figure 5.   The profile diagram.

foundation deformation d should be less than 10 mm. For achieving the goal of deformation control, the technological process of pre-grouting is used as shown in Figures 4 and 5.

3   NUMERICAL SIMULATION

3.1   *Modeling*

For checking the reliability of construction design, a three-dimensional numerical simulation is applied. All modeling parameters are determined on the basis of that actual engineering

Figure 6.   Full view of the ground.

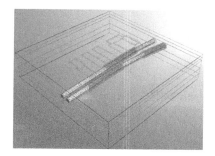

Figure 7.   Full view of the tunnels.

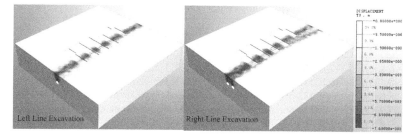

Figure 8.   Vertical displacement contour of the ground.

mentioned above. In particular, the soil property is taken as elasto-plasticity model using Mohr Coulomb yield criterion (Li, 2004) and the modulus of elasticity E is equal to the one of compression E_s according to previous research experiences. The typical finite element meshes are shown in Figures 6 and 7.

There are three key points directly related to the quality of simulation. They are boundary conditions definition, initial geostatic equilibrium and load distribution. For this numerical calculation, boundary conditions are such that the bottom of the soil mesh is fixed in all three coordinate directions and the circumference is fixed in radial direction (Tao, 2003). It is worth mentioning that the hardening behavior of initial tunnel lining is also realized by changing boundary conditions here. Since only the relative deformation of ground after tunneling is cared, the initial geostatic equilibrium must be calculated with the vertical loads of that 4 existing residential buildings. In MIDAS GTS, load distribution can be conveniently simulated by seting up the parameter named *LDF* in different construction stage (MIDAS IT, 2007). According to the results of previous field tests, 0.5, 0.25 and 0.25 are assumed in the excavation stage, lining construction stage and lining hardening stage respectively.

3.2   *Results*

Indicated in Figure 8 is the vertical displacement contour of ground at different construction stage. To be specific, the maximum settlement of ground is about 6.6 mm after left line tunneling and the value after right line tunneling is increasing to 7.6 mm. Both of them are less than 10.0 mm, which is the maximum of d. That is to say the construction design is reliable normally and the 4 buildings are safe in the course of metro construction.

4   MEASUREMENT VERIFICATION

After final completion, a series of leveling surveying work is carried out to evaluate the validity of numerical simulation method. The arrangement plan of observation points is shown in Figure 9.

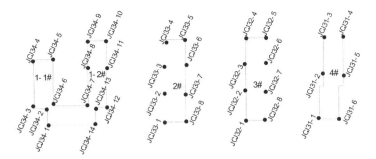

Figure 9. The arrangement plan of observation points in Mengjiacun.

Table 2. Summary of the settlement value at observation point.

Serial number	Settlement	Serial number	Settlement	Serial number	Settlement	Serial number	Settlement
JCJ31-1	0.5 mm	JCJ32-3	6.0 mm	JCJ33-3	8.1 mm	JCJ34-6	1.1 mm
JCJ31-2	7.2 mm	JCJ32-4	8.7 mm	JCJ33-4	8.4 mm	JCJ34-7	1.3 mm
JCJ31-3	8.0 mm	JCJ32-5	8.5 mm	JCJ33-5	8.3 mm	JCJ34-8	6.0 mm
JCJ31-4	7.9 mm	JCJ32-8	0.5 mm	JCJ33-8	0.8 mm	JCJ34-9	8.3 mm
JCJ32-1	0.4 mm	JCJ33-1	0.7 mm	JCJ34-4	4.0 mm	JCJ34-10	8.4 mm
JCJ32-2	2.2 mm	JCJ33-2	2.0 mm	JCJ34-5	3.9 mm	JCJ34-12	2.0 mm

Compared with the initial state, the settlement at observation point can be measured by using *Trimble DiNi03* and summarized in Table 2. It is shown that the ground deformation over tunnels is more remarkable than the other. The maximum value is 8.7 mm (<10.0 mm), which is much similar to the one of numerical simulation. In other words, both the validity of numerical simulation method and the safety of construction design are verified powerfully.

5 CONCLUSIONS

1. With reasonable calculation parameters, three-dimensional numerical simulation method has an outstanding performance on checking the reliability of metro construction design, not only in qualitative analysis but also in quantitative computation.
2. Considering the importance of adjacent buildings, strict supervision of reinforcement construction and deformation monitoring must run through the whole process of metro construction to make sure d is under control.

ACKNOWLEDGEMENT

The project is supported by the foundation (No. 2013-K3-22) of Ministry of Housing and Urban-Rural Development of China.

REFERENCES

Chen, X. 2011. Control standards for settlement of ground surface with existing structures due to underground construction of crossing tunnels. *Journal of engineering geology*. 19(1): 103–108. (in Chinese).
Li, D. 2004. 3-D numerical simulation of frozen construction of a connected aisle in metro. *Rock and soil mechanics*. 25 (supp. 2): 472–474. (in Chinese).
MIDAS IT, 2007. *MIDAS/GTS 03 analysis reference*. Beijing: MIDAS Information Technology CO., LTD.
Tao, L. 2003. Numerical simulation of ground settlement due to constructing metro-station based on shield tunneling. *Journal of China university of mining & technology*. 32(3): 236–240. (in Chinese).

Hydraulic Engineering III – Xie (Ed)
© 2015 Taylor & Francis Group, London, ISBN 978-1-138-02743-5

Analysis of temperature influence on steel strut based on FLAC3D

Zhi Zhou, Chaoyang Heng & Xiyuan Sun
Institute of Foundation Engineering, China Academy of Building Research, Beijing, China

ABSTRACT: Steel strut plays an important role in controlling deformation and maintaining the pit stability in strutted retaining structure. Axial force in the steel strut is influenced by temperature obviously because of the steel thermal characteristic. By simulation in FLAC3D of an actual project, the axial force in steel struts is analyzed when temperature rises. The origin axial forces that may influence the force changes are discussed. The way that axial force changes approximate linearly with temperature rises in the numerical model is basically consistent with the monitoring. This conclusion could provide a useful reference for engineering and construction in similar deep excavation projects.

1 INTRODUCTION

Steel strutted retaining structure is one of the common forms widely used in construction in deep foundation pit because of its convenient installation and disassembly. Lots of economic benefits are linked with the reuse of steel struts. Since steel strut can work immediately to effectively reduce the displacement of excavation after installation, priority could be given to the use of this form of supporting in case of buildings adjacent around the foundation pit or small deformation of the pit required (Bin, 2012 & Wang, 1998).

In this structure system, the steel struts that support the horizontal force caused by soil pressure, groundwater, ground load, pre-pressure and temperature changes play an important role in controlling deformation and maintaining the stability. Because of the steel thermal characteristic, axial force in the steel strut is influenced by temperature obviously. Monitoring results indicates that the axial force in steel strut changes with temperature significantly, therefore, this changing should be considered in engineering and construction. In this paper, the axial force in steel struts is analyzed when temperature rises by simulation in FLAC3D of an actual project, and some factor that influence the force changes are discussed also.

2 PROJECT INTRODUCTION

2.1 Project overview

Baishiqiaonan station is the 12th station of Beijing Metro Line 9, the station is located in the northwest corner of intersection of Shoutinan Road and Chegongzhuang Street, and it is an interchange station with Metro Line 6 also (Fig. 1). Standard region of the pit is about 18.2 m in width, excavation depth is about 18 m; transfer node pit is about 38.5 m in width, excavation depth is about 25.5 m. Steel strutted retaining structure is used in this pit. Bored piles are 1000 mm in diameter by spacing 1400 ~ 1600 mm. The steel strutted system uses steel tube brace. The first layer braces which are 800 mm in diameter, 12 mm in thickness are installed on the cap beam; the other layer braces which are 800 mm in diameter, 12~16 mm in thickness are installed on steel purlins which are made by the combination of I45b section steel (Fig. 2) (Zhou, 2013).

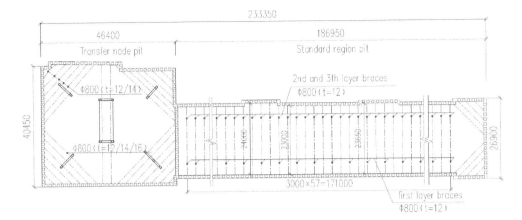

Figure 1. Baishiqiaonan station steel strutted retaining structure plan.

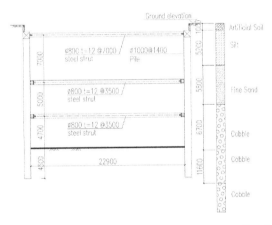

Figure 2. Baishiqiaonan station standard region retaining structure profile with stratigraphic section.

2.2 Geological conditions

According the formation lithology and the physical and mechanical property, there are basically six stratums in this station construction site in turn: filling ①, silt ③, find sand ④, cobble ⑤, cobble ⑦, cobble ⑨(Fig. 2). And there is no groundwater within the excavation region.

3 NUMERICAL MODEL AND PARAMETER

3.1 Model foundation of simulation

Figure 3-a shows standard region pit simulation model is founded in FLAC3D as analysis object. the excavation is 23 meters in width and 19 meters in depth as the actual project. The model is 223 meters totally in width and 90 meters in depth, the calculation length is 9 meters in longitudinal direction of the pit. The gridpoints along the bottom boundary are fixed in the z-direction. Roller boundaries are placed on four sides of the model expect the top and the bottom. In the standard region pit, soldier pile is 1 meter in diameter spacing 1.4 meters. The steel strut is 0.8 meters in diameter, and 12 mm in thickness. Steel struts are created with SEL beam, soldier pile walls are created with SEL liners which could model pile walls for which both normal-directed compressive/tensile interaction and shear-directed frictional interaction with the soil, as be shown in Figure 3-b. The whole model contains 25920 zones, 29965 grid-points and 2544 structure elements in total.

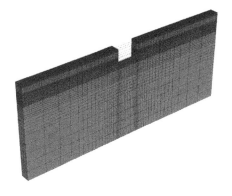

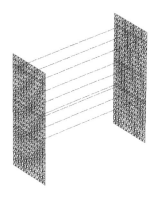

a. Overall excavated model   b. Steel strutted retaining structure model

Figure 3.   Numerical model.

Table 1.   Material properties of the soil for the analysis.

Soil	Density (kg/m³)	Cohesion (c) (kPa)	friction angle (φ) (°)	Young's modulus (MPa)
Filling-1	1800	5	10.0	30
Silt-3	1950	24	26.1	48
Find sand-4	1900	0	28.0	72
Cobble-5	1900	0	38.0	210
Cobble-7	1900	0	40.0	300
Cobble-9	1900	0	42.0	360

3.2 Soil and structure parameter

The soil is considered as Mohr-Coulomb material. Table 1 presents the material properties of the soils used in this analysis.

The beam and pile are modeled as elastic materials. The strut has the following properties: Young's modulus, E 2.06E5 MPa; Poisson's ratio, v 0.3; Cross-sectional area, 0.029707 m²; polar moment of inertia, 0.004612 m⁴ (Chen. 1994). The solider pile walls are modeled as walls which has the same second moment in unit length with the following properties: Young's modulus, E 2.5e4 MPa; Poisson's ratio, v 0.22; thickness, 0.716 m; Density, ρ 2500 kg/m³ (Liu. 1989). The modulus g of the gravity vector may be approximated as 9.8 m/s². No groundwater influence is involved because of the low groundwater level.

3.3 Analysis stages

The analysis is divided into two stages. In the first stage, the pit is excavated as the actual construction steps. After create the piles, the soil above the elevation which below the first layer strut 0.5 m is excavated using NULL command. Then struts are installed with prestress. With this excavation and struts installation procedure cycles, a total of four sequential excavation and support steps are performed.

In the second stage, a FISH program executes to simulate the temperature of the struts rises from 0°C to 40°C every 5°C. Here the 0°C means the struts installation temperature, not actual temperature. In this simulation, temperature influence on the struts axial forces are the main research object, so the axial forces of supports are monitored throughout the construction. These values are recorded at the end of each step and stored in files.

4 RESULTS ANALYSIS

4.1 Simulation results

The axial forces in struts versus temperature rising are shown in Figure 4-a. Figure 4-a illustrates what happens to the axial forces when the temperature is changed. As the temperature increases, the axial forces in every layer struts increases almost linearly. The axial force in the first layer struts increases almost 37 kN with 10°C increase in temperature, also 76 kN in the second layer struts and 195 kN in the third layer struts. This increase maybe reaches 40% of the origin axial force with just 30°C increase which is very common in summer between day and night. This means temperature change influences the strut axial force so significant that the influence can't be ignored in engineering and construction especially in the building site where the temperature varies greatly within short periods.

4.2 Comparison with monitoring

Figure 4-b shows the monitoring results of axial force of a first layer strut in two days with temperature change. The way that axial force increases linearly with temperature is basically consistent with the simulation results. The axial forces increase with 10°C increases in temperature from simulation and a monitoring site in standard region pit are summarized in Table 2. There's about 30% difference between simulation and monitoring. This difference is mainly caused by difference between the real soil properties and the assumption of soil material properties. This difference could be decreased with the accumulation of calculation experience. So it's feasible to analyses temperature influence on steel strut based on FLAC3D.

4.3 Origin axial force influence

In an object, most factors that may influence the force in struts increases with temperature rising are always decided by the site condition and the engineer, such as the soil, the pile or the steel strut itself. Origin force difference caused by installation prestress error or

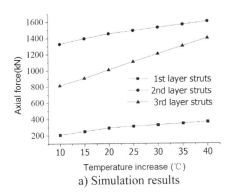

a) Simulation results

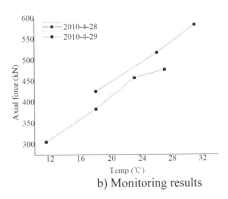

b) Monitoring results

Figure 4. Axial force versus temperature increase.

Table 2. Axial force increase pre 10°C increase.

Results	1st layer (kN)	2nd layer (kN)	3rd layer (kN)
Simulation	37	76	195
Monitoring	55	109	176

346

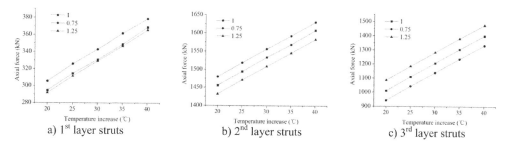

a) 1ˢᵗ layer struts   b) 2ⁿᵈ layer struts   c) 3ʳᵈ layer struts

Figure 5.   Axial force versus temperature increase.

installation temperature is almost the only factor might influence causes the way of force increase caused by construction. In order to make sure that whether the slope of the force increase line is changes with difference origin axial force, two more simulation are executed. Prestress is changed only in those two models, that is, one with 75% prestress applying and 125% prestress is applied in the other. Axial forces in different layers with different prestress are shown in Figure 5. The figure indicates that origin axial force has no influence on the way that axial force in struts changes with temperature rising. This can also be seen in Figure 4-b, although the origin forces are different in different monitoring days, the force still increase with temperature rising linearly in the same slop. The slope of axial force increases with temperature rising linearly, therefore, is only decided by the geotechnical conditions and retaining structure. This character is useful for construction and monitoring. By using two axial force monitoring dates, the slope of axial force increase with temperature rising can be easily calculated. With this slope, the value of axial force in the future can be forecasted, and an appropriate installation temperature can be determined, so that the value of axial force should always under the steel struts capabilities even if a large temperature rises. With this slope, monitoring engineers can calculate the value of origin axial force without temperature influence, this value can help engineers judge the state of pit with truly steel struts axial forces, and the false high value can be corrected.

5   CONCLUSION

The way that axial force increases linearly with temperature rises in the numerical model is basically consistent with the monitoring. It's feasible to analyses temperature influence on steel strut based on FLAC3D.

Temperature change influences the strut axial force so significantly that the influence should be paid more attention to in engineering and construction especially in the building site where the temperature varies greatly within short periods.

The slope of axial force increases with temperature rising linearly is only decided by the geotechnical conditions and retaining structure. Origin axial force has no influence on the value of slope of axial force increase with temperature rising. The slope can be easily calculated by using two axial force monitoring dates. This conclusion could be used by construction and monitoring engineers to guide the project.

The simulation in FLAC3D can get a regular result basically consistent with the monitoring in this project. But more work should be done in parameter choosing, so that the simulation could be closer to the reality more.

ACKNOWLEDGEMENT

This paper is supported by the foundation (No. 2013-K3-22) of Ministry of Housing and Urban-Rural Development of the People's Republic of China.

REFERENCES

Bin, Yang. 2012. *Technical Specification for Retaining and Protection of Building Foundation Excavations.* Beijing: China Architecture & Building Press.

Jiwang, Wang. 1998. *Code for Technique of Building Foundation Pit Engineering*. Beijing: Beijing Sunyi Xinghua Press.

Shaofan, Chen. 1994. *Steel Structural*. Beijing: Beijing Fusheng Press.

Zhaopei, Liu & Wenmei, Zhang. 1989. *Structural Mechanics*. Tian Jin: Tianjin University Press.

Zhi, Zhou & Xiyuan, Sun & Chaoyang, Heng. 2013. Temperature influence to the steel brace in piles in row-steel bracing system. *Advances in underground space development*: 325.

Hydraulic Engineering III – Xie (Ed)
© 2015 Taylor & Francis Group, London, ISBN 978-1-138-02743-5

Comprehensive evaluation of cracking characteristics of box girder high performance concretes

Beixing Li, Xianglin Jiang & Zhigang Zhu
State Key Laboratory of Silicate Materials for Architecture, Wuhan University of Technology, Wuhan, China
Office of Jiujiang Yangtze River Expressway Bridge Construction Project, Transportation Department of Jiangxi Province, Jiujiang, China

ABSTRACT: An experimental investigation was carried out to evaluate the early age cracking tendency of box girder high performance concretes containing supplementary cementitious materials by tracking the development of thermal, physical and deformation properties. The test variables included the type and the amount of supplementary cementitious materials (fly ash and ground granulated blast-furnace slag). Portland cement was replaced with FA up to 30% and blends of FA and GGBS up to 30%. The results confirmed that FA performs better than the combinations of FA and GGBS for the decrease of hydration heat, adiabatic temperature rise, autogenous shrinkage and plastic shrinkage cracking. However, the ternary mixture containing FA and GGBS performs better than the binary mixture containing FA to increase the cracking stress and reduce the cracking temperature. The comprehensive results show that 25% of FA replacement was found to be an optimum level and demonstrated excellent performance in crack resistance.

1   INTRODUCTION

High-Performance Concretes (HPCs) are being widely used in large-span prestressed concrete continuous girder bridges. However, the early age cracking of HPCs has been mentioned frequently in recent years and confirmed in engineering practice (Zhang, 2006; Zhao, 2006; Zhu, 2006). In view of the fact that in HPCs of low w/b ratios, the early age cracking can be due to three processes occurring simultaneously, autogenous shrinkage and thermal effects (Bentz, 2008; Holt, 2008), coupling with large dry shrinkage if inappropriate moisture curing (Qin, 2001). Thus, some products for minimizing shrinkage cracking were specifically developed, such as fibers and shrinkage reducing admixtures.

For producing HPCs, it is well recognized that the use of SCMs such as Silica Fume (SF), ground Granulated Glass Blast-Furnace Slag (GGBS) and Fly Ash (FA), are necessary. These materials, when used as Supplementary Cementitious Materials (SCMs) in HPCs, can improve either or both the strength and durability properties of concretes (Elahi, 2010).

Crack resistance is of significant performance relevant to various properties of concrete structures, but the influence of these SCMs on crack resistance of HPCs, hard to acknowledge, is rarely studied (Bentz, 2004; Haque, 1998; Jiang, 2004; Bentur, 2003). Thus, a variety of qualitative cracking tests were used to evaluate the sensitivity to early age shrinkage cracking of HPCs containing different SCMs. The overall objective of the research was to attempt to establish the evaluation systems of cracking characteristics of box girder HPCs.

2 EXPERIMENTAL PROGRAMME

2.1 *Mix proportions*

The C55 strength grade HPC mix proportions were designed to apply in the super-wide prestressed concrete box girder of Jiujiang Yangtze River Highway Bridge. The final mix proportions were arrived at after having done many trials so as to have a slump ranging from 210 to 240 mm at a constant water-binder ratio (w/b) of 0.305, sand-total aggregate ratio (s/a) of 39% and a constant total binder content of 495 kg/m³. The slump was adjusted by adding different dosages of the superplasticiser. The mix proportions are summarized in Table 1. One control mix incorporating only Portland Cement (PC) and five other mixes containing SCMs were prepared.

2.2 *Materials*

A type-II 42.5 grade Portland cement complying with the Chinese National Standard GB 175-2007, C-type first grade FA satisfied with GB/T 1596-2005 and S95 grade GGBS manufactured according to GB/T 18046-2008 were used as binders. Both the fine and coarse aggregates were obtained from local sources in Jiujiang. The fine aggregate used was medium graded river sand with a fineness modulus of 2.6, whereas, the coarse aggregate used was crushed limestone with a nominal maximum size of 20 mm. The superplasticiser (SP) used was a set retarding-type polycarboxylate containing 29.7% solids, commercially branded as MAPEI SX-C18. The water content in the mixes was adjusted to account for the water in the SP, whilst retaining the constant w/b of 0.305.

2.3 *Experimental methods*

2.3.1 *Workability*
The workability of concrete was measured in terms of slump and slump flow, noted immediately after manufacturing the concrete. The slump and slump flow tests were carried out in accordance with the Chinese National Standard GB/T 50080-2002. These results are reported in Table 3. It can be seen from this table that the measured slump was within the target range of 210–240 mm.

2.3.2 *Compressive strength*
The compressive strength of the concrete was determined by crushing three cubes of 150 mm size at the age of 7 and 28 days for each mix. The test was carried out according to GB/T50081-2002.

Table 1. Mix proportions.

Mix ID	Mass of ingredients (kg/m³)						
	PC	FA	GGBS	Water	Fine aggregate	Coarse aggregate	SP
HPC-PC	495	0	0	151	723	1131	6.435
HPC-FA20	396	99	0	151	723	1131	5.445
HPC-FA25	371	124	0	151	723	1131	5.445
HPC-FA30	346	149	0	151	723	1131	5.445
HPC-FA15 + BS10	371	75	49	151	723	1131	5.445
HPC-FA15 + BS15	346	75	74	151	723	1131	5.445

Notes: The numeral in mix designation corresponds to the percentage of SCMs present in the binder. For example HPC-FA25 means that the high performance concrete was prepared using FA (25%) and the rest was cement. For instance, in HPC-FA15 + BS10, the FA content was 15%, GGBS content was 10% and the rest was cement.

2.3.3 Chloride diffusion

The non-steady state diffusion test was carried out according to GB/T50082-2009. Tests were conducted using 100 mm diameter and 50 mm length concrete discs, and a non-steady-state diffusion coefficient of rapid migration (DRCM) was obtained.

2.3.4 Hydration heat of binder

To measure the hydration heat release process of binder, the auto isothermal calorimeter of TAM Air was used. Tests were conducted using binder of 10 g and w/b ratio of 0.40 and the test temperature was 25 °C.

2.3.5 Adiabatic temperature rise test

The JR-2 adiabatic temperature rise tester was used to measured the adiabatic temperature rise of actual concrete mixes with hydration time according to the Chinese Electric Power Trade Standard DL/T 5150-2001.

2.3.6 Autogenous shrinkage

To measure the autogenous volume deformation of the concrete under the condition of no moisture exchange between specimen and the ambient, the magnetic suction type multi-channel sensor test device was used. Tests were conducted using 100 mm × 100 mm × 515 mm concrete prisms at the ages of 1, 2, 3, 4, 5, 7, 14 and 28 days for each mix.

2.3.7 Plate test for plastic shrinkage crack

Plastic shrinkage crack test was conducted according to the Chinese Civil Engineering Society Standard CCES 01-2004 and reference (Soroushian, 1993). In order to induce cracks in the concrete, a steel plate of 600 mm × 600 mm × 63 mm with 14 bolts of 10 mm in diameter, mounted regularly at each four sides of the rigid frame, was used to restrict possible drying shrinkage of concrete. Concrete was placed in the steel form and cured with a cover of acrylic board for 2 hours under a constant temperature of 30 °C and relative humidity of 60%. Subsequently, it was exposed to an air flow with a velocity of 8 m/s under a constant temperature of 30 °C and relative humidity of 60%. Then, the age of crack initiation, number of cracks, crack length and crack width were measured up to the age of 24 hours. Parameters representing crack propagation characteristics at the age of 24 hours may be the average crack area, number of cracks per unit area and total crack area per unit area.

2.3.8 Temperature and stress tests for evaluation of thermal cracking

The uniaxial restrained and adjusted Temperature and Stress Testing Machine (TSTM) was used evaluate the cracking sensitivity of concretes and the thermal stresses developed in adiabatic conditions (Hu, 2007; Zhang, 2002). Fresh concrete was placed in the 150 mm × 150 mm × 1500 mm frame. The temperature of the fresh concrete should be 25 ± 2 °C. There was no moisture exchange between specimen and the ambient, thus no dry shrinkage took place and in consequence, the deformation of specimen consisted of only autogenous shrinkage under full restraint and thermal deformation at the same time.

3 RESULTS AND DISCUSSION

3.1 Compressive strength and chloride diffusion

The results of compressive strength are presented in Table 2. From the compressive strength results at different ages with different FA replacement levels, the strength development of the concretes at all the FA replacement levels was consistently lower than that of the control PC mix at all the ages. Table 2 demonstrate that not only all the FA mixes had a lower strength compared to the control mix, but also an increase in FA content further reduced the strength. As all the mixes were cast with the same w/b and the slump was similar, the reduction in the strength of FA mixes is considered to be the result of slower pozzolanic reaction of the FA in

Table 2. Workability, strength and resistance to chloride permeability of concretes.

Mix ID	Slump (mm)	Slump flow (mm)	Compressive strength (MPa)		28d D_{RCM} (10^{-12} m²/s)
			7d	28d	
HPC-CC	210	510	72.6	83.9	3.92
HPC-FA20	215	590	64.0	81.3	3.24
HPC-FA25	240	615	60.9	77.7	3.09
HPC-FA30	240	630	56.4	70.1	3.30
HPC-FA15 + BS10	235	600	63.1	80.8	2.14
HPC-FA15 + BS15	200	480	59.6	79.4	2.53

Table 3. Hydration heat of binders in concretes.

Mix ID	Maximum rate of hydration heat release (KJ/Kg·h)	Time of maximum hydration heat rate (h)	Hydration heat (KJ/Kg)				
			1d	2d	3d	5d	7d
HPC-PC	10.44	9.55	132.09	191.81	228.67	274.20	290.40
HPC-FA20	8.88	7.32	136.30	173.16	195.98	223.15	244.50
HPC-FA25	7.80	10.14	102.93	151.81	188.42	221.78	242.96
HPC-FA30	7.95	7.64	123.72	157.96	176.27	202.78	218.46
HPC-FA15 + BS10	8.67	10.28	134.28	165.6	200.61	236.85	259.35
HPC-FA15 + BS15	9.83	10.91	131.37	164.23	196.05	233.94	257.87
HPC-FA25 + SP1.1	2.27	37.25	44.99	92.92	125.20	172.16	205.17

Note: SP—Superplasticiser. The numeral in SP corresponds to the percentage of SP present in the binder.

these mixes. The compressive strength of ternary mixes of PC, FA and GGBS were observed to be greater than binary mixes of PC and FA, which is considered to be due to the contribution made by GGBS. The rate of hydration of GGBS is considerably higher than that of FA. As a result, GGBS starts to contribute to the strength development at an early stage in hydration compared to FA.

The chloride diffusion test yielded an effective diffusion coefficient (D_{RCM}), the D_{RCM} values of various mixes at 28 days are shown in Table 2. All the binary and ternary mixes showed an decrease in D_{RCM} from 17.5% to 45.4% smaller than the control PC mix. Amongst the FA mixes, the lowest D_{RCM} was obtained for FA 25, and the addition of GGBS decreased the D_{RCM} of the FA mixes. It is normally considered that the addition of FA or GGBS improves the microstructure due to its filling ability and pozzolanic reaction.

3.2 *Hydration heat*

The hydration heat results of the binders in concretes are presented in Table 3. The results shows that with the increase of FA replacement levels, the hydration heat release rate was slowed down and the hydration heat quantity was gradually reduced. The reduction in hydration heat is attributed to the decrease of the hydration particle numbers in per binder paste with the addition of FA. The rate and quality of hydration heat of both the 25% FA and 30% FA mixes were lower than those of the ternary mixes of FA and GGBS at all the test ages. It can be concluded that the addition of FA is better than GGBS for controlling hydration heat of HPCs.

Furthermore, the addition of 1.1% retarding polycarboxylic-acid SP to the 25% FA mix resulted in not only a large decrease of the release rate and quality of hydration heat, but also a great delay of the occurrence time of exothermic peak, especially a significant reduce of hydration heat at early ages. Accordingly, the composite of FA and set retarding-type SP is very beneficial to the control of temperature gradient and thermal stress of HPCs.

3.3   Adiabatic temperature rise

Figure 1 shows the adiabatic temperature rise development of the concretes at different ages with different SCMs. The temperature rise development of the binary and ternary mixes was consistently lower than that of the control mix at all the ages. In Figure 1, the addition of 25% FA was found to decrease the temperature rise at 7 days by 10.6 °C compared to the control (100% PC). When the 10% FA was replaced with 10% GGBS, the associated decrease in temperature rise was 8.5 °C. Therefore it could be considered that there is more benefit when adding FA than GGBS.

Moreover, there is a rapid growth periods for adiabatic temperature rise between 12 hours and 36 hours, the temperature rise tends to be stable after 36 hours. Thus, the initial 36 hours should be considered as the key periods of temperature control of HPCs, and during HPCs construction the effective measures should be taken to reduce the temperature peak.

3.4   Autogenous shrinkage

Figure 2 shows the autogenous shrinkage development at different ages with different SCMs. The autogenous shrinkage values of all the HPCs develop rapidly in the initial periods of 7 days, and become stable when up to 28 day ages. The partial replacement of PC with 25% FA or both 15% FA and 10% GGBS inhibited the autogenous shrinkage ratio at all the ages. The FA25 mix showed a reduce in autogenous shrinkage ratio by 14.2% and 10.9% at 7 days

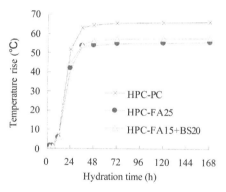

Figure 1.   Adiabatic temperature rise of binary and ternary mixes containing FA and GGBS.

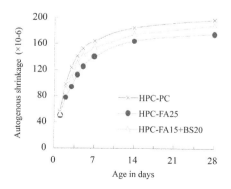

Figure 2.   Autogenous shrinkage of binary and ternary mixes containing FA and GGBS.

and 28 days compared to the control PC mix. Improvements in autogenous shrinkage of concrete mixes containing FA can be explained by the low reactivity rate of FA in at early ages. The decrease of the early hydration speed, equivalent to increasing effective water-cement ratio (w/c), can reduce the rate and degree of self-desiccation inside the concrete (An, 2001), resulting in a reduce of autogenous shrinkage of the binder pastes.

3.5 Plastic shrinkage cracking test

Table 4 shows the crack propagation parameters of three concrete mixes in the wind condition in the end of test at 24 hours. Compared with the control mix, the FA binary mix reduced 66.6% in the crack numbers of unit area, and decreased 77.5% in the average crack area, and the level of crack resistance is increased to II grade. However, the inclusion of GGBS in the FA binary mixes considerably reduced the crack resistance. Hence, there is a increased risk of plastic cracking of HPCs with addition of GGBS.

3.6 Temperature-stress cracking test

The temperature-stress machine test mainly yielded five parameters, viz. the maximum temperature rise, the 2nd zero stress temperature, room temperature stress, cracking temperature and cracking stress. These are reported in Table 5. It can be seen from Table 6 that the temperature rise of hydration heat, the room temperature stress and the cracking temperature of the FA25 mix are lower than those of the control mix, and the stress reserve is higher. Improvements

Table 4. Plastic shrinkage cracking parameters.

Mix ID	Area of cracks (mm²)	Number of cracks (N)	Maximum width of cracks (mm)	Average crack area (mm²/N)	Number of cracks per unit area (N/m²)	Total crack area per unit area (mm²/m²)	Crack evaluation grade
HPC-PC	53.80	12	0.20	2.24	33	73.98	III
HPC-FA25	12.10	4	0.06	1.51	11	16.64	II
HPC-FA15 + BS15	98.00	17	0.10	2.88	47	135.47	III

Table 5. Cracking parameters by temperature and stress test machine.

Cracking parameters	Mix ID		
	HPC-PC	HPC-FA25	HPC-FA15 + BS15
Temperature of fresh concrete (°C)	26.5	27.2	27.4
The first zero stress temperature (°C)	34	27.9	30.2
Maximum compressive stress (MPa)	0.324	0.274	0.412
Temperature at maximum compressive stress (°C)	52.5	50.2	53.6
Maximum expansion value (μm)	46.6	100.7	118.1
Maximum temperature (°C)	63.2	56.1	58.6
Temperature rise (°C)	36.7	28.9	31.2
Time for temperature rise (h)	16.48	25.57	26.34
The second zero stress temperature (°C)	55.6	51.8	47.8
Room temperature stress (MPa)	−2.29	−1.73	−1.7
Maximum shrinkage value (μm)	145.2	101.8	97.5
Cracking stress (MPa)	−2.95	−2.76	−2.87
Cracking temperature (°C)	14.8	8.7	6.5
Stress reserves (%)	22.4	37.3	40.8

in cracking tendency of the concrete mix containing 25% FA can be explained by two effects due to the incorporation of FA, viz. low hydration heat and less shrinkage. The addition of FA reduced the rate and quality of hydration heat and increased the concrete ductility, leading to a decrease of the second zero stress temperature. The inclusion of FA lowered the concrete autogenous shrinkage and dry shrinkage as well as tensile stress due to temperature gradient, resulting in a decrease of the room temperature stress. Although the development of early tensile strength of the FA mix is slower, its crack resistance is still perform better.

Compared with the FA binary mix, although the ternary mix of FA and GGBS had higher temperature rise of hydration heat, there was a more apparent reduce in the internal tensile stress generated by temperature shrinkage, autogenous shrinkage and dry shrinkage, which resulted in the decreases of the 2nd zero stress temperature and the room temperature stress. In addition, the ternary mix of FA and GGBS exhibited rapid development of the tensile strength and high cracking stress. Hence, the ternary mix of FA and GGBS exhibited lower cracking temperature and higher stress reserve, resulting in a slightly better crack resistance.

4   CONCLUSIONS

On the basis of the results obtained from this research work, the following conclusions have been drawn:

1. With the w/b kept constant at 0.305, the compressive strength was detrimentally affected by the replacement of PC with both FA and GGBS at all ages. However, workability and resistance of chloride diffusion were improved significantly. The ternary mixes of FA and GGBS performed better for strength and resistance of chloride diffusion than the FA binary mixes.
2. A comprehensive evaluation system was introduced to determine the early cracking tendency of box girder HPCs with inclusion of SCMs by tracking the development of thermal, physical and deformation properties, which included hydration heat test, adiabatic temperature rise test, autogenous shrinkage test and plastic shrinkage test as well as Temperature and Stress Machine test, et al.
3. There was a decrease in hydration heat at all ages for all the HPC mixes with SCMs, and FA was more pronounced in decreasing the hydration heat than GGBS. In addition, the inclusion of set retarding-type polycarboxylate superplasticizer further decreased the hydration heat release rate and could delay the occurrence time of exothermic peak.
4. The results of the tests carried out on the HPC concretes containing FA and GGBS clearly illustrated that the FA binary mix is superior to the ternary mix of FA and GGBS for various cracking index presented in this paper. The addition 25% FA was more pronounced in decreasing the adiabatic temperature rise, autogenous shrinkage ratio and plastic shrinkage cracking of concrete than the composite addition of 15% FA and 10% GGBS. Both the FA binary mix and the ternary mix of FA and GGBS can reduce cracking temperature and increase stress reserve of the bridge HPCs, but the latter showed the better performance in improving the cracking temperature and cracking stress.

ACKNOWLEDGEMENTS

The financial supports under Natural Science Foundation of China (grant number 51372185) and the Science and Technology Planning Item of Transportation Department of Jiangxi Province (grant number 2010C00004) are gratefully acknowledged.

REFERENCES

An Mingzhe, Zhu Jinquan and Qin Weizu. Measures to restrain the autogenous shrinkage of high performance concrete. *Concrete* 2001; (5):41–46.

Bentz D.P and Jensen O.M. Mitigation strategies for autogenous shrinkage cracking. *Cement and Concrete Research* 2004; 26(6): 603–762.

Bentz D.P and Peltz M.A. Reducing thermal and autogenous shrinkage contributions to early-age cracking. *ACI Materials Journal* 2008; 105(4): 414–420.

Bentur A. and Kolver K. Evaluation of early age cracking characteristics in cementitious systems. *Materials and Structures* 2003; 36(4): 183–190.

Elahi A, Basheer P.A.M, Nanukuttan S.V. and Khan Q.U. ZM. Mechanical and durability properties of high performance concretes containing supplementary cementitious materials. *Construction and Building Materials* 2010; 24 (3): 292–299.

Haque M.N and Kyaali O. Properties of high-strength concrete using fine fly ash. *Cement and Concrete Research* 1998; 28(10):1445–1452.

Holt E and Leivo M. Cracking risks associated with early age shrinkage. *Cement and Concrete Composites* 2008; 23(2): 263–267.

Hu Shuguang, Chen Jing and Zou Zhifeng. Control system of uniaxial restrained and adjusted temperature-stress testing machine. *Journal of Wuhan University of Technology* 2007; 29(1):55–57, 61.

Jiang Zhengwu, Sun Zhenping and Wang Peimi. Study on self-desiccation effect of high performance concrete. *Journal of Building Materials* 2004; 7(1):19–24.

Qin Weizu. Shrinkage and cracking of concrete and its evaluation and prevention. *Concrete* 2001; (7):3–7.

Soroushian P, Mirza F and Alhozaimy A. Plastic shrinkage cracking of polypropylene fiber reinforced concrete. *ACI Materials Journal* 1993; 92(5):553–560.

Zhang Shihai, Qin Weizu and Zhang Tao. Evaluation on cracking performance of concrete at early age-the advances of uniaxial restrained test method. *China Concerete and Cement Products* 2002; (3):13–16.

Zhang Hua, Liu Zhao and Wang Libo. Discussion on temperature crack control of concrete box girder. *Construction Technology* 2006; 35(7):112–113.

Zhao Qilin, Zhou Wangjin and Jiang Kebing. Analysis and repair of soleplate cracks of pre-stressed concrete box girder during construction. *Journal of Highway and Transportation Research and Developmen* 2006; 23(6), 85–88.

Zhu Hanhua, Chen Mengchong and Yuan Yingjie. *Cracks Analyses and Control of Prestressed Concrete Continuous Box-girder Bridge.* Beijing: the Chinese people's traffic press; 2006. p37–84.

Hydraulic Engineering III – Xie (Ed)
© 2015 Taylor & Francis Group, London, ISBN 978-1-138-02743-5

Elementary physical model to study the monogranular cohesive materials

V. Pasquino
Department of Hydraulic Engineering, University of Naples Federico II, Italy

E. Ricciardi
Department of Structures for Engineering and Architecture, University of Naples Federico II, Italy

ABSTRACT: A well-know model good for cohesionless monogranular materials is considered: it is provided with inter-central elasto-plastic braces simulating, in harmony with the physical behavior, the cohesive brace. Such a model, already examined for an initial dynamic study of the monogranular cohesive media under micro-earthquakes, is more profoundly studied as regards Mohr's curve and the curve of the heaviest stresses; also post-critical behaviours for big deformations are deduced, from which it follows that the material is, in such a field, prevalently unstable, since the diagram $\varphi - \tau$ presents a negative gradient through the whole interval $[\varphi(\tau_{max}); 60°]$.

1 INTRODUCTION

A cohesive soil constituted by elements whose dimensions aren't very variable is considered; idealizing this soil in a set of spherical elements, joined by elasto-plastic braces, a very acceptable Mohr's curve is drawn. The model permits to obtain the disgregation's bond between the frequency and the amplitude of a sinusoidal given motion, i.e. the conditions under which the cohesion disappears.

The soil is idealized as a set of spherical elements having the same radius R and the same unit weight, touching one another in horizontal strata. The center C of the generical sphere (Fig. 1), and the centers A and B of the spheres on which it rests, lie in a vertical plane. Considering the set of spheres contained between two parallel vertical planes, whose distance is 2 R, the cartesian orthogonal axes (x, y, z) are fixed so that x is vertical and downward directed, and the centers of the spheres contained in the plane (y, z).

The static range can be examined fixing the spheres A and B, and charging the sphere C by a radial vertical load *N* and by a radial horizontal load T parallel to z axis. The system is plane, and the C point is obliged on two circular lines, whose radius in *R*, and centers are B if $\varphi > 0$, A if $\psi > 0$; we have a one freedom's degree system, but the line on which C is obliged

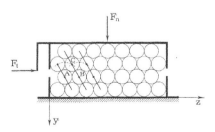

Figure 1. Geometrical model.

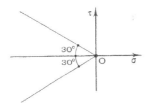

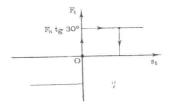

Figure 2. Coulomb's bilateral.   Figure 3. Rigid-plastic behavior.

isn't regular in $\varphi = \psi = 0$. Let N be fixed and T increasing from zero in $R+$; if the surface are smooth, the sphere C doesn't move until:

$$-T R\cos 30° + N R sen 30° \geq 0 \tag{1}$$

That is until $T \leq N \, tg30°$. If $T = N \, tg30°$, the position $\varphi = 0$ is a position of equilibrium, bit this is unstable; if a little displacement in induced, the sphere C steps over the sphere B and all be spheres of the same stratum. According that we can write $\sigma = N/4R^2$, $\tau = T/4R^2$, the Coulomb's bilateral is reported in Figure 2 and the diagram τ, γ, for a fixed value of σ, shows a classic rigid-plastic behavior (Fig. 3).

2   MOHR'S CURVE IN PRESENCE OF COHESION

The presence of cohesion c can be simulated, in the model reported in Figure 4, by means of the presence of two elastic braces connecting the center C to the two lower centers A and B (Fig. 5); let the behavior of these braces be ideally elastoplastic, with elastic stiffness k and elastic strain limit ε_l.

Depending on φ, we can write (Fig. 4), for $\varphi \in [0,60°]$,

$$v_c = -2R\sin(60° + \varphi) + 2R\sin 60° = -2R[0,5\sin\varphi - 0,866(1 - \cos\varphi)] \tag{2}$$

$$w_c = 2R\cos 60° - 2R\cos(60° + \varphi) = 2R[0,5(1 - \cos\varphi) + 0,866\sin\varphi] \tag{3}$$

See too

$$\overline{AC'} - \overline{AC} = R[\{8(1 - 0,5\cos\varphi + 0,866\sin\varphi)\}^{0.5} - 2] \tag{4}$$

The strain energy in the brace AC is

$$L = 0.5k(\overline{AC''} - \overline{AC})^2$$

and the total potential energy is given by

$$\begin{aligned} E = &\, 2NR[0,5\sin\varphi - 0,866(1 - \cos\varphi)] - 2TR[0,5(1 - \cos\varphi) + 0,866\sin\varphi] \\ &+ 0.5kR^2[8(1 - + 0,5\cos\varphi + 0,866\sin\varphi) + 4 - 4\{8(1 - 0,5\cos\varphi + 0,866\sin\varphi)\}^{0.5}] \end{aligned} \tag{5}$$

Thus, the equilibrium condition

$$\frac{dE}{d\varphi} = 0 \tag{6}$$

can be written as

358

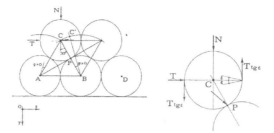

Figure 4.   Geometrical model.

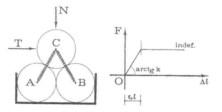

Figure 5.   Geometrical model with elastic braces.

$$N(0,5\cos\varphi-0,866\sin\varphi)-T(0,5\sin\varphi+0,866\cos\varphi)$$

$$+2kR\left[0,5\sin\varphi+0,866\cos\varphi-\frac{0,5\sin\varphi+0,866\cos\varphi}{\sqrt{2(1-0,5\cos\varphi+0,866\sin\varphi)^{\frac{1}{2}}}}\right]=0 \qquad (7)$$

From equation (7) we have

$$T=N\frac{0,5\cos\varphi-0,866\sin\varphi}{0,5\sin\varphi+0,866\cos\varphi}+2kR\left[1-\frac{1}{\sqrt{2(1-0,5\cos\varphi+0,866\sin\varphi)}}\right] \qquad (8)$$

T (in equation 8) is defined on an interval $]0,\ \varphi_1]$ of φ where φ_1 corresponds to the expansion limit ε_l. From equation (4), it is

$$\varepsilon_l=\frac{R\left[\sqrt{2(1-0,5\cos\varphi_1+0,866\sin\varphi_1)}-1\right]}{R} \qquad (9)$$

from which

$$0,5\cos\varphi_1-0,866\sin\varphi_1=1-0,5(\varepsilon_l+1)^2 \qquad (10)$$

For $\varepsilon_l=0,2\rightarrow\varphi_1=13,5°$; for $\varepsilon_l=0,3\rightarrow\varphi_1=21°$.
On this interval, the following cases can be possible (Fig. 7):

a. $T(\varphi)$ has a local maximum at an interior point φ_m; $T(\varphi_m)$ is the limit value T_0 for T, the equilibrium is stable for $\varphi\in[0,\ \varphi_m[$, unstable for $\varphi\in]\varphi_m,\ \varphi_1]$;
b. $T(\varphi)$ is strictly increasing and has a maximum at $\varphi=\varphi_1$; the equilibrium is always stable and $T(\varphi_1)$ is the limit value T_0 for T, reached when the brace is plasticized;
c. $T(\varphi)$ is strictly decreasing and has a maximum at $\varphi=0$; the equilibrium is always unstable on $[0,\ \varphi_1]$ and $T(0)$ is the limit value T_0 for T.

Let us observe that for $\varphi=0$ the equilibrium values for T are all those included on the interval $[0,\ T(0)]$ but only $T(0)$ satisfies (7), because this condition is verified in $]0,\ \varphi_1]$.

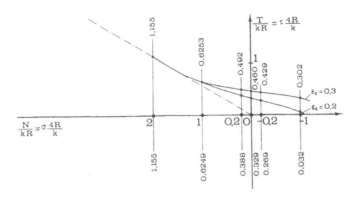

Figure 6.   T-N diagram.

Case c) occurs for high values of N with respect to $2\,kR$; case b) occurs for small values of N. Figure 6 shows the values assumed by T in function of N, in the two hypotheses of $\varepsilon_l = 0,2$ and $\varepsilon_l = 0,3$; we can observe that, for $N/kR > 2$, the effect of cohesion is not felt anymore, and the curve coincides with the 30° sloped line with respect to the N axis.

Equation (8), for $N = 0$ (case b) gives:

$$T_0 = T_\infty = 2kR \cdot \left[1 - \frac{1}{\sqrt{2(1 - 0,5\cos\varphi_1 + 0,866\sin\varphi_1)}} \right] \tag{11}$$

Since

$$c = \frac{T_\infty}{4R^2} \tag{12}$$

we have, for $\varepsilon_l = 0,2$ ($\varphi_1 = 13,5°$)

$$T_\infty = 0,329kR; \quad c = 0,0822\frac{k}{R}; \quad k = 12,17cR$$

and for $\varepsilon_l = 0,3$ ($\varphi_1 = 21°$)

$$T_\infty = 0,460\,kR; \quad c = 0,115\frac{k}{R}; \quad k = 8,69cR \tag{13}$$

Thus, from the values of cohesion c and of the predominant particle size R and assuming an ε_l, we can obtain the value for the stiffness k. Let us here observe that, multiplying the coordinates of the diagram in Figure 4 by $k/4\,R$, we obtain the (σ, τ) diagram.

3   DEGRADATION BOND

In order to simulate the earthquake and its effects on cohesion, let us study the sketch in Figure 7, where the three-spherical model lies on a shaking table. As all the mass is involved in the movement of the soil, we must assume that, impositions which are different from those of rest, the action of the strata located above the level in question is felt by means of a force N directed along $C'B$ if $\varphi > 0$, along $C'A$ if $\varphi < 0$. Hence, the potential energy of N does not vary during the motion.

Let us choose, as Lagrangian coordinate, the angle CBC' if clockwise ($\varphi > 0$) while the angle CAC' if counterclockwise ($\varphi < 0$); the value of φ and its sign define the position of the

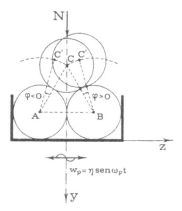

Figure 7. Three-spherical model on a shaking table.

upper sphere with respect to two lower ones. Let the table be subject to a sinusoidal motion directed along z axis:

$$w_p = \eta \sin \omega_p t \tag{14}$$

for point C, we can write:

$$v = R(-\varphi + 0,866\varphi^2)$$
$$w = R(1,732\varphi + 0,5\varphi^2) + \eta \sin \omega_p t$$

whence

$$\dot{v} = -R\dot{\varphi} + 1,732\varphi\dot{\varphi}$$
$$\dot{w} = 1,732R\dot{\varphi} + \varphi\dot{\varphi} + \eta\omega_p \cos \omega_p t$$

Therefore, neglecting the terms of higher order, we obtain:

$$E = 1.5k\varphi^2 R^2$$
$$C = \frac{m}{2}(\dot{v}^2 + \dot{w}^2) = \frac{m}{2}\left(4R^2\dot{\varphi}^2 + \eta^2\omega_p^2 \cos^2 \omega_p t\right)$$

Thus, we obtain

$$\frac{d}{dt}\frac{\partial E}{\partial \varphi} = \frac{m}{2}(8R^2\dot{\varphi} + 1,732\eta\omega_p \cos(\omega_p t)) = 4R^2 m\ddot{\varphi} - 0,866 Rm\eta\omega_p^2 \sin \omega_p t$$
$$\frac{\partial U}{\partial \varphi} = -\frac{\partial E}{\partial \varphi} = -3kR^2\varphi$$

which can be rearranged to obtain the equation of motion, as follows

$$4R^2 m\ddot{\varphi} + + 3kR^2\varphi = 0,866 Rm\eta\omega_p^2 \sin \omega_p t \tag{15}$$

Equation (15) can be written as

$$\ddot{\varphi} + \omega^2\varphi = a\sin \omega_p t \tag{16}$$

where

$$\omega^2 = 0,75\frac{k}{m} \tag{17}$$

361

$$a = 0{,}2165 \frac{\eta \omega_p^2}{R} \tag{18}$$

A particular integral for (16) is

$$\ddot{\varphi} = \frac{a}{\omega^2 - \omega_p^2} \sin \omega_p t$$

hence its complete integral is

$$\varphi = A \sin \omega t + B \cos \omega t + \frac{a}{\omega^2 - \omega_p^2} \sin \omega_p t.$$

From the conditions

$$t = 0 \rightarrow \varphi = \dot{\varphi} = 0$$

we obtain

$$B = 0$$

$$A\omega + a \frac{\omega_p}{\omega^2 - \omega_p^2} = 0$$

whence

$$A = -a \frac{\omega_p}{\omega} \cdot \frac{1}{\omega^2 - \omega_p^2}.$$

therefore it is

$$\varphi = -\frac{a}{\omega^2 - \omega_p^2} \left(\frac{\omega_p}{\omega} \sin \omega t - \sin \omega_p t \right) \tag{19}$$

The motion is the sum of two harmonic motions, whose angular frequency are ω and ω_p. From equations:

$$k = \alpha c R$$

$$m = \frac{\gamma}{g} \cdot \frac{4}{3} \pi R^3$$

we obtain

$$\omega^2 = 0{,}179 \frac{\alpha c g}{\gamma R^2} \tag{20}$$

Equation (19) shows that φ can reach the value

$$\varphi_m = \frac{a}{\omega^2 - \omega_p^2} \tag{21}$$

by equating this value to φ_1 we obtain, for every ω_p, the value of η that nullifies cohesion, i.e. the disaggregation bond. This bond is therefore given by equations (18) and (20)

Table 1. Experimental results.

f_p (sec^{-1})	ω_p (sec^{-1})	η (cm)
1	6,28	3686,88
2	12,57	920,90
3	18,85	408,68
4	25,13	229,40
5	31,41	146,42
10	62,83	35,78
20	125,66	8,12
30	188,49	3,00
40	251,33	1,22
50	314,15	0,38
	365,46	0
	366	−0,003

$$\varphi_1 = 0,2165\frac{\eta\omega_p^2}{R}\cdot\frac{1}{0,179\dfrac{\alpha c g}{\gamma R^2}-\omega_p^2}=\frac{0,2165\eta\omega_p^2\gamma R}{0,179\alpha c g-\gamma R^2\omega_p^2}$$

whence

$$\eta = \varphi_1\frac{0,179\alpha c g-\gamma R^2\omega_p^2}{0,2165\gamma R\omega_p^2}\qquad(22)$$

For $\varepsilon_1 = 0,2$ $\varphi_1 = 13,5° = 0,236$ rad and $\alpha = 12,17$.

The graph of the function $\omega_p \rightarrow \eta$ (equation 22) is shown in the table with the following data:

$c = 0,1$ kg cm^{-2}; $\gamma = 0,0016$ kg cm^{-3}; $R = 1$ cm; $g = 981$ cm sec^{-2}

We can immediately observe that the proper harmonic of the earthquake cannot affect the cohesion since it is accompanied by values of η which are much lower than those needed to verify equation (25). Instead, the danger may exist for microseisms, characterized by very high frequencies and by displacements of the order of millimeters, which can satisfy equation (25).

4 CONCLUSIONS

The proposed model provides Mohr's curves which are very close to the real ones. In particular, all these curves have an asymptote at Coulomb's line, passing through origin and inclined at 30°, in accordance with recent results (Leonards, 1962), (Hess and Stoll, 1966), (Wai-Fah Chen, 1975), (Adriani, Franciosi, Pasquino, 1980). For a complete variational formulation, see e.g. (De Anglis, 2000) and (De Angelis, 2007). For analysis of time-dependent material behavior see e.g. (De Angelis, 2012) and (De Angelis, 2013). For dynamic nonlinear considerations see among others (Cancellara and De Angelis, 2012). For a nonlinear analysis under building effects see e.g. (Cancellara and De Angelis, 2012).

I was thus led to study, through this model, the behavior of a coherent soil under the effect of earthquakes and microseisms, also driven by a purely mechanical interest in such a model, whose constraint curve is non-regular at a point. The danger that the microseisms might frustrate the cohesive bond has therefore been confirmed, at least from a qualitative point of view, and we have tried to quantify, for monogranular soils, the range of danger in the amplitude-frequency plane. Such danger can have a particular weight in all those cases in which long-lasting and uniform vibrations (traffic, motors, mechanical engines, etc.) affect the foundations. Experimental tests, necessary to confirm or deny equation (40), are being organized at the Department of Structures for Engineering and Architecture using contributions provided by the National Research Council (CNR); a shaking table of modest size will be used, subject to multi-directional pulses whose frequency and amplitude vary with continuity. If there is an experimental confirmation, it will be my intention to study, on a similar model to the proposed one, the disaggregation bond for mortars, given the actuality of this problem in relation to the damages of old masonry buildings due to the increase of traffic in town centers and, above all, to the construction of subways.

REFERENCES

Adriani L., Franciosi V., Pasquino M. (1980)—Un modello fisico per lo studio degli agglomerati monogranulari coesivi (*in italian*). Proceedings of the 5th AIMETA congress, Palermo.

Cancellara D., De Angelis F. (2012)—A nonlinear analysis for the retrofitting of a RC existing building by increasing the cross sections of the columns and accounting for the influence of the confined concrete, Applied Mechanics and Materials, Vol. 204–208, pp. 3604–3616.

Cancellara D., De Angelis F. (2012)—Dynamic nonlinear analysis of an hybrid base isolation system with viscous dampers and friction sliders in parallel, Applied Mechanics and Materials, Vol. 234, pp. 96–101.

Cancellara D., De Angelis F. (2012)—Hybrid base isolation system with friction sliders and viscous dampers in parallel: comparative dynamic nonlinear analysis with traditional fixed base structure, Advanced Materials Research, Vols. 594–597, pp. 1771–1782.

Cancellara D., De Angelis F. (2012)—Seismic analysis and comparison of different base isolation systems for a multi-storey RC building with irregularities in plan, Advanced Materials Research, Vol. 594–597, pp. 1788–1799.

Cancellara D., De Angelis F. (2012)—Steel braces in series with hysteretic dampers for reducing the seismic vulnerability of RC existing buildings: assessment and retrofitting with a non-linear model, Applied Mechanics and Materials, Vol. 204–208, pp. 2677–2689.

De Angelis F. (2000)—An internal variable variational formulation of viscoplasticity, Computer Methods in Applied Mechanics and Engineering, Vol. 190, Nos. 1–2, 35–54.

De Angelis F. (2007)—A variationally consistent formulation of nonlocal plasticity, Int. Journal for Multiscale Computational Engineering, Vol. 5 (2), pp. 105–116, New York.

De Angelis F. (2012)—A comparative analysis of linear and nonlinear kinematic hardening rules in computational elastoplasticity, Technische Mechanik, Vol. 32 (2–5), pp. 164–173.

De Angelis F. (2012)—On the structural response of elasto/viscoplastic materials subject to time-dependent loadings, Structural Durability & Health Monitoring, Vol. 8, No. 4, pp. 341–358.

De Angelis F. (2013)—Computational issues and numerical applications in rate-dependent plasticity, Advanced Science Letters, Vol. 19, Number 8, pp. 2359–2362.

Hess M.S., Stoll R.D. (1966)—Interparticle Sliding in granular materials. Columbia University, Burm. Lab. Soil Mech. Res. Rep. n. 1.

Leonards G.A. (1962)—Foundation Engineering, Mc Graw Hill, Book Company New York.

Wai-Fah Chen (1975)—Limit Analysis and Soil Plasticity (p. 27). Elsevier Publishing Company, Amsterdam.

Hydraulic Engineering III – Xie (Ed)
© 2015 Taylor & Francis Group, London, ISBN 978-1-138-02743-5

Connection design of Concrete-Filled Twin Steel Tubes column

Y.F. Zhang
Department of Civil Engineering, North China University of Technology, Beijing, P.R. China

Z.Q. Zhang
School of Civil Engineering, Chang'an University, Xi'an, P.R. China

W.H. Wang
College of Civil Engineering, Huaqiao University, Xiamen, P.R. China

ABSTRACT: This paper concentrates on Concrete-Filled Twin Steel Tubes (CFTST) columns with an ideal section of inner circular tube and outer square tube. The column can take advantage of not only the good performance of circular steel tubes to confine concrete well but also the convenient connections between square steel tubes and beams. Two types of connections were introduced between CFTST column and steel beams, reinforced concrete beams respectively. The formation and feature of these connections were described. Then some experimental data including typical failure modes, load-carrying capacity, ductility, etc., were shown in order to express the good aseismatic behavior of these two types of connections. The behaviors influenced by the main connection components were also analyzed. Both these connection systems can easily achieve the anti-seismic design principle, and could be applied to CFTST structure especially in seismic region.

1 INTRODUCTION

Concrete-Filled Steel Tubular (CFST) has been applied widely in civil engineering for many years. Up to today, there have been a number of studies conducted to investigate the behavior of CFST columns (Ou et al. 2011). The main benefit of using CFST is that it utilizes the advantages of both steel and concrete, viz. steel members have high tensile strength and ductility, whereas concrete members are advantageous in compressive strength and stiffness (Roeder et al. 2010). With the development of CFST structure, recently there occurs Concrete-Filled Twin Steel Tubes (CFTST) column, which consists of an inner tube and an outer tube fully filled with concrete.

2 CFTST COLUMNS

CFTST column is a typical member in the entire Concrete-Filled Steel Tubes (CFST) family. As its cross-section is shown in Figure 1, the column can take advantage of not only the

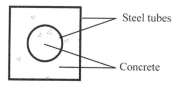

Figure 1. Cross section of the column.

good performance of circular steel tubes to confine concrete but also the convenient connection between square steel tubes and beams (Zhang et al. 2010). Similar to common CFSTs, CFTST columns exhibit excellent structural and constructional benefits. Most important is that this type of columns has good fire resistance. Test results (Zhang et al. 2013; Cai and Jiao 1997) suggest that CFTSTs exhibit better seismic resistance than common CFSTs, particularly when subjected to high axial compressive loads. Furthermore, this kind of columns, without needing to consider the limit of axial compression force ratio, can be used in large span bridges and columns of ultra-high-rise buildings. For example, this composite column with no more than 600 mm in diameter has been used in Wupertal city building in German. It resolved the problem of one fire-resisting and overloading column with the axial capacity of 8000 kN load.

It has a theoretical significance and application value to adopt CFTST column in practice, so the beam-column connection should be solved firstly. Much research (Han and Li 2010; Nie et al. 2008) has been conducted in order to better understand the behavior of CFST moment frames, as well as to develop new connection types that demonstrate adequate structural performance under seismic loading conditions; however, very little research has been done to investigate the connection of CFTST structure. This paper presented two new types of connections between the CFTST columns and RC beams, steel beams respectively.

3 CONNECTED WITH RC BEAM

3.1 *Detail of the connection system*

The ring-beam joint between CFST column and RC beam is typically through-beam connections, the beams directly passing through the panel zone because of the discontinuous outer tube. The connection details are shown in Figure 2. In the ring-beam joint with discontinuous outer tube, the inner tube can be continuous to provide the column with enough load bearing capacity and stability, while the outer tube is interrupted and multiple lateral and vertical reinforcements, making up a steel cage, are used to confine the core concrete in the joint zone. Vertical reinforcements are inserted through the outer tube with enough anchorage length in the concrete above and below the joint. Shear studs are welded on the inner wall of the discontinued outer tubes close to the joint zone to ensure a more uniform transfer of stresses between the concrete and the outer steel tube. Since the outer tube was interrupted at the joint area, there was expected stress concentration at the vicinity of the interruption. In order to ensure the enough bonding between the outer tube and the concrete, shear studs were welded upon the inner wall of the outer tube.

Because the inner circular tube has nearly the same width with the RC beam, the beam's reinforcements can wrap around the continuous inner tube to achieve the through-beam joint. To ensure the CFTST column's continuity at the joint, an octagonal ring beam with multi-layer annular reinforcements and radiation stirrups is located outside the joint core to improve the compression area of the column. The key of this new connection system is that the continuous RC beams are achieved by the interruption of the outer steel tube. The transferring of moments and shear forces can be ensured by the longitudinal reinforcements of the RC beam; and the reduced stiffness of the column due to the interruption of the outer steel tube can be compensated by the confinement of the steel cage anchored inside the joint zone and the improvement on the compression area of the octagonal ring beam located outside the column. The outer concrete of the composite column and the concrete of the ring beam form one united entity to provide the joint with good integrity and continuity.

3.2 *Test results*

The behavior of this joint under seismic loads was studied through testing four beam-column assemblage specimens. The changing parameters include different transverse reinforcement ratios for the ring beam and different diameter for the inner tube. The tests were used to comprehend the force transfer mechanisms between the CFTST columns and RC beams

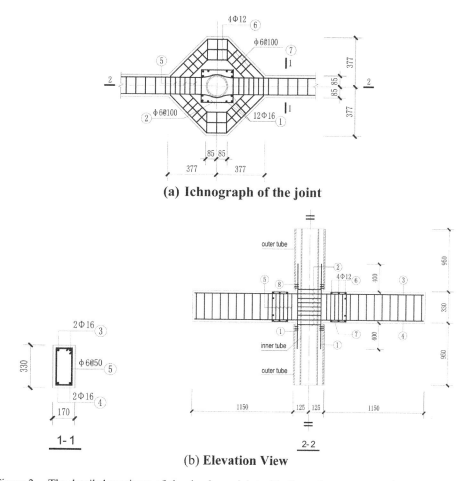

(a) Ichnograph of the joint

2Φ16 ③

Φ6@50 ⑤

2Φ16 ④

1-1

2-2

(b) Elevation View

Figure 2.   The detailed specimen of the ring beam joint with discontinuous outer tube.

(Zhang et al. 2012). In general, according to the design criteria of beam failures, the plastic hinge formed as expected in the beam end near the connection, and the design requirement of "strong joints" could be achieved. Based on these investigations, the conclusions are summarized as follows:

1. The connections belong to the through-beam connection family in which the outer steel tube is interrupted in order to make the longitudinal reinforcement of RC beams continuous. The beam moment and shear force can easily be transmitted by the RC beam. Details such as the vertical reinforcements, welded shear studs, and ring beam ensured the integrity of the joint to compensate the decreased stiffness of the composite column due to the interruption of the outer steel tube. Meanwhile, the continuous inner CFST played an important role in the rigidity and integrity of the joint, especially the shear resistance. This through-beam detail provided the most effective detail to achieve the ideal rigid connection condition.

2. The ring beam joint with discontinuous outer tube, when designed with suitable parameters, performed quite well in carrying the cyclic loads as they could generally attain their strength and deformation capacity. because the moment was mainly transmitted through the vertical rebars in this connection, more reinforcement with a sufficient anchorage length can achieve a high load capacity.

3. The reinforcement ratio of the ring beam showed an obvious influence on the load carrying capacity of the joints, and the ductility at the maximum load was improved with the improvement of the inner tube. Meanwhile, the anchorage length l_a and cross-section ratio

367

P_v of the vertical reinforcement are the critical parameters in the behavior analysis of ring beam joints with a discontinuous outer tube.

4 CONNECTED WITH STEEL BEAM

4.1 Detail of the connection system

The vertical stiffener joints between CFST column and H-shaped steel beam consists of mainly vertical stiffeners and embedded anchorage plate. The key of this new connection system is that vertical stiffeners are served as the primary force-transferring mechanism. Meanwhile anchorage plate is groove welded on inner steel tube, going through the outer steel tube, and welded and bolted connected with the steel beam web. The connection Ichnography details for CFTST column are shown in Figure 3, where extension of the vertical stiffener is 80 mm. Vertical stiffener is groove welded on the column flange, and its extension was welded to the end plate sides, so the extension is same long as the widened end plate. In order to avoid the stress concentration problem, so the radius-cut section is the used for the end plate. The end plate as wide as the column flange was butt welded to the beam flange.

It can be realized that the part of beam flange near the column flange will weaken the column; thereby the widened end plate induces the location of plastic hinge formation inside of the beam in order to prevent the brittle fracture of the columns. The application of anchorage plate (No. ⑤ in Fig. 4) embedded in the concrete between inner tube and outer tube, can improved workability and globality of the joint. Because the anchorage plate embedded inside the concrete is higher than beam web, anchorage plate can be chosen as two types including with triangle ribs (see SBJ2-1 in Fig. 4(b)) and without ribs (see SBJ1-1 in Fig. 4(a)). As shown in Figure 4(b), in the ribbed connection, the end plate and outer tube are slotted and double bevel groove welded to the ribbed anchorage plate. Therefore, this fully restrained connection is commonly characterized by: (1) vertical stiffener; (2) anchorage plate; and (3) reduced beam section. All continuity plates were welded to columns using partial-penetration double bevel groove welds. Complete-penetration single-bevel groove welds were used to connect the beam flanges to the column flange in all specimens.

4.2 Discussion of test results

The behavior of the new joints under seismic loads was studied through testing six beam-column assemblage specimens (Zhang et al. 2013). The variables in the tested specimens include extension of the vertical stiffener (same length as the widened end plate), anchorage plate with or without ribs, width–thickness ratio of square steel tube, and axial compression ratio. As a result, the capacity of the various elements of the connection detail can be predicted. In general, the plastic hinge formed as expected in the reduced beam section where the end plate was cut as curvature. It is confirmed that the destruction of the joint is plastic hinge damage of the steel beam.

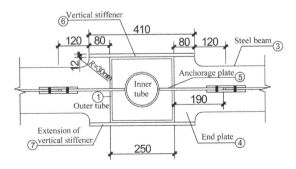

Figure 3. Ichnography details of CFTST column-beam joint.

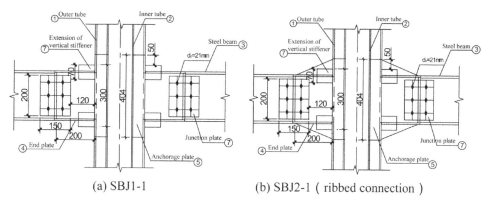

(a) SBJ1-1   (b) SBJ2-1 (ribbed connection)

Figure 4.   Elevation view of CFTST column-beam joint.

According to the tests, initial stiffness and maximum bearing capacity have improved signifi-cantly for the specimens with ribbed anchorage plates. The comparable improvement in load capacity and flexural failure prove that the applicability of ribbed anchorage plate can evidently improve the joint stiffness and strength. In case of the specimens with different vertical stiffener length, the main significant difference exists in the failure mode because the plastic hinge can be removed farther from the surface of the column when the extension is longer. Therefore, the extension of the vertical stiffener can effectively protect the panel zone of the joint. Therefore, it can be found that increasing the extension of vertical stiffener will increase the anti-shear resist-ance of the connection to protect the panel zone. The maximum bearing capacity of the con-nections depends on whether the anchorage plate is ribbed or not, so the best way to improve the bearing capacity of the composite connections is to use ribbed anchorage plates.

Meanwhile, it was also concluded that high quality welding procedures allow a frame speci-men to be exercised well beyond elastic limits in order to obtain sufficient ductility. The compari-son of the composite specimens with different tube thickness suggests that using thicker tube or increasing local thickness of the steel tube can provide some benefit to the joint performance.

5   CONCLUSIONS

CFTST columns exhibit excellent structural, constructional benefits and good fire resistance. Two types of new connections were described in this paper for the steel beam and reinforced concrete beam respectively, which were devised to cater for the seismic requirements of "strong column-weak beam" philosophy.

All the analysis demonstrates that both the new-types of connections have the advantage of transferring load reliably, simple construction details, thus they have wider foreground in CFTST structures especially in seismic region.

ACKNOWLEDGMENTS

The authors would like to acknowledge the support provided by the Chinese National Science Foundation (Grant No. 51478004 and No. 51008027). Meanwhile, the financial support pro-vided by the Beijing Innovation Platform of Scientific research base construction in China is also appreciated.

REFERENCES

Cai S.H. & Jiao Z.S.1997. Behavior and ultimate load analysis of multi-barrel tube-confined concrete columns. *Journal of Building Structures* 18(6):20–25 (in Chinese).

Han L.H. and Li W. 2010. Seismic performance of CFST column to steel beam joint with RC slab: Experiments. *Journal of Constructional Steel Research* 66(11):1374–1386.

Nie J.G., Qin K. and Cai C.S. 2008. Seismic behavior of connections composed of CFSSTCs and steel-concrete composite beams-experimental study. *Journal of Constructional Steel Research* 64(10): 1178–1191.

Zhang Y.F. 2010. Study on axial compressive behavior of the composite CFST and seismic research on RC beam-column connections. Doctor Thesis, Chang'an University, Xi'an, China.

Zhang Y.F., Zhao J.H., Cai C.S. 2012. Seismic behavior of ring beam joints between concrete-filled twin steel tubes columns and reinforced concrete beams. *Engineering Structures* 39(6):1–10.

Zhang Y.F., Zhao J.H. and Yuan W.F. 2013. Study on compressive bearing capacity of concrete filled square steel tube column reinforced by circular steel tube inside. *Journal of Civil Engineering and Management* 19(6):787–795.

Zhang Y.F., Zhang D.F., Zhao J.H. 2013. Experimental study on seismic behavior of connection between composite CFST column and H-shaped steel beam, *Journal of Building Structures* 34(9):40–48 (in Chinese).

Hydraulic Engineering III – Xie (Ed)
© 2015 Taylor & Francis Group, London, ISBN 978-1-138-02743-5

The improved Simplified Bishop's Method considering the Difference of Inter-slice Shearing Force

Nianchun Xu, Wenjing Xia, Baoyun Zhao & Tongqing Wu
*Department of Civil Engineering and Architecture, Chongqing University of Science and Technology,
Chongqing, China*

ABSTRACT: The Bishop's method is widely used in slope stability analysis because its derivation is simple, its concept is clear and the calculated result is easily convergent. The stability factor calculated through traditional Simplified Bishop's Method (SBM) is close to K_J calculated one through Janbu's method, which is one of strictly slice method, for slopes with small dip angle. But the error of the factor gets bigger for steep slopes due to the absence of the Difference of Inter-Slice Shearing Force (DISF). The improved SBM obtains the distribution of shearing stress through finite elements methods for slopes with varying dip angles, uses linear function and quadratic function to describe the average shearing stress's varying under the slope's top face and incline face respectively, develops the formula to calculate the DISF based on the functions, then gets the anti-sliding force contributed by DISF. Using the improved SBM, a stability factor, which in closer to K_J, can be got for slopes with small dip angle, and the stability factor is obviously increased for steep slopes.

1 INTRODUCTION

Limit equilibrium method is still the most widely used method to analyze slopes' stability. It includes Fellenius's method (or Sweden method), Simplified Bishop's Method (SBM), Janbu's method, Sarma's method, Morgenstern & Price's method and Spencer's method, et al. Among these methods, Fellenius's method and SBM are the most popular method in engineering field.

The Fellenius's method ignores any interaction between two slices, and gets the factor of stability directly based on entirely stress analysis. The Bishop's method considers the action of inter-slice vertical shear force and the playing degree of soil's anti-shear strength, get the finally stability factor through iterative algorithm. Although the Bishop's method doesn't meet the force balance in horizontal direction and torque balance, and traditional SBM eventually neglects the difference of inter-slice shearing force, but the SBM have always been praise highly in geotechnical engineering field, the reasons follow as: its derivation is simple, its concept is clear; the calculated result is easily converge and has higher accuracy.

However, when the traditional SBM is applied to analyze steep slopes (the dip angle > 60°), the error of the stability factor is some big. This article will conduct research on this, and propose an improved SBM with consideration of the difference of inter-slice shear force.

2 THE BISHOP'S METHOD TO CALCULATE SLOPES' STABILITY FACTOR

A slope is shown in Figure 1 and the radius of circular sliding surface is R. The sliding mass has been divided into vertical slices. For the slice No. i, the horizontal width is b_i, the underside's dip angle is θ_i, the weight is W_i, the slice No. i−1 acting a horizontal lateral pressure E_i and vertical shear force X_i at the right side, the slice No. i+1 acting a horizontal lateral

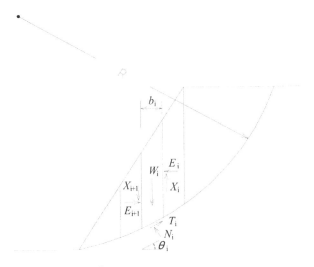

Figure 1.   The forces acting on the slice.

pressure E_{i+1} and vertical shear force X_{i+1} at the left side, at its bottom surface there are normal counter-force N_i and parallel anti-shear force T_i. The sliding surface's anti-shear parameters are c_i (cohesion) and ϕ_i (internal friction angle). Bishop A.W. supposed the playing degree of the soil's anti-shear strength is equal for the entire sliding surface, that is $1/K$, where K is stability factor.

Based on Mohr-Coulomb criterion, Equation (1) can be got.

$$T_i = N_i \tan\phi_i / K + c_i b_i \sec\theta_i / K \tag{1}$$

3   GETTING STARTED

3.1   *The template file*

The force balance in vertical direction can be described as Equation (2).

$$N_i \cos\theta_i + T_i \sin\theta_i = W_i + X_{i+1} - X_i \tag{2}$$

Equation(3), (4) can be got through combination of Equation (1), (2).

$$N_i = (W_i + X_{i+1} - X_i - c_i b_i \tan\theta_i / K)/m_i \tag{3}$$

$$T_i = [(W_i + X_{i+1} - X_i)\tan\phi_i / K) + c_i b_i / K]/m_i \tag{4}$$

where $m_i = \cos\theta_i + \tan\phi_i / K \sin\theta_i$.

Considering the moment bout the center of the sliding surface, for the E_i and X_i exist in pairs (with their counter-force), the sum of their moments is zero, the acting line of N_i pass through the center, so the final moment balance can be written as Equation (5).

$$\sum T_i R = \sum W_i R \sin\theta_i \tag{5}$$

Equation (6) can be obtained if Equation (4) is put into Equation (5).

$$K = \frac{\sum (c_i b_i + (W_i + X_{i+1} - X_i)\tan\phi_i)/m_i}{\sum W_i \sin\theta_i} \tag{6}$$

The traditional SBM neglects the difference of inter-slice shearing force, so Equation (6) can be written as Equation (7).

$$K = \frac{\sum (c_i b_i + W_i \tan \phi_i)/m_i}{\sum W_i \sin \theta_i}$$

(7)

4   THE CALCULATING EXAMPLES USING TRADITIONAL SBM

Using the Equation (7) to calculate the slopes with different dip angle, the stability factors K' can be got as shown in Table 1. In Table 1, K_J (calculated with Janbu's method) and K_F (calculated with Fellenius's method) also are listed. Seen from Table 1, when the slope's dip angle α is small, K' and K_J are very close, but with the increase of the dip angle, the difference of them is increased. When the slope angle is increased to 50°, the difference is obvious. Due to divergency, the value of K_J for slopes with $\alpha \geq 60°$ dip angle cannot be obtained with Janbu's method, but it can be found that $K' < K_F$ when $\alpha \geq 70°$.

The Fellenius's method does not take into account any interaction between the slices, the slices will be considered as isolated bodies, the calculated stability factor K_F will be small, the results are somewhat safe. So we can judge from that: when α is bigger, the stability factor calculated with traditional SBM is too small, and the error is larger. To eliminate the error, the following of this paper will propose an improved SBM which takes the difference of inter-slice shear force into consideration.

5   THE FEM ANALYSIS OF SHEAR STRESS DISTRIBUTION IN SLOPES

To obtain the distribution of shear stress in slopes, the Finite Element Method (FEM) is the best choice, the general-purpose finite element program Ansys10.0 has been used for numerical analysis.

During the finite element analysis, the parameters of slopes be taken as h (height) = 10 m, E (elastic modulus of the soil) = 20 MPa, v (Poisson's ratio) = 0.32, γ (unit weight) = 18 kN/m³, α (dip angle) = 30°–80°. In order to obtain the shear stress distributions under the slope's top face and under the slope's incline face respectively, seven stress paths have been defined in slope, their position shown in Figure 2.

Through FEM analysis, the shear stress S_{xy} along the seven stress paths can be obtained. Dividing S_{xy} with path's length can get the average shear stresses AS_{xy}. AS_{xy} be conducted nondimensionalization through dividing it with γh, then the variety of $AS_{xy}/\gamma h$ with x-axis representing h_i/h under the slope's incline face and variety of $AS_{xy}/\gamma h$ with x-axis representing x_i/h under the slope's top face can be got, as shown in Figure 3 and 4 respectively.

Table 1.   The calculated result of stability factor K' through traditional SBM*.

No.	Height (h/m)	Dip angle (α/°)	T. SBM (K')	Janbu's (K_J)	Fellenius's (K_F)
1	10	20	2.360	2.381	2.223
2	10	30	1.764	1.790	1.670
3	10	40	1.431	1.461	1.368
4	10	50	1.198	1.241	1.166
5	10	60	1.019	—**	1.013
6	10	70	0.864	—**	0.885
7	10	80	0.721	—**	0.767

* $c = 15$ kPa, $\phi = 25°$, $\gamma = 18$ kN/m³.
** means hard to give a precise result due to bad convergences.

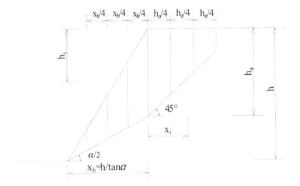

Figure 2. The position of seven stress paths during FEM analysis.

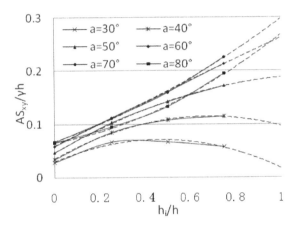

Figure 3. Variety of average shearing stress under the slope's incline face.

Using quadratic function to fit the six curves in Figure 3 respectively, the fitting curves are displayed with broken lines which also be shown in Figure 3. It can be seen that the solid curves and broken lines have a good fitting result. Based on the fitting curves, the distribution equation of average inter-slice shear stress under the slope's incline face can be got which can be described as Equation (8).

$$F\left(\frac{h_i}{h}\right) = aa\left(\frac{h_i}{h}\right)^2 + bb\frac{h_i}{h} + cc \qquad (8)$$

where aa, bb, cc represent the varying coefficients of shearing force, as shown in Table 2.

Seen from the Figure 4, the shear stresses only exist on part of the section. So line function can be used to describe the distribution of the AS_{xy}. The distribution equation of average inter-slice shear stress under the slope's top face shown as Equation (9).

$$\begin{cases} G\left(\dfrac{x_i}{h}\right) = kk\dfrac{x_i}{h} + cc & \dfrac{x_i}{h} \leq -cc/kk \\[2mm] G\left(\dfrac{x_i}{h}\right) = 0 & \dfrac{x_i}{h} > -cc/kk \end{cases} \qquad (9)$$

where cc is identical to cc in Table 2, kk represent the varying range, $-cc/kk$ represent the distributing zone. The cc and kk are shown in Table 3, they jointly describe the distributing characteristic of shear stress under the slope's top face.

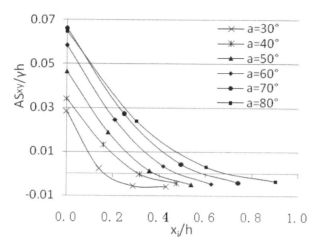

Figure 4.   Variety of average shearing stress under the slope's top face.

Table 2.   Varying coefficients of the shearing force under the slope's incline face.

No.	Dip angle ($\alpha/°$)	Coefficients		
		aa	bb	cc
1	30	0.188	0.175	0.030
2	40	−0.174	0.236	0.035
3	50	−0.101	0.243	0.047
4	60	−0.006	0.209	0.059
5	70	0.083	0.149	0.067
6	80	0.127	0.076	0.065

Table 3.   Characteristic coefficients of the shearing force under the slope's top face.

No.	Dip angle ($\alpha/°$)	Characteristic coefficients	
		kk	−cc/kk
1	30	−0.189	0.159
2	40	−0.113	0.310
3	50	−0.133	0.353
4	60	−0.129	0.457
5	70	−0.110	0.609
6	80	−0.092	0.707

6   THE IMPROVED SBM

Based on Equation (8), the difference of inter-slice shear force under the slope's incline face can be written as Equation (10).

$$X_{i+1} - X_i = F\left(\frac{h_{i+1}}{h}\right) \cdot \gamma h \cdot h_{g(i+1)} - F\left(\frac{h_i}{h}\right) \cdot \gamma h \cdot h_{gi} \tag{10}$$

where h_{gi}, $h_{g(i+1)}$ represent the height of slice No. i and i+1 respectively, as shown in Figure 5.

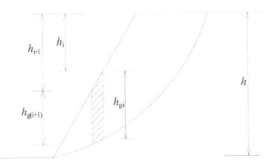

Figure 5. The height of the slice.

Table 4. The stability factors through improved SBM*.

No.	Height (h/m)	Dip angle ($\alpha/°$)	T. SBM (K')	I. SBM (K)	$(K–K')/K' \times 100\%$
1	10	30	1.764	1.772	0.45%
2	10	40	1.431	1.445	0.98%
3	10	50	1.198	1.221	1.92%
4	10	60	1.019	1.049	2.94%
5	10	70	0.864	0.911	5.44%
6	10	80	0.721	0.774	7.35%

* $c = 15$ kPa, $\phi = 25°$, $\gamma = 18$ kN/m³.

Putting Equation (10) into Equation (6), the anti-slide force R_1 can be got which induced by difference of shear force under the slope's incline face. R_1 can be written as Equation (11).

$$R_1 = \sum \left[F\left(\frac{h_{i+1}}{h}\right) \cdot \gamma h \cdot h_{g(i+1)} - F\left(\frac{h_i}{h}\right) \cdot \gamma h \cdot h_{gi} \right] \tan\phi_i / m_i \qquad (11)$$

Based on Equation (9), the difference of inter-slice shear force under the slope's top face can be written as Equation (12).

$$X_{i+1} - X_i = G\left(\frac{x_{i+1}}{h}\right) \cdot \gamma h \cdot h_{g(i+1)} - G\left(\frac{x_{i+1}}{h}\right) \cdot \gamma h \cdot h_{gi} \qquad (12)$$

Putting Equation (12) into Equation (6), the anti-slide force R_2 can be got which induced by difference of shear force under the slope's top face. R_2 can be written as Equation (13).

$$R_2 = \sum \left[G\left(\frac{x_{i+1}}{h}\right) \cdot \gamma h \cdot h_{g(i+1)} - G\left(\frac{x_i}{h}\right) \cdot \gamma h \cdot h_{gi} \right] \tan\phi_i / m_i \qquad (13)$$

R_1 and R_2 added together are resultant anti-slide force induced by difference of shear force, the stability factor K of improved SBM can be calculated finally if the resultant anti-slide force be added into traditional SBM's anti-slide force.

The results of stability factor calculated with improved SBM of the six slopes given by Table 1 are listed in Table 4. It can be seen from Table 4 that the K are greater than K', the bigger the dip angle, the bigger of the increasing magnitude, when $\alpha = 70°$, the increasing magnitude is 5.44%, when $\alpha = 80°$, it is 7.35%.

Comparing the data in Table 4 and Table 1, it can be seen that for slopes of $\alpha = 30–50°$, the value K and K_1 are closer the K' and K_J. For slopes of $\alpha \geq 70°$, the $K > K_F$ in Table 4.

Table 5. The stability factors for steep slopes*.

No.	c (kPa)	ϕ (°)	T. SBM (K')	I. SBM (K)	$(K–K')/K' \times 100\%$
1	10	10	0.548	0.579	5.66%
2	20	10	0.950	0.984	3.58%
3	10	20	0.694	0.742	6.92%
4	20	20	1.099	1.165	6.01%
5	10	30	0.840	0.905	7.74%
6	20	30	1.266	1.352	6.79%
7	10	40	0.996	1.052	5.62%
8	20	40	1.449	1.556	7.38%

*$h = 6$ m, $\alpha = 80°$, $\gamma = 18$ kN/m³.

For steep slopes, with h = 6 m, $\alpha = 80°$, varied anti-shear parameters, the stability factors have been calculated with T. SBM and I. SBM respectively, the results are listed in Table 5.

Seen from Table 5, all of the K have obvious increasement. The increasing magnitude locates in 3.58–7.74%.

7 THE DISCUSSION ON RESEARCH METHOD AND CONCLUSIONS

7.1 The discussion on research method

During FEM analysis, the bottoms of seven stress paths form two straight lines, while the actual sliding face is circular. Although the error is inevitable from this, but there are three elements to guarantee the method is reasonable and the error is acceptable: 1) the position of the dangerous circular sliding surface is adjacent to the two lines and the changing of angle is similar; 2) the FEM analysis is used to obtain the average shear stress but not the resultant of shear force; 3) the anti-slide force R is induced by the difference of inter-slice shear force.

During FEM analysis, the soil's elastic modulus E be taken as 20 MPa. In actually, the value of E just change the slope's deformation, and have no effect on slope's stress distribution.

During FEM analysis, the Poisson's ratio v be taken as 0.32. The value of stresses in slope changing with the value of v, but the variation is very small.

The distribution of shear stress in slopes is mainly decided by dip angle. As the shear stresses have already been conducted nondimensionalization, the data in Tables 2 and 3 can be applied to slopes with any h, E, v and γ. For slopes with layered soil, a comprehensive γh can be calculated based on every layer's γ_i and depth. When there is a uniformly distributed load q on the slopes' top face, the γh in Equation (12), (14) should be replaced by $\gamma h + q$.

7.2 Conclusions

Through the previous analysis and discussion, it can be seen that the improved SBM proposed by this article has substantial theory basis and ideal calculated result. Although there is some limitations in analysis method and parameter selection, but the inducted error is very small. The improved SBM can be used effectively to analyze most slopes.

ACKNOWLEDGEMENTS

This work was funded by the Scientific and Technological Research Program of Chongqing Municipal Education Commission (KJ1401304), by the Chongqing municipal commission of urban-rural development (2011-84) and by Chongqing Research Program of Basic Research and Frontier Technology (cstc2013 jcyjA90004).

REFERENCES

Bishop A.W. 1955. The use of the slip circle in the stability analysis of slopes. *Geotechnique* 5(1):7–17.

Chen Xiaoping. 2008. Soil mechanics and foundation engineering. Beijing: China Hydraulic and hydropower Press.

Dai Z., Shen P. 2002. Numerical solution of simplified Bishop method for stability analysis of soil slopes. *Rock and Soil Mechanics,* 23(6):760–764.

Fellenius W. 1936. Calculation of the stability of earth dams. *Trans 2nd Cong Large Dams* 4–445.

Janbu N. 1973. Slope stability computations, embankment dam engineering. New York: John Wiley and Sons 47–86.

Jiang B., Kang W. 2008. Analytical formulation of Bishop's method for calculating slope stability [J]. *Journal of China University of Mining & Technology* 37(3):287–290.

Zhang L., Zheng Y. 2004. An extension of simplified Bishop method and its application to non-circular slip surface for slope stability analysis [J]. *Rock and Soil Mechanics* 25(6):927–930.

Zhao C. 2008. The discussion on convergence of Janbu's method and the realization of limit equilibrium's generalization [Sci. M. Thesis]. Xi'an: Chang'an University.

Zhu B., Zhou Y., Zhu D. 2003. A modified algorithm for calculating slope stability using Janbu's generalized procedure of slices [J]. *Journal of Disaster Prevention and Mitigation Engineering* 23(4):56–60.

Zhu D., Deng J., Tai J. 2007. Theoretical verification of rigorous nature of simplified Bishop method. *Chinese Journal of Rock Mechanics and Engineering* 26(3):455–458.

Hydraulic Engineering III – Xie (Ed)

Study on effects of straight arch with two articulations structure system caused by knee

Wenyi Cui, Zhixin Zhang & Junchao Huang
Civil Engineering of Yanbian University, Yanji, Jilin Province, China

ABSTRACT: Previous studies found that when straight arch with two articulations bears loads, the turning point formed a complex stress state, which made the mechanical properties of the straight arch change greatly. In this paper, mechanical analysis was developed on both ends fixed hinge support straight arch, using the method of plane link system analysis and three-dimensional finite element analysis, which both use ANSYS finite element analysis software. By analyzing the different mechanical data derived from the two methods this paper proposed a theory of integrated effects of straight arch with two articulations mechanical structure system caused by break point.

1 INTRODUCTION

Straight arch with two articulations is supported on reinforced concrete pillars or reinforced concrete ring beam of architecture; it is a new form of roof beam structure with the development of light steel roof structure system. Currently, this form of structure is generally applied in the light steel roof structure system, but the current code for design of steel structures and regulations has not been clearly put forward the calculation principle and design method. In practical engineering application, the calculation theory and design method have disputed at present (Cui. 2012). This structural system (names are herringbone beam, inclined arches beam, yamagata beam, etc. (Li. 2007)) seem simple, but steel structure system under load generates the larger horizontal thrust and the larger bending moment, bearing is hard to form a fixed end constraint, So the whole structural system is different from the type of door rigid frame structure system. At present, some domestic researchers carry out a more extensive study on mechanical analysis of straight arch with two articulations and put forward their views, such as using a two-hinged arch principle make mechanical calculations and structural analysis; research the stress of the structure system from the horizontal thrust structural system and so on. Traditional mechanics analysis make the structure of bar section shrink into a straight line along the stem axis (Cai & Chen. 2009). This analysis method has some limitations. The actual structure of the various components of the system is the three-dimensional shape, usually straight arch with two articulations is H section. The turning point of straight arch with two articulations formed a complex stress state due to the geometry of the mutation in internal force transmission and deformation process under the action of load, which have a greater impact on entire structural system of force distribution. Traditional mechanical analysis (section shrink into the rod axis) can not explain this state. In this paper, mechanical analysis was developed on straight arch with two articulations, using the method of plane link system analysis (traditional mechanical analysis) and three-dimensional finite element analysis, which both use ANSYS finite element analysis software. By analyzing the different mechanical data derived from the two methods, this paper proposed a theory of integrated effects of straight arch with two articulations mechanical structure system caused by break point.

2 CALCULATION MODEL

Calculation model chosen both ends fixed hinge support of the single span straight arch, that combined with practical engineering applications, as shown in Figure 1.

To facilitate the finite element analysis software and future structure of the test specimen, the span of straight arch with two articulations taken 3000 mm, the section of straight arch with two articulations taken HN150X75X5X7. According mansard beams in practical engineering applications, double-sided slope chosen six cases as 1/10, 1/9, 1/8, 1/7, 1/6, 1/5, considering structure test in the future, load use in beam span third-class equinoctial position under the action of two concentrated load (uniformly distributed load is difficult to put in the laboratory), load size according to the theory calculation take 18 kN and choose Q235 steel.

3 ANSYS SOFTWARE ANALYSIS

According to the established mechanical calculation model, using the ANSYS finite element analysis software, using the method of plane link system analysis and three-dimensional finite element analysis, under the section size of selected and the load, Six different slope states of straight arch with two articulations are analyzed in stress. Plane frame calculation model shown in Figure 1, three-dimensional calculation model shown in Figure 2.

Two different methods of analysis, reaction force, mid-span vertical displacement and mid-span moment of the results show Tables 1–3.

Figure 1.   Straight arch with two articulations calculation diagram.

Figure 2.   Three-dimensional computational model.

Table 1.   Plane frame calculation results.

Slope	Vertical force/kN	Horizontal force/kN	Displacement/mm
1/10	18	104	4.37
1/9	18	99.7	3.78
1/8	18	94.1	3.18
1/7	18	87.1	2.59
1/6	18	78.6	2.03
1/5	18	68.6	1.51

Table 2. Three-dimensional calculation results.

Slope	Vertical force/kN	Horizontal force/kN	Displacement/mm
1/10	18	47.6	2.21
1/9	18	48.9	2.03
1/8	18	49.8	1.84
1/7	18	49.96	1.67
1/6	18	49.02	1.42
1/5	18	46.5	1.19

Table 3. Two analysis methods span bending moment.

Slope	Plane frame/kN·m	Three-dimensional/kN·m
1/10	−2.40	−10.86
1/9	−1.35	−9.83
1/8	−0.31	−8.64
1/7	+0.64	−7.31
1/6	+1.65	−5.75
1/5	+2.58	−4.05

4 THE DATA ANALYSIS

According to the analysis of the results in Table 1 and Table 2 and Table 3, mapping the cylindrical coordinates about Reaction force, mid-span vertical displacement and mid-span moment, as follows Figures 3–5.

That can be observed from the above figure, the Plane frame calculation result of Reaction force and mid-span vertical displacement are bigger than Three-dimensional calculation, the plane frame calculation result of mid-span moment is smaller than Three-dimensional calculation. Compared to two kinds of calculation results: the same of material, section size, load and support constraint condition, the calculation results of reaction force, mid-span vertical displacement and Mid-span moment are not equal, magnitude difference is bigger also. From the change of the slope, when the slope is small, reaction force, mid-span vertical displacement and Mid-span moment are bigger, with the increase of slope, reaction force, mid-span vertical displacement and mid-span moment become small. If the two methods of structure analysis are exact reaction structure system of the actual stress state, so two kinds of analysis and calculation results should be equal, But actual analysis and calculation results are not equal, so two calculation methods of structure analysis are many differences in reflecting the structure system of the actual stress state. The first, the biggest characteristic of Plane frame analysis method is that the structure of bar section along the stem axis shrink into a straight line, And it has own section stiffness. When loading, the turning point of mansard beams reflect only normal constraints and internal force transfer case in specify the stiffness, which can not reflect the reaction of other constraints. The second, the biggest characteristic of Three-dimensional analysis method is that the calculation model set up according to the actual three-dimensional shape straight arch with two articulations beam section, which The beam element was divided into finite, continuous three-dimensional finite element are used to analyze and calculate by Using the method of finite element mesh division. So this analysis method can accurately reflect the actual stress state in internal different parts of the three-dimensional cross section and the integrated effect. Through the above analysis, the Plane frame analysis and calculation cannot reflect the actual stress state in internal different parts of the three-dimensional cross section and the integrated effect, So the result and the actual state has a gap, but finite element analysis method exactly reflect the actual stress state of structure system.

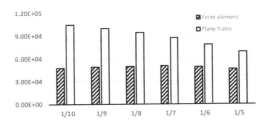

Figure 3.　Horizontal thrust under different slope/N.

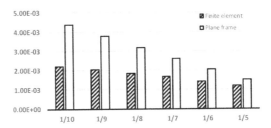

Figure 4.　Mid-span vertical displacement/m.

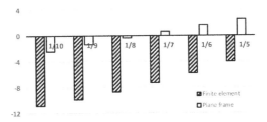

Figure 5.　Mid-span moment under different slope/kN·m.

5　MECHANICAL ANALYSIS

A large gap exists between the plane truss analysis results and the three-dimensional finite element analysis results. Plane linkage analysis cannot reflect the exact force state of the structure system. There are serious flaws when plane linkage analysis is applied to straight arch with two articulations. Serious flaw is reflected in that because the geometry of the straight arch with two articulations (bearing fixed at both ends hinged H-shaped cross-section beams) changes after a certain slope mansard folded, large changes occur in the delivery status of internal forces and stress state of the turning point. Three-dimensional finite element analysis method can exactly reflect the changes, however plane linkage analysis cannot reflect the changes. At present, people use traditional mechanics analysis method to analyze beam structure system. If we analyze the curved beam, using plane frame analysis method, comprehensive analysis of additional effect caused by the break point should be added to mechanical analysis process. In order to analyze deeply comprehensive mechanical effects caused by straight arch with two articulations fold point position, we put the calculation diagram 1 into calculation diagram 6 (shear force is zero at mid-span). By the force balance we know that the vertical reaction force is a fixed value. Learned from Table 1 to Table 3 calculation results, changes happened in horizontal thrust and span bending moment. By comparative analysis of the results of two equal state we know that the results of three-dimensional finite element analysis is less than the results of linkage analysis in horizontal thrust bearing (large difference), however, in terms of mid-span moment calculations, the results of linkage analysis is less than three-dimensional finite element analysis. This shows that the combined effect of the complex stress state at turning point of straight arch with two articulations and the change of straight arch with two articulations geometry caused these changes, which reflect in offsetting the effect of large horizontal thrust, and a substantial increase in cross-site moment carrying capacity. To explain the combined effect, Figure 1 is changed to Figure 7 (assumed calculation diagram).

Figure 7 is a assumed calculation diagram, which in order to reflect combined effect at break point, pull rod is added at the lower flange of mid-span fold point on the basis of plane linkage calculation diagram. We assume that the tension of the rod equals to the value plane truss analysis state minus the three-dimensional finite element analysis state, as well as the corresponding moment the tension produced to the folded position. In order to facilitate mechanical analysis, Figure 7 is broken down into Figure 8. According to Figure 8 assumed calculation diagram and Figure 9 cross-sectional analysis of internal force diagram, the following equation can be established.

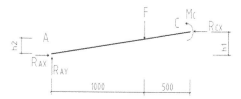

Figure 6.   Girder section calculation diagram.

Figure 7.   Assumed calculation diagram.

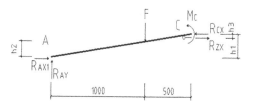

Figure 8.   Assumed calculation diagram.

$R_{ZX} = R_{CX} - R_{CX}$, where R_{AX1} = Three-dimensional finite element horizontal thrust; R_{CX} = Horizontal thrust of plane frame; R_{ZX} = horizontal thrust difference between plane frame and three-dimensional finite element.

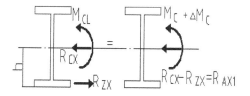

Figure 9.   Cross-sectional analysis of internal force.

383

In plane truss analysis condition, the stress state generated by comprehensive effects of adding tie rod is same to that of finite element analysis. So according to the mechanical equilibrium principle, the following equation is established.

$$M_{CL} = M_C + \mu\left[R_{ZX} \cdot h_3\right] \tag{1}$$

$$R_{CX} = R_{AX1} + R_{ZX} \tag{2}$$

where M_{CL} = span moment of three-dimensional finite element; M_C = span moment of planar truss; h_3 = distance from pull line of rod to the cross-sectional center; μ = moment coefficient of the combined effect.

The key in the equation is to reasonably explain the effect of additional bending moment $\mu[R_{ZX} \cdot h_3]$. $[R_{ZX} \cdot h_3]$ is the moment calculation formula in the normal state of stress. However, the object of our analysis is the mansard architecture, as well as the bending effect at turning point caused by the comprehensive stress state. Therefore, the combined effect coefficient μ is added.

6   COMPREHENSIVE ANALYSIS

6.1   *Analysis of the additional moment effect*

From the actual stress state of the straight arch with two articulations system, there is no tensile connection at the lower flange of mid-span fold point. The purpose of additional pull rod is to reflect the comprehensive effect caused by complex stress condition at the fold point. Since the comprehensive effect exists, pull rod is placed at the lower flange of mid-span fold point, as shown in Figure 9.

$$R_{ZX} = R_{CX} - R_{AX1} \tag{3}$$

$$\Delta M_C = M_{CL} - M_C = \mu\left[R_{ZX} \cdot h_3\right] \tag{4}$$

In the type:

$$h_3 \approx h = 0.15m/2 = 0.075m$$

$$\mu = \Delta M /\left[R_{ZX} \cdot h_3\right] = h_1 / h_3 \tag{5}$$

Additional bending moment embodies arch effect.

6.2   *Analysis of stress effect at fold point*

To explain the combined effect due to the formation of the turning point for the sake of stress distribution, ANSYS post-processing function is used to arrange horizontal stress vertical turning point cutting plane, as shown in Figures 10 and 11 (slope of 1/8).

From the stress distribution, we know that section stress is linear change state, the whole section is in the state of compression. Thus, theory of elasticity can be applied to the analysis.

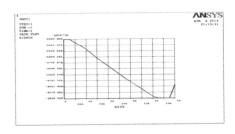

Figure 10.   Stress distribution map of fold point.

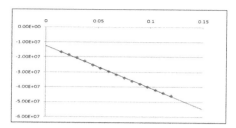

Figure 11. Stress distribution (1/8).

Table 4. All kinds of influence coefficient table.

Slope	Factor μ	Factor φ	The deflection ratio
1/10	2.0	5.3	0.51
1/9	2.2	4.5	0.54
1/8	2.5	4.1	0.58
1/7	2.86	3.4	0.64
1/6	3.3	3.1	0.70
1/5	4.0	2.4	0.79

The data in Figure 11 is organized into stress distribution diagram shown in Figure 12. The section stress is decomposed into three parts: uniform compressive stress caused by horizontal force R_{AX1}; Stress caused by a plane linkage effect moment; M_c stress caused by the additional bending moment ΔM_C; The formation of ΔM_C and the resulting stress effect, which is not the normal steady state, is due to the entire line beam arch effect and complex stress state of break point position. Comprehensive stress factor φ should be added to the mechanical analysis process.

$$\sigma_{max} = \frac{R_{AX1}}{A} + \frac{M_C}{W_X} + \frac{\Delta M}{\varphi W_X}$$

$$\varphi = \Delta M \bigg/ \left(\sigma_{max} - \frac{R_{AX1}}{A} - \frac{M_c}{W_X} \right) W_X \qquad (6)$$

Equation (6), in addition to comprehensive stress factor, all other data can get through cross-sectional characteristics and Table 1–Table 3, and ANSYS finite element stress data after processing. Therefore, the impact coefficient can be drawn in case of different slope.

6.3 Span deflection analysis

The axial deflection and bending deformation occur at the same time during the process of generating deformation (deflection) when straight arch with two articulations are under load. From Table 1, Table 2, we can see that the plane truss analysis results span deflection is greater than the three-dimensional finite element analysis results, and the magnitude of difference is large. Plane truss analysis method, cannot reflect the comprehensive effect of straight arch with two articulations, so the deflection value is smaller. Difference between the results reflects that the comprehensive effect does exist and great influence. In order to facilitate a comprehensive analysis the bending effect the stress effect and deflection ratio (three-dimensional/plane frame) are listed in Table 4.

From Table 4, it can be observed that additional moment effect increases with increase of the slope, stress effect increases gradually decreases with the slope increases, deflection ratio gradually increases with increase of the slope (difference less small).

7 CONCLUSION AND OUTLOOK

1. When straight arch with two articulations is analyzed, traditional mechanical analysis methods (plane linkage analysis) can not reflect the comprehensive effect of complex stress state due to generating fold point and arching.
2. Three-dimensional finite element analysis method can exactly reflect the comprehensive effect of complex stress state due to generating fold point and arching.
3. The specific impact of the combined effect is reflected in greatly reducing bearing horizontal thrust and span deflection, and improving mid-span moment greatly.
4. Currently, in engineering application, the calculation principle of design calculation is more use of plane linkage. Span deflection and horizontal force bearing of such calculations would be too large, especially horizontal force. This will have great influence on the lower part of the steel beam structure (concrete column), which need to pay attention to.
5. There are many factors (support restraint, span, section size, load, slope) related to the comprehensive effect. Deeper study is needed in order to provide a theoretical basis for correct design calculations of mansard steel girder structure system.

REFERENCES

Cai, Y.Y & Chen, Y.Q. 2009. The problems of design and construction of steel inclined arch supported on concrete columns. Steel Construction (10):32–35.

Chen, J.L. 2007. Analysis of mixed structure design of concrete column beam. Steel Construction (7):26–28.

Cui, W.Y. 2012. Research on stressed status of breakpoints at tortuous linear steel shape beam. Shanxi Architecture 17(32):44–46.

Hong, F.Y. 2009. Structure design of single-storey building with light steel roof on concrete Column. Fujian Construction Science & Technology 5:22–24.

Hong, X.H. 2013. Discussion on Column internal force and beam-column joint design of concrete column and light steel roof structure. Fujian Construction Science & Technology 2:31–33.

Leng, J.B. 2013. Design of steel roof and concrete substructure. The World of Building Materials 34(1):65–67.

Li, J. 2007. Thrust analysis on orographic structure of concrete columniation and steel beam system. Steel Construction (11):31–34.

Wan, C.Y. 2007. Approach to structural system of reinforced concrete column and steel beam for single story industrial mill building. Industrial Construction 37:387–390.

Wei, C.W. & Chen, Y.Q. 2005. Design of Straight Steel Arch with Hinge-supported on Concrete Columns. Building Structure 35(2):31–33.

Xu, S.J. 2008. Effect of gradient on thrust at springer of two-hinged arch. Steel Construction 12(23):19–20.

Hydraulic Engineering III – Xie (Ed)
© 2015 Taylor & Francis Group, London, ISBN 978-1-138-02743-5

A gradually-varied Hoek–Brown disturbance factor for analyzing an axisymmetrical cavern

S.P. Yang, J.H. Yu & Z.B. Zhao
Sichuan Water Conservancy Vocational College, Chengdu, China

W.X. Fu
State Key Laboratory of Hydraulic and Mountain River Engineering, College of Water Resource and Hydropower, Sichuan University, Chengdu, China

ABSTRACT: In this study the disturbance factor in the general Hoek–Brown (HB) criterion is considered to be a gradually-attenuated variable from the excavation surface to the deep surrounding rocks. The elasto-plastic analytical solution is formulated for an axisymmetrical cavern model in which there exist a supported pressure at the wall of tunnel and a far-field pressure at infinity. The presented analytical model can well reflect the disturbance of the HB rock mass triggered by drilling and blasting excavation.

1 INTRODUCTION

Although the Hoek–Brown (HB) criterion is an empirical criterion, it is widely accepted in the field of geotechnical engineering and has been applied to many rock engineering projects (Hoek et al. 2002). In the general HB criterion the disturbance factor D is used for describing the excavation disturbance of the surrounding rocks, subjected by blast damage and stress relaxation. Also, the general HB criterion gives the guidelines for roughly estimating the D-value from the appearances of the rock mass near the excavation surface (Hoek et al. 2002).

When employing the drilling and blasting technique to excavate a cavern however, the rock mass near excavation inevitably suffers more serious blast damage than those remote from the excavation surface. In addition, during the process of the stress relaxation caused by removal of the rock mass within the cavern, occurrence of the plastic zone around the cavern extends progressively until the eventual equilibrium state reaches. Clearly, it is appropriate to consider that the disturbance degree gradually attenuates from the excavation surface to the deep rock mass.

However, the D-value is often fixed to be a constant in the existing analysis for the underground engineering projects in the HB rock mass (Chen & Tonon 2011, Fraldi & Guarracino 2010, Li et al. 2009, Park & Kim 2006, Shen et al. 2010, Zhong et al. 2009, Zhou & Li 2011). In this study D is treated as a variable. To formulate the elasto-plastic analysis solution for an axisymmetrical cavern, a linear function is chosen to quantitatively describe D. Compared with the elastic perfectly-plastic and elastic-brittle-plastic results, the present analysis can objectively reflect the excavation disturbance of the surrounding rocks.

2 DESCRIPTION OF ANALYTICAL MODEL

When the ratio of the excavation radius of a deeply-buried circular cavern to its axial line length is small enough, the analyzed problem can be transformed to a problem of plane strain. The analytical sketch is shown in Figure 1. The compressive stress is positive as well as the outward radial displacement.

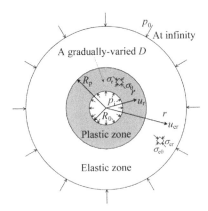

Figure 1. The analytical sketch for an axisymmetrical cavern.

For an axisymmetrical cavern with radius R_0, the radial and tangent stresses σ_r and σ_θ correspond to the minor and major principal stresses σ_3 and σ_1 respectively, the shear stress $\tau_{r\theta}$ in the r–θ plane is zero, and the out-of plane stress σ_z is an intermediate principal stress.

The present work mainly focuses on the influence of D on the redistribution stresses and radial displacements around an axisymmetrical cavern. The D-value is taken to be gradually decreased from the elastic-plastic interface to the excavation surface; but it is fixed to be 0 within the elastic zone. Under the condition without any internal pressure, the D-value is 1 near excavation and 0 at the elastic-plastic interface.

The radial stress σ_r at the elastic-plastic interface equals the critical pressure p_c. The D-value within the plastic zone is considered to be a linear function with respect to σ_r. Therefore, together with the conditions of $D = 0$ for $\sigma_r = p_c$ and $D = 1$ for $\sigma_r = 0$, D can be evaluated by the expression as follows:

$$D(\sigma_r) = [(1 - p_i / p_c)(\sigma_r - p_c)]/(p_i - p_c) \tag{1}$$

Replacing D with $D(\sigma_r)$ we can rewrite the general HB criterion as the yield function form as follows:

$$F_{HB} = \sigma_\theta - \sigma_r - \sigma_{ci}[m_b(D(\sigma_r)) \cdot (\sigma_r / \sigma_{ci}) + s(D(\sigma_r))]^a = 0 \tag{2}$$

where $m_b(D(\sigma_r)) = m_i e^{(GSI - 100)/[28 - 14D(\sigma_r)]}$, $s(D(\sigma_r)) = e^{(GSI - 100)/[9 - 3D(\sigma_r)]}$, $a = (1/2) + (e^{-GSI/15} - e^{-20/3})/6$, σ_{ci} is the uniaxial compressive strength of intact rock, m_i is a material constant that can be determined by laboratory triaxial compression testing for intact rock samples, GSI is an abbreviation of the Geological Strength Index.

When determining the radial displacements in the plastic zone, a plastic potential needs to be specified in advance. However, different-form plastic potentials have significant influences on dilatant plastic deformations (Zienkiewicz et al. 1975). In this study the dilatant plastic deformations are assumed to be related to stress levels. A non-linear non-associated flow rule is employed (Clausen & Damkilde 2008):

$$g_{HB} = \sigma_\theta - \sigma_r - \sigma_g[m_g(\sigma_r / \sigma_g) + s_g]^{a_g} \tag{3}$$

where σ_g, m_g, s_g and a_g in Equation 3 are material constants.

If σ_r and σ_θ are replaced with σ_3 and σ_1 respectively, the above analytical model can be extensively applied to underground excavations in a complicated geological condition.

388

3 ELASTO-PLASTIC ANALYTICAL SOLUTION

3.1 Critical internal pressure

The classical Lame's elastic solutions for the axisymmetrical cavern in Figure 1 are:

$$\begin{cases} \sigma_{\text{er}} = p_0[1-(R_0/r)^2] + p_i(R_0/r)^2 \\ \sigma_{\text{e}\theta} = p_0[1+(R_0/r)^2] - p_i(R_0/r)^2 \\ u_{\text{er}} = -[(1+v)/E](R_0^2/r)(p_0 - p_i) \end{cases} \tag{4}$$

where E and v are the Young's modulus and Poisson's ratio respectively.

Assuming that p_c acts at $r = R_0$ and the D-value is $D(p_c) = 0$, and substituting the elastic limit stresses of $\sigma_{\text{er}} = p_c$ and $\sigma_{\text{e}\theta} = 2p_0 - p_c$ at $r = R_0$ lead to:

$$2p_c + \sigma_{\text{ci}}[m(0)p_c/\sigma_{\text{ci}} + s(0)]^a - 2p_0 = 0 \tag{5}$$

where $m(0) = m_i e^{(GSI-100)/28}$ and $s(0) = e^{(GSI-100)/9}$.

The p_c-value in Equation 5 can be solved by iteration, e.g. the Newton–Raphson method.

3.2 Solution in the elastic zone

For the eventual equilibrium state with p_c acting at $r = R_p$ as shown in Figure 1, the radial and tangent stresses and radial displacement at $r \geq R_p$ can be directly determined from Equation 4:

$$\begin{cases} \sigma_{\text{er}} = p_0[1-(R_0/r)^2] + p_i(R_0/r)^2 \\ \sigma_{\text{e}\theta} = p_0[1+(R_0/r)^2] - p_i(R_0/r)^2 \\ u_{\text{er}} = -[(1+v)/E](R_0^2/r)(p_0 - p_i) \end{cases} \tag{6}$$

3.3 Solution in the plastic zone

1. Stress solution

Due to axisymmetry of Figure 1, the equilibrium equation at $r < R_p$ is expressed by:

$$\sigma'_r + (\sigma_r - \sigma_\theta)/r = 0 \tag{7}$$

where $\sigma'_r = d\sigma_r/dr$.

The tangent stress σ_θ in Equation 7 is evaluated by Equation 2. The correct stress solutions evaluated by Equation 7 must satisfy any one blow:

$$\begin{cases} \sigma_r|_{r=R_0} = p_i \\ \sigma_\theta|_{r=R_0} = p_i + \sigma_{\text{ci}}[m(D(p_i)) \cdot p_i/\sigma_{\text{ci}} + s(D(p_i))]^a \end{cases} \tag{8}$$

To attain correct solutions for the elastic and plastic zones, R_p must be determined first. By use of dichotomy technique, together with numerical integration for Equation 7 and any boundary condition in Equation 8, we can approximately estimate R_p.

Substituting σ_θ in Equation 2 into Equation 7, we transform Equation 7 as the first-order differential equation with respect to σ_r, which is written as the following ordinary form:

$$y' = Y(r, y) \tag{9}$$

where y' is the first-order derivative of σ_r with regard to r, y is σ_r, $Y()$ is a function symbol.

The radial stress σ_r in Equation 7 can be found by numerical integration techniques, *e.g.* the Runge–Kutta equation as follows:

$$y_{j+1} = y_j + hk_2 \quad (j = 1, 2, \ldots, n) \tag{10}$$

where $k_1 = Y(r_j, y_j)$, $k_2 = Y(r_j + h/2, y_j + hk_1/2)$, $h = (R_0 - R_p)/n$, and n is the division number.

In the dichotomy, one can take a large initial upper limit for R_p, *e.g.* $6R_0$, and can directly take the lower limit for R_p as R_0. During iteration for R_p, σ_r can be simultaneously estimated by Equation 7, and σ_θ is determined by Equation 2. When R_p is estimated, σ_{er}, $\sigma_{e\theta}$ and u_{er} in the elastic zone are calculated by Equation 6. In addition, care must be taken if $[m(D(\sigma_r)) \cdot \sigma_r/ \sigma_{ci} + s(D(\sigma_r))] \leq 0$ in Equation 2 during iteration for R_p. For this case, since the real value for σ_θ evaluated by Equation 2 does not exit, R_p must be set again in a new dichotomy.

2. Displacement solution

When R_p, σ_r and σ_θ are determined, the radial displacements can be found by the plasticity theory (*i.e.* Park & Kim 2006). The principal plastic strains at $r < R_p$ can be evaluated by:

$$\begin{cases} \varepsilon_{pr} = \lambda\left(\left[\int_{R_p}^r (\partial g/\partial \sigma_r)dr\right] \bigg/ \int_{R_p}^r dr\right) = \lambda\left(\left[\int_{R_p}^r (\partial g/\partial \sigma_r)dr\right] \bigg/ (r - R_p)\right) \\ \varepsilon_{p\theta} = \lambda\left(\left[\int_{R_p}^r (\partial g/\partial \sigma_\theta)dr\right] \bigg/ \int_{R_p}^r dr\right) = \lambda\left(\left[\int_{R_p}^r (\partial g/\partial \sigma_\theta)dr\right] \bigg/ (r - R_p)\right) \end{cases} \tag{11}$$

where λ is a positive plastic scale coefficient, ε_{pr} and $\varepsilon_{p\theta}$ are the radial and tangent plastic strains respectively.

The total strains can be decomposed as the elastic and plastic parts:

$$\begin{cases} \varepsilon_r = \varepsilon_{pr} + \varepsilon_{er} \\ \varepsilon_\theta = \varepsilon_{p\theta} + \varepsilon_{e\theta} \end{cases} \tag{12}$$

where ε_r and ε_θ are the total radial and tangent strains respectively, ε_{er} and $\varepsilon_{e\theta}$ are the radial and tangent elastic strains respectively.

According to the Hooke's law, the elastic parts in Equations 12 can be written as:

$$\begin{cases} \varepsilon_{er} = [(1+v)/E][(1-v)(\sigma_r - p_0) - v(\sigma_\theta - p_0)] \\ \varepsilon_{e\theta} = [(1+v)/E][(1-v)(\sigma_\theta - p_0) - v(\sigma_r - p_0)] \end{cases} \tag{13}$$

Substituting Equations 11 and 13 into Equation 12 and eliminating λ lead to:

$$\varepsilon_r - \bar{A}_1\varepsilon_\theta = [(1+v)/E][\bar{A}_2(\sigma_r - p_0) - \bar{A}_3(\sigma_\theta - p_0)] \tag{14}$$

where $\bar{A}_1 = \int_{R_p}^r \dfrac{\partial g/\partial \sigma_r}{\partial g/\partial \sigma_\theta} dr \bigg/ (r - R_p) = \int_{R_p}^r (\partial g/\partial \sigma_r)dr \bigg/ (r - R_p) \approx \sum_{j=1}^k [-1 - a_g m_g (m_g \sigma_{rj}/\sigma_g + s_g) \cdot h_j$

and $k = 1, 2, \ldots, n$; $\bar{A}_2 = 1 - v + \bar{A}_1 v$, $\bar{A}_3 = v + \bar{A}_1 - \bar{A}_1 v$.

Substituting the geometry equations $\varepsilon_r = -du_r/dr = u_r'$ and $\varepsilon_\theta = -u_r/r$ into Equation 14 leads to:

$$u_r' - \bar{A}_1 u_r / r = -[(1+v)/E][\bar{A}_2(\sigma_r - p_0) - \bar{A}_3(\sigma_> - p_0)] \tag{15}$$

According to the compatibility of deformations, the boundary condition of Equation 23 is determined from u_{er} of Equation 6:

$$u_r(R_p) = u_r\big|_{r=R_p} = -R_p[(1+v)/E](p_0 - p_c) \tag{16}$$

Equation 16 can also be written as the ordinary form in Equation 9. Accordingly, the u_r-value at $r \leq R_p$ can be calculated from the Runge–Kutta equation in Equation 10.

390

4 RESULTS AND COMPARISON

Under the situation that the disturbance factor is independent of the stress state in the plastic zone, *i.e.* a fixed D-value, the presented analytical model can be degenerated to an elastic perfectly-plastic or elastic-brittle-plastic model. The corresponding analytical solutions of the elastic perfectly-plastic or elastic-brittle-plastic model can be easily derived. We only give the analytical results of the elastic perfectly-plastic and elastic-brittle-plastic models for comparison with the present analytical model.

The primary parameters are $R_0 = 5$ m, $p_i = 0.5$ MPa, $p_0 = 10$ MPa, $\sigma_{ci} = 75$ MPa, $m_i = 15$, and $GSI = 30$ in three models with an identical plastic potential. The constants in Equation 3 are $\sigma_g = \sigma_{ci}$, $m_g = m(0)/4$, $s_g = s(0)$, and $a_g = a$, and a. In the elastic perfectly-plastic model, the D-value is taken to be 0.5 for the peak and residual envelopes. The D-value in the elastic-brittle-plastic model is fixed to be 0 for the peak envelope and 0.5 for the residual envelope. The HB envelopes used for analyzing the three elasto-plastic models and the respective application conditions are shown in Figure 2.

For the presented model, the D-value is evaluated by Equation 1, the pre-estimated interval for R_p is $[R_0, 6R_0]$ during iteration, and the division number n in $h = (R_0 - R_p)/n$ is 2000 over $[R_p, R_0]$. In addition, the convergence tolerances for both of p_c and R_p are controlled to be 1.0e-6.

The analytical results evaluated by the three models are given in Figure 3. To check the analytical solutions, the results based on the finite element calculation with the convergence value of the ratio of unbalance force 1.0e-6, are plotted in Figure 3 as well.

Observation for Figure 3 shows that the analytical solutions of the three models are in good agreement with the corresponding FEM results. In addition, comparison of the results

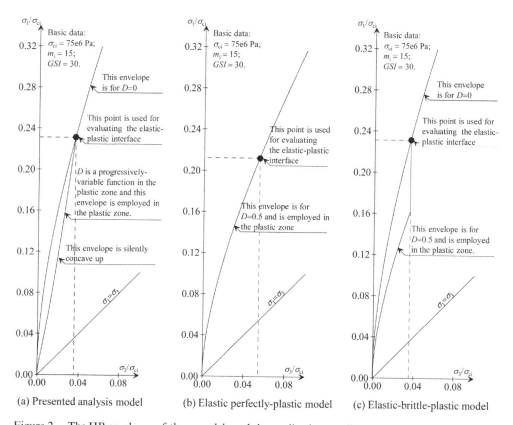

(a) Presented analysis model   (b) Elastic perfectly-plastic model   (c) Elastic-brittle-plastic model

Figure 2. The HB envelopes of three models and the application conditions.

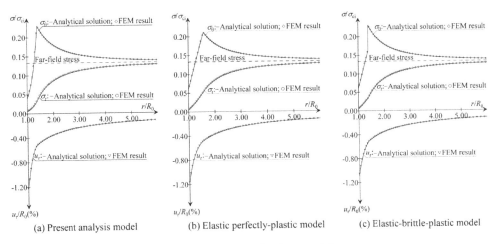

Figure 3.   The analytical solutions and FEM results for three models.

based on the three models shows an obvious difference in the distribution of the stresses in the plastic zone. For the tangent stress evaluated by the elastic-brittle-plastic model, there is a sudden drop at the elastic-plastic interface; but the stress gradient of the plastic zone calculated by the elastic perfectly-plastic model is less than that of the plastic zone calculated by the present model.

5   CONCLUSIONS

In practice of rock engineering projects, the generalized HB criterion is suited to a homogeneous, isotropic and massive rock with few discontinuities or to a heavily jointed rock mass. When a cavern or cavern group is constructed in the heavily jointed rock mass, it has been observed that the surrounding rock failure depends on mutual separation and slip of discontinuities. For this case in the elasto-plastic analysis, it can be accepted that the three parameters of σ_{ci}, m_i and GSI in the generalized HB criterion are fixed to be constants, but D is treated as a variable related to the confining pressure σ_3.

The presented analytical method can objectively evaluate the stress and displacement distribution and the plastic extent around a cavern. However, the quantitative evaluation of D, disturbance influence on the deformation parameters of the HB rock mass, and time-dependent disturbance response *etc.*, are also worth further investigating.

REFERENCES

Chen R., Tonon F., 2011. Closed-form solutions for a circular tunnel in elastic-brittle-plastic ground with the original and generalized Hoek–Brown failure criteria. Rock Mechanics and Rock Engineering, 44(2): 169–178.

Clausen, J. & Damkilde, L. 2008. An exact implementation of the Hoek–Brown criterion for elasto-plastic finite element calculations. International Journal of Rock Mechanics & Mining Sciences 45(6): 831–847.

Fraldi, M. & Guarracino, F. 2010. Analytical solutions for collapse mechanisms in tunnels with arbitrary cross sections. International Journal of Solids and Structures 47(2): 216–223.

Hoek, E., Carranza-Torres, C. & Corkum, B. 2002. Hoek–Brown failure criterion—2002 edition. In Hammah, R., Bawden, W., Curran, J. & Telesnicki, M. (ed.), the 5th North American symposium—NARMS-TAC 2002, Mining Innovation and Technology, 2002, Toronto (downloadable for free at Hoek's corner, http://www.rocscience.com).

Li, S.S, Qian, Q.H, Zhang, D.F & Li, S.C. 2009. Analysis of dynamic and fractured phenomena for excavation process of deep tunnel. Chinese Journal of Rock Mechanics and Engineering, 28(10): 2104–2112. (in Chinese).

Park, K.H., & Kim, Y.J. 2006. Analysis solution for a circular opening in an elastic-brittle-plastic rock. International Journal of Rock Mechanics & Mining Sciences, 43(4): 616–622.

Shen, Y.J., Xu, G.L., Zhang, L. & Zhu, K.J. 2010, Research on characteristics of rock deformation caused by excavation disturbance based on Hoek–Brown criterion. Chinese Journal of Rock Mechanics and Engineering, 29(7): 1355–1362. (in Chinese).

Zhong, Z.Q., Peng, Z.B. & Peng, W.X. 2009. Excavation response of tunnel in stratified rock mass based on Hoek–Brown criterion. Journal of Central South University (Science and Technology), 40(6): 1689–1694. (in Chinese).

Zhou, X.P. & Li, J.L. 2011. Hoek–Brown criterion applied to circular tunnel using elastoplasticity and in situ axial stress. Theoretical and Applied Fracture Mechanics 56(2), 95–103.

Zienkiewicz, O.C., Humpheson, C. & Lewis, R.W. 1975. Associated and nonassociated viscoplasticity in soil mechanics, Géotechnique, 25(4): 671–689.

Initial establishment and application of cell membrane chromatography with mouse macrophage

Yuqiang Wang, Xiaoli Zhou, Wen Tang, Yiming Zhou & Xu Chen
School of Perfume and Aroma Technology, Shanghai Institute of Technology, China

ABSTRACT: Mouse macrophage, cultured in vitro, was used to prepare its cell membrane stationary phase, using silica gel as carrier. Its surface and chromatographic characteristics were studied after the establishment of mouse macrophage model. The mouse macrophage cell membrane chromatography model was applied to identify existence of affinity adsorption. Results revealed mutual interaction between vitro ligand and mouse macrophage in cell membrane chromatography model. Flavonoids can selectively act on mouse macrophage in chromatographic conditions. This showed that mouse macrophage cell membrane chromatography could be used as screening model to study and find certain bioactive molecule.

1 INTRODUCTION

Cell membrane chromatography was a novel affinity chromatography method to study mutual interaction between receptor and drugs. This method, combining high performance liquid chromatography, cell biology and receptor pharmacology, successfully conducted dynamic simulation for some reactions which occurred in vivo in chromatographic column by taking advantage of specific affinity between active ingredient and membrane receptor. CMC was to fix active membrane on the surface of certain carrier, making Cell Membrane Stationary Phase (CMSP) (He L.C et al., 1996; He L.C et al., 2001). Using buffer solution as mobile phase and active compounds as sample, mutual interaction between active substrate and cell membrane of stationary phase was studied by HPLC. Bioactive compounds can directly be screened in certain CMC model without extraction and separation. CMC method had such merits as easy operation, stability and reliability and high sensibility. CMC model had been used to conduct basic research for natural plant and Chinese medicine materials.

Recently, researches had proved that flavonoids showed significant antioxidant effect on many cells such as mouse macrophage. Flavonoids had obvious physiological and pharmacological activity and anti-inflammatory activity (Li D et al., 2010; Xue C Y et al., 2005).

In this paper, retention characteristics of buckwheat flavonoids and its main ingredients which were rutin and quercetin were studied.

2 MATERIALS AND METHODS

2.1 *Material and chemicals*

Tartary buckwheat (Fagopyrum tataricum (L.) Gaertn) was cultivated in Shanxi province of China. Mouse macrophage was purchased from Shanghai Institute of Biochemistry and Cell biology. Circular spherical macroporous silica gel was purchased from Welch Materials in Shanghai. All other chemicals were purchased from Sinopharm Chemical Reagent (Shanghai, China).

2.2 Extraction of flavonoids from tartary buckwheat

Extraction of flavonoids was estimated by previous study with slight modifications (Zoggel H et al., 2012). Firstly, buckwheat seeds were washed in 0.9% (w/v) NaCl solution and then immersed in deionized water for 10 h. Secondly, germinated seeds were hulled and freezed drying under −20 °C. Finally, these freeze-dry seeds were crushed into powder to dissolve in 70% ethanol for further flavonoids extraction.

2.3 Cell culture and activation of silica gel

Mouse macrophage was maintained in RPMI-1640 medium consisting of 10% FBS, 100 IU/mL penicillin and 100 ug/mL streptomycin. Those cells were kept in a moist environment containing 5% CO_2 at 37 °C with its culture medium replaced every 2–3 days. When cells overgrew, they were added in 0.25% (w/v) trypsin and 0.02% (w/v) EDTA for 2 min and replaced to a new flask.

An amount of macroporous spherical silica gel was added into 100 ml 1.0 M hydrochloric acid, followed by 15 min ultrasonic processing and counter-flow for 2 h in a heating and stirring condition. Then, silica was transferred to beaker with distilled water for a few hours. Silica gel was placed at 120 °C for 7 h when its supernatant were dumped (He L C., 1996).

2.4 Preparation for CMSP, standard sample and flavonoids sample

Mouse macrophage, smashed to pieces after ultrasonic treatment, was mixed with some gel in a reaction tube for adsorption in low temperature, vacuum and agitation conditions. Then, physiological saline was added to dilute the solution, waiting for cell phospholipid bilayer fusion until the occurrence of even membrane in the surface of silica. After centrifugation at 2000 r/min to remove supernatant, silica gel carrier cell membrane was washed by Tris-HCl buffer 2–3 times in order to eliminate unconjugated cell membrane (He L C., 2001).

0.1 g rutin and 0.1 g quercetin was dissolved with 25% methanol phosphate to 20 mL respectively, obtaining 5 mg/mL rutin and quercetin standard solution. Then, equivalent rutin and quercetin standard solution was both mixed with 3 g silica gel and 3 g CMSP to full reaction at room temperature. The above solutions were centrifuged at 3000 r/min for 10 min to obtain supernatant respectively (Shu Y X et al., 2002).

Flavonoids extracting solution was mixed with 3 g CMSP to get full reaction and then centrifuged at 3000 r/min for 10 min to acquire supernatant.

2.5 HPLC analysis

The above three supernatants of two standard samples and flavonoids sample were filtered through 0.45 μm nylon membrane and then injected into a Shimadzu LC-10 AVP Plus system (Shimadzu corporation). The column used in this experiment was Symmetry C18 (4.6 mm × 150 mm, 5 μm) from WondaSil. Mobile phases consisted of solvent A: acetonitrile and solvent B containing 0.05% phosphoric acid in water. Both mobile phases A and B were filtered through a 0.45 μm nylon membrane and ultrasonic degassed before any analytical run. Method to identify rutin in samples was first eluted by a binary gradient of 90% solvent B for 2 min, followed by a 60% B for 8 min at a flow rate of 0.8 ml/min. Method to determine quercetin in samples was first eluted by a binary gradient of 50% B for 10 min, 60% B for 10 min, followed by 70% B for 10 min at a flow rate of 0.8 mL/min. The separation was monitored by a diode array detector at a wavelength of 254 nm.

2.6 Statistical analysis

The statistical analysis of the results was carried out by origin 8.0 software.

3 RESULTS

3.1 *Surface characteristics of mouse macrophage cell membrane stationary phase*

Under the above experimental condition, mouse macrophage cell membrane would absorbed on the surface of silica, which was irreversible. Completely covered cell membrane stationary phase was achieved when fragments of mouse macrophage got mutual fusion in a certain temperature. CMSP of mouse macrophage was tested by electron microscopy technique and surface energy spectrum technology.

Figure 1 showed that of surface of mouse macrophage CMSP was mostly covered with mouse macrophage by comparing pure silica gel.

Figure 1(c) showed that there were strong energy spectrum of oxygen and silicon. Figure 1(d) showed that energy spectrum of carbon increased and silicon decreased, which demonstrated that the surface of silica gel was covered by cell membrane and CMSP of mouse macrophage would show some characteristics of cell membrane.

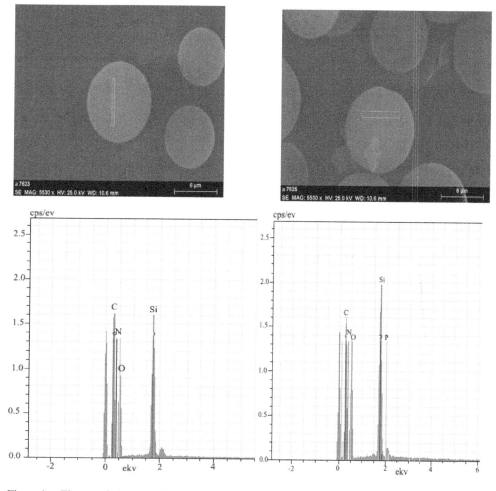

Figure 1.   Figures of electron microscope and the analysis of the spectrum of surface energy for pure silica gel and CMSP of mouse macrophage. (a)   Pure silica gel carrier (magnified 3000 times); (b) CMSP of mouse macrophage (magnified 3000 times); (c) figure of surface energy for pure silica gel; (d) CMSP of mouse macrophage.

3.2 Results of HPLC

According to the absorption of rutin and quercetin on the UV spectra, 254 nm was selected as detection wavelength, where rutin and quercetin had maximum absorption.

Figure 2(a) showed that pure rutin had the unique highest response at 2.5 min. Figure 2(c) showed that two substances had high absorbance at 2 min and 2.5 min which meant there was reaction occurring between rutin and CMSP. Figure 2(e) demonstrated several substances had high absorbance at 254 nm at 1 min, 2 min, 2.5 min and 4.5 min, meaning rutin in flavonoids also had reaction with CMSP. At the same time, Figure 2(b) demonstrated that pure quercetin had the unique highest response at 7.1 min. Figure 2(d) showed that there

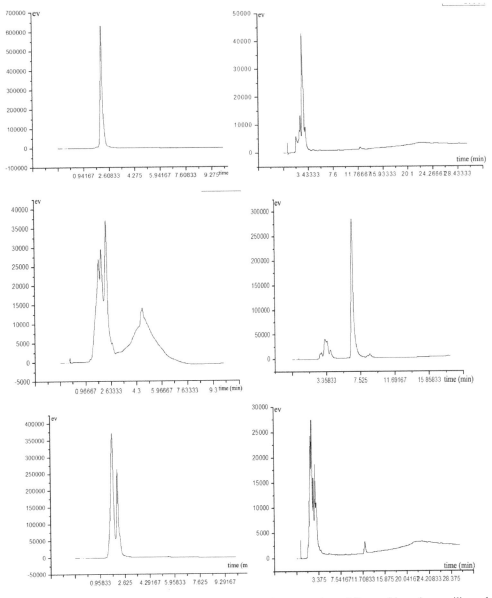

Figure 2. Results of HPLC after reactions between rutin, quercetin and flavonoids and pure silica gel, CMSP and CMSP, respectively.

was several substances at about 3.4 min which meant there was reaction occurring between quercetin and CMSP. Figure 2(f) showed some matters had high response at about 3.4 min, meaning quercetin in flavonoids also had reaction with CMSP.

4 CONCLUSIONS

In conclusion, mouse macrophage chromatography model had similar biological activity and chromatographic separation, so it can imitate mutual interaction between active ingredient and mouse macrophage cell membrane. This model can be used to study mutual reaction between drug molecule or active molecule and targets. Results just meant that there was reaction there, as for what kind of biological effect would happen demanded further validation in vitro experiment.

ACKNOWLEDGMENT

This study was supported by Sino-German Cooperation Project (GZ 727).

REFERENCES

He L.C, Geng X.D. 1996. Cell membrane chromatography—a new method to study drug and receptor. *New advances in biomedical chromatography*, 3: 8–9.

He L.C, Wang S C, Geng X.D. 2001. Coating and fusing cell membrane onto a silica surface and their chromatographic characteristics. *Chromatographia*, 54: 71–76.

He L.C, Wang S.C. 2001. Establishment and preliminary application of vascular endothelial cell membrane chromatography model. *Chromatographia*, 56: 60–64.

He L.C, XinDu G. 1996. Enzyme activity and chromatographic characteristics of cell membrane fixed in the surface of silica gel. *New advances in biomedical chromatography*, 3: 80–85.

He L.C. 1999. Evaluation of drug-muscarinic receptor affinities by using cell membrane chromatography. *Science bulletin*, 44: 632–636.

Li D, Ding X.L. 2010. Screening, analysis and in vitro vasodilatation of effective componets from Ligus ticum Chuanxiong. *Food Science*, 26: 113–116.

Ren X, Zhang X, Li Y. & Wang Z. 2000. Angiogenesis in cancer and other diseases. *Life science*, 32: 1314–1317.

Shu Y.X, RuL B. 2002. A key marker defining haemopoietic stem cells. *Methodology of pharmacological experiment*, 33: 393–394.

Xue C.Y, Zhang Y.H, Liu Y.H, Zhang R.X, Jing H.J & Zhang Y. 2005. Protein measurement with the Folin phenol reagent. *Chinese Journal of Clinical Rehabilitation*. 35: 108–111.

Zoggel H, Carpentier G, Dos S.C, Hamma-Kourbali J, Courty M. & Amiche J. 2012. Inhibitory effects of the alkaloids from Radix caulophylli on the proliferation of human vascular endothelial cell. *Chinese Journal of Clinical Rehabilitation*. 6: 124–127.

Hydraulic Engineering III – Xie (Ed)
© 2015 Taylor & Francis Group, London, ISBN 978-1-138-02743-5

Risk of accident investigation and correlation analysis of the foundation pit

De-Hua Zhu & Qiang Zhang
Shanghai Jianke Engineering Consulting Co. Ltd., Shanghai, P.R. China

ABSTRACT: According to relevant documents and payment in the form of questionnaires, 224 cases of the domestic excavation accident in past two decades are collected in this paper. Based on processing and analysis of accident cases, the common risk events and factors are determined in the construction of deep excavation. Meanwhile the relativity between risk of quality and risk of safety and risk of investment is analyzed according to causes and losses of the accident, and the graph of macroscopic logical relation established between the risk of three objective dimension, which provides a reference for comprehensive risk management excavation construction process and baseline data for the theoretical analysis of risk.

1 INTRODUCTION

In recent years, the characteristics of foundation pit engineering become "big, deep, tight, near". The foundation pits with area among 10000~50000 sqm are more and more big and the foundation pit depth can reach 20~30 m since there scales are more and more big. The locations of foundation pit engineering are compact, which are often in areas concentrated with buildings and lifeline engineering, besides some were close to the red line. The surrounding environment of the foundation pit is complex, other than buildings or metro tunnels, there are also water or gas pipes and other major pipelines (CHEN Zhong-han, 2002). Since the 1990s, though the increasingly advancement of the foundation pit engineering design theory, the development of the construction technology, and the persistent improvement of the calculation method, there were still many accidents occur in foundation pit construction due to complex strata, combined with the bad management of design and construction, which caused huge economic losses, casualties, and even delayed the construction period, resulted in a bad social impact. Such as on July 21st, 2005, in Guangzhou, the Haizhu Square foundation pit collapse killed 3 people and wounded 3 people. On November 15th, 2008, Hangzhou metro foundation pit collapsed, 21 people were killed and more than two dozen people were injured, the direct economic losses were more than 50 million Yuan.

From the above, many unforeseen factors in the process of construction of foundation pit, will affect the implementation of the foundation pit engineering, which will increase the risk of the project in various ways, embodied in: the quality and safety accidents, and investment cost overruns, etc. Meanwhile these different types of risk are not independent or unrelated, instead they influenced and closely linked to each other. Literature (TANG Ye-qing, 1999), (YANG Li-jun, 2003), (HE Shou-ye, 2003) and (LIU XIANG, 2007) summarized the cause of the accident and the disposal measures via analyzing the common accident in foundation pit construction; Literature (HUANG Zhong-huan, 2012) studied the preventive measures for foundation pit engineering accident; and Literature (HUANG Hong-wei, 2005) studied risk management in the process of foundation pit construction. The previous studies mostly focused on the construction quality and safety, without taking investment, quality and safety into consideration simultaneously, let alone studied the relationship among the quality, safety, and investment risk. The thesis indicates the cause of the accident and performs an analysis on the relationship among the quality risk accident, safety risk accident and investment risk accident.

Table 1. List of foundation pit risk accident cases.

No.	Project name	Form of retaining structure	Excavation method	Support form	Excavation depths	Accident type	Accident cause	Casualty	Additional investment
1	An underground mall in Qingdao	Soil nailing wall	Open excavation method	Hybrid support	12 m	Local instability	Inadequate exploration	0	0.2 million
2	A square in Guangzhou	Soil nailing wall	Open excavation method	Steel support	17 m	Overall instability	Over excavation/ overload/ timeout	3 people died, five people seriously injured	30 million
...... 224	A metro station in Shanghai	Diaphragm retaining wall	Open excavation method	Hybrid support	14.9 m	Diaphragm retaining wall leakage	Improper construction	0	0.3 million

2 RISK ACCIDENT INVESTIGATION OF THE FOUNDATION PIT

Construction engineering is an activity throughout the development process of human society. We human moved into skyscraper from the cave by accumulating successful experience and we can also prevent the accidents from recurring via the analysis of previous failures to master relevant laws.

The study collected 224 cases of the domestic typical foundation pit accidents mainly by referring to relevant literature and issuing questionnaires, yet it could not cover all cases limited to all aspects of objective reason, as shown in Table 1 (due to the limited space, not enumerate). From the viewpoint of statistics, the greater the sample size, the more reliable the analysis results are. And case collection is a continuous process, therefore, in the follow-up work we will enrich the cases via different approaches, timely update and optimize the analysis result.

3 RISK ACCIDENT DATA ANALYSIS OF THE FOUNDATION PIT

3.1 Accident type list of foundation pit accident

As shown in Table 2, in all kinds of accident types, local instability and overall instability accounted for the largest proportion, respectively is 37% and 23%; followed by piping erosion and leakage of retaining structure. The outcome is connected with the district the foundation pit in and the enclosure type it belongs to.

Table 3 indicates that local instability and overall instability have higher probability in bored pile and soil nailing wall, because the distance between the bored piles was two large, furthermore the grouting range and overlap width were insufficient; yet soil nailing wall has a great horizontal displacement that it could not exert supporting capacity on the soil that tend to creep. Therefore both the two enclosure types caused local Instability and overall Instability commonly. It also explains why local instability and overall instability accounted for a larger proportion in all accidents. All types of accidents have occurred in bored piles due to its widely application.

3.2 Cause list of foundation pit accident

Table 4 and Table 5 illustrate that unreasonable design caused the most local instability accidents and overall Instability accidents, the ratio respectively are 28% and 33%, therefore it is necessary to conduct checking calculation for Overall Stability and Local Stability of the foundation pit as a successful design proposal. On account of any negligence or mistake in design could lead to serious foundation pit engineering accident, the proposal must be exactly right, what's more it is necessary to conduct checking calculation for the retaining the strength of structure, Overall Stability and Local Stability of the foundation pit, structure and ground Deformation, and the effects that local reinforcement of soft soil layer exerted on adjacent buildings, in addition, putting forward preventive measures for potential accidents is also needed.

As shown in Table 3 former, accidents caused by Retaining Structure leakage make up a great part of the whole in Bored Pile and Diaphragm Retaining Wall. The distance between the bored piles was two large, furthermore the grouting range and overlap width were

Table 2. Accident type distribution list of foundation pit.

Accident type	Local instability	Overall instability	Retaining structure leakage	Supporting instability	Piping erosion	Soil landslide inside the pit	Pile broken	Surrounding environment disruption
Qty	82	51	23	19	26	1	16	6
Ratio	37%	23%	10%	8%	12%	0%	7%	3%

Table 3. Accident type distribution list of foundation pit.

	Local instability	Overall instability	Retaining structure leakage	Supporting instability	Piping erosion	Soil landslide inside the pit	Pile broken	Surrounding environment destruction
Soil nailing wall	17	11	1	0	2	0	1	1
Bored pile	36	15	11	9	14	1	7	3
Diaphragm retaining wall	6	5	11	5	2	0	1	0
Steel sheet pile	4	7	0	0	3	0	0	0
Step-slope	3	4	0	0	1	0	0	0
Manual hole digging pile	4	0	0	0	1	0	2	0
Precast concrete pile	3	2	0	3	0	0	0	1
Cement-soil mixing pile	0	1	0	0	1	0	0	1
Ejector anchor	9	6	0	1	2	0	0	0

Table 4. Cause distribution list of local instability.

Accident cause	Over-excavation	Insufficient dewatering and drainage	Wrong exploration	Unreasonable design	Improper construction	Unqualified construction quality
Qty	5	15	5	23	23	11
Ratio	6%	18%	6%	28%	28%	14%

Table 5. Cause distribution list of overall instability.

Accident cause	Over-excavation	Insufficient dewatering and drainage	Wrong exploration	Unreasonable design	Improper construction	Unqualified construction quality
Qty	7	10	5	17	9	3
Ratio	14%	19%	10%	33%	18%	6%

insufficient; whereas for Diaphragm Retaining Wall, the slot may have deviation or distortion, the steel reinforcement cage was hard to be put into the slot or prone to come-up, the joints was to difficult to handle, thus any carelessness in the construction may result in Retaining Structure leakage. It explained why in Table 6 improper construction and unqualified construction quality account for large proportion of the accident causes.

Table 7 shows that insufficient dewatering and drainage caused most Piping Erosion and the ratio reached up to 31%. The existence of groundwater increased the difficulty of foundation pit project once improper control or disposal would lead to engineering accidents. As the retaining structure was lack of water proof curtain or coupled with waterproof curtain but

Table 6. Cause distribution list of retaining structure leakage.

Accident cause	Insufficient dewatering and drainage	Unreasonable design	Improper construction	Unqualified construction quality
Qty	2	3	7	11
Ratio	9%	13%	30%	48%

Table 7. Cause distribution list of piping erosion.

Accident cause	Insufficient dewatering and drainage	Wrong exploration	Unreasonable design	Improper construction	Unqualified construction quality
Qty	8	3	7	6	2
Ratio	31%	11%	27%	23%	8%

Table 8. Cause distribution list of pile broken.

	Over-excavation	Insufficient dewatering and drainage	Wrong exploration	Unreasonable design	Improper construction	Unqualified construction quality
Qty	1	4	1	6	2	2
Ratio	6%	25%	6%	37%	13%	13%

had defect. When dewatering heavily, the foundation soil within some range would come to an uneven settlement and cause the pavement and underground pipeline to cracking or even destruction, etc. Therefore when dewatering and draining of foundation pit, monitoring and supervision should be strengthen, it is necessary to observe and record the displacement of retaining structure, dewatering depth and influence scope of water level change, pit wall stability and seepage, and the settlement of adjacent buildings, so as to get hold of information and take effective measures to solve the problems in order to avoid accidents.

In pile design, puny thinking of the load, improper parameters selection, unreasonable material selection and layout design all these could lead to pile broken, as shown in Table 8, unreasonable design accounted for largest pile broken and the ratio reached 37%.

4 CORRELATION ANALYSIS OF QUALITY RISK ACCIDENT, SAFETY RISK ACCIDENT AND INVESTMENT RISK ACCIDENT OF FOUNDATION PIT ENGINEERING

Previous Table 1 presented that many of the foundation pit quality accidents caused casualties and increased investments, in other words, foundation pit quality accidents tend to trigger safety accidents and investment accidents. Via arranging and analyzing the collected accident cases, the thesis finds a tentative logical relationship among the quality accidents, safety accidents and investment accidents of foundation pit engineering, as shown in Figure 1, the solid line means supported by cases whereas the dashed line means lack of cases but may have a practical correlativity.

This section highlights studying the correlativity of quality accidents, safety accidents and investment accidents aiming at the risk recognition and assessment of foundation pit construction.

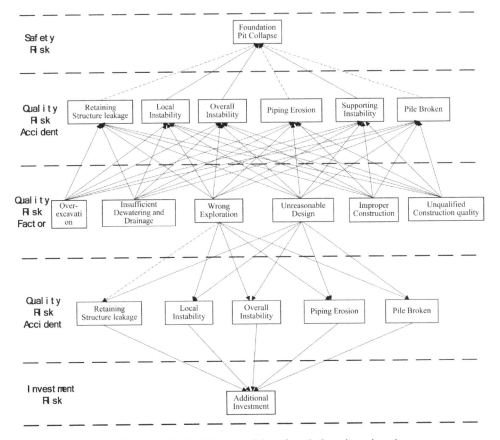

Figure 1. Three risk dimensions logical diagram of deep foundation pit engineering.

4.1 *Correlativity between quality accidents and safety accidents*

Via screening the collected foundation pit accident cases, we figure out the amount of safety accidents triggered by quality accidents, as shown in Table 9. It reveals that safety accidents triggered by quality accidents in foundation pit engineering concentrated in foundation pit collapse, of which, the proportion cause by overall instability reached 82%. Furthermore, the two common form of foundation pit collapse induced by overall instability presented as collapse due to retaining structure breakdown and collapse due to support structure destruction.

4.1.1 *Foundation pit collapse due to retaining structure breakdown*

This kind of destruction mode were mainly retaining structure shear failure or fracture as a result of its insufficient strength owing to the unreasonable designer improper construction, which lead to unstable failure of the whole foundation pit, for example, retaining wall shear destruction. As it plugged in solid soil layer or lack of supporting while the earth pressure behind it was heavy, the flexible retaining wall broke down as a result of an overload structural stress accompanying with the foundation pit fall down inward, as shown in Figure 2.

4.1.2 *Collapse due to support structure breakdown or tension anchor destruction*

Such failure were mainly support structure or tension anchor destruction by reason of its insufficient strengthen and result in overall buckling collapse of the foundation pit, as shown in Figure 3.

Table 9. Relationship between quality accidents and safety accident.

Quality risk accidents	Foundation pit collapse	Ratio
Local instability	8	13.1%
Overall instability	50	82.0%
Supporting instability	3	4.9%

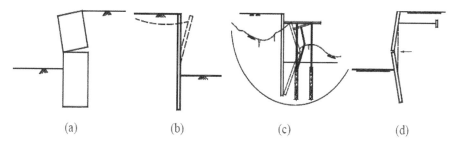

(a)	(b)	(c)	(d)

Figure 2. Failure of retaining structure.

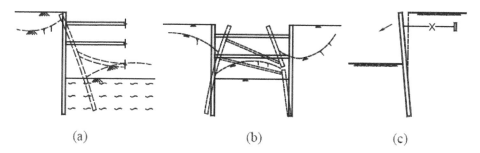

(a)	(b)	(c)

Figure 3. Destruction of support structure or tension anchor.

Table 10. Relationship between quality accidents and investment accident.

Risk factors	Quality accidents	Cost claims	Ratio
Wrong exploration	Local instability	5	20%
	Overall instability	5	
	Retaining structure leakage	0	
	Piping erosion	3	
	Pile broken	1	
Unreasonable design	Local instability	23	80%
	Overall instability	17	
	Retaining structure leakage	3	
	Piping erosion	7	
	Pile broken	6	

4.2 Correlativity between quality accidents and investment accidents

Via screening the collected foundation pit accident cases, we figure out the amount of safety accidents triggered by quality accidents, as shown in Table 10. It reveals that investment accidents triggered by quality accidents in foundation pit engineering mainly presented as cost claim. The risk factors that induce investment accidents mainly come from exploration and design. The unreasonable design caused quality accidents following by cost claim and this situation held a highest proportion up to 80%. Table 2 also presents that the cost claim cases due to local instability are the most numerous, of which, cause by wrong exploration were 5, caused by unreasonable design were 23, totally were 28.

5 CONCLUSION AND DISCUSSION

Quality risk accidents, safety risk accidents and investment risk accidents of deep foundation pit are not independent from each other, when consider one kind of risk accidents and its factors separately we may leave out some inner relation of them and then get a distorted risk recognition result. Via combing accident cases of deep foundation pit, the article carried out an analysis on the relationship among quality risk, security risk and investment risk and then established a macroscopic logic diagram of the influence rule of the three objective dimensions, reached a tentative conclusion as follows:

1. Quality accident types of deep foundation pit engineering mainly include: local instability, overall instability, retaining structure leakage, supporting instability, piping erosion, soil landslide inside the pit, pile broken, surrounding environment disruption, Among which, local instability and overall instability share higher probabilities, in particular the foundation pit instability of bored pile holds the maximum probability. Unreasonable design and improper construction take the blame for all kinds of foundation quality accidents for the most part while insufficient dewatering and drainage is also an important factor.
2. The two common form of foundation pit collapse are retaining structure breakdown and support structure destruction. Foundation pit collapse induced by foundation pit instability have a very high proportion of which overall instability contribute 82%. Unreasonable design undertake the leading responsibility for foundation pit collapse, in addition, unreasonable design refer to: inaccurate calculation parameter selection, defected supporting scheme, incorrect supporting design and insufficient design safety margin, etc.
3. Correlativity between quality risk and investment risk manifested as cost claim due to mistaken exploration and unreasonable design of which the latter contributes the most. Since the semi-empirical and semi-theoretical foundation pit design, any negligence or mistake in design could result in serious foundation pit engineering accident.

This thesis aims at offering a reference for overall risk management of deep foundation pit construction, in addition, collecting basic data for risk theoretical analysis and study. Furthermore it is more important to learn a lesson from failure cases and accumulate experience in order to provide reference for the subsequent deep foundation pit construction.

REFERENCES

Chen Zhong-han. "Deep Foundation Pit Engineering" [M]. Beijing: China Machine Press, 2002.

He Shou-ye, Yao Yang-ping. Analysis of Loess Foundation Pit Engineering Accidents Caused by Water Immersion [C]. Chinese Civil Engineering Society ninth Soil Mechanics and Geotechnical Engineering Conference. 2003, 904–909.

Huang Hong-wei, Bian Yi-hai. Risk Management in the Construction of Deep Excavation Engineering [J]. Chinese Journal of underground space and Engineering. 2005, 4.

Huang Zhong-huan, Li Lu-fa. Prevention Measures for the Foundation Pit Engineering accident [J]. Construction Security. 2012, 2.

Liu Xiang, Zhang Chen, Zhao Xiang. Analysis of a soft Soil Foundation Pit Accident caused by seepage [J]. Construction Technique. 2007, 36(9):72–79.

Tang Ye-qing. Analysis and Treatment of an Accident in a Deep Foundation Construction [M]. Beijing: China Architecture & Building Press, 1999.

Yang Li-jun, Zhou Wei-dong. Common Problems and Treatment Measures of Deep Foundation Pit Engineering [J]. Western Exploration Engineering. 2003, 15(8):58–59.

Author index

Printed and bound by CPI Group (UK) Ltd, Croydon, CR0 4YY

18/10/2024

01776219-0008